Networking, Transport, and Quality of Service in Vehicular Networks

Truong Khang Nguyen
Van Lang University, Vietnam

Devasis Pradhan
Acharya Institute of Technology, India

IGI Global
Scientific Publishing
Publishing Tomorrow's Research Today

Vice President of Editorial — Melissa Wagner
Managing Editor of Acquisitions — Mikaela Felty
Managing Editor of Book Development — Jocelynn Hessler
Production Manager — Mike Brehm
Cover Design — Phillip Shickler

Published in the United States of America by
IGI Global Scientific Publishing
701 East Chocolate Avenue
Hershey, PA, 17033, USA
Tel: 717-533-8845
Fax: 717-533-7115
Website: https://www.igi-global.com E-mail: cust@igi-global.com

Library of Congress Cataloging-in-Publication Data

LCCN: 2024053105 (CIP Data Pending)
ISBN13: 979-8-3693-6422-2
Isbn13Softcover: 979-8-3693-6423-9
EISBN13: 979-8-3693-6424-6

British Cataloguing in Publication Data
A Cataloguing in Publication record for this book is available from the British Library.

Table of Contents

Detailed Table of Contents

M. Clement Joe Anand, Mount Carmel College, Bengaluru, India
V. A. Anjana, Mount Carmel College, Bengaluru, India
Manokamana Agarwal, Silver Oak College of Engineering and Technology, Silver Oak University, India
Utpal Saikia, Silapathar College, India
B. Ranjitha, Mohan Babu University, India
Chitra Ramaprakash, Dayananda Sagar College of Engineering, India

This chapter explores the integration of Neutrosophic Queuing Theory (NQT) and Neutrosophic Network Analysis (NNA) into the realm of vehicular communication at traffic signals to address the uncertainties and complexities of traffic flow, including variable arrival rates, unpredictable service times, and dynamic network conditions. Using real-world traffic data, a neutrosophic queueing model is developed to optimize traffic signal timings specifically for enhancing vehicular communication by determining key traffic parameters such as traffic intensity, the probability of crossing the signal, and various performance measures emphasizing the expected waiting time and queue length. Also proposes a novel framework that integrates NNA into the design and optimization of communication protocols for vehicles at traffic signals with the help of the score function and modified Kruskal's algorithm. The results demonstrate significant improvements in communication reliability, reduced latency, and enhanced system throughput, highlighting the effectiveness of the proposed approach in enhancing vehicular communication efficiency.

Decision-making on the selection-based problem of the location of Roadside units (RSU) in vehicular networks is complicated. This intricate problem shall be resolved with the intervention of Explainable Artificial Intelligence (XAI) and fuzzy-based multi-criteria decision-making methods (MCDM). This research proposes the integrated decision-making approach in making optimal location selection considering different criteria and economic aspects. The decision-making problem discussed in this chapter considers location selection of roadside units as the choice-making of the location is very significant in enhancing vehicular efficiency. This decision problem considers the criteria of coverage area, traffic density, accident frequency, infrastructure availability, environmental impact, energy efficiency, and cost components such as cost effectiveness, revenue competence, and investment returns. The integrated approach is leveraged to make an optimal choice of the RSU locations.

The most promising system-proposed solution is to benefit real-time traffic density; congestion can be minimized through dynamic fine-tuning of signal timings. Making use of PIR sensors and LED countdown timers, the system effectively enhances traffic flow and driver awareness. However, future upgrades for greater performance optimization might consider broadening the data collection scale through an enlarged sensor network, advanced traffic modeling techniques, and integration with other urban infrastructural systems. Additional work on the machine learning algorithms probably allowed the system to learn from historical traffic patterns for tuning the responses accordingly. Addressing such aspects further strengthens the system as an effective tool for traffic management in urban areas.

The varied types of Vehicular Networks Communication models (VNCM) circumscribe the decision-makers into a chaotic decision-making environment. The choice-making of VNCM is purely dependent on the attributes of traffic volume, vehicle density, road type, weather conditions, and application demands. However, the attribute of management techniques with the attribute values of dynamic routing, load balancing, and congestion control is more significant in choosing the desirable communication model. This challenging decision-making problem shall be resolved by applying the machine learning approach. This chapter proposes a matching model with supervised data that associates VNCM with management techniques and other core attributes. Supervised machine learning algorithms are applied to develop a matching model that best associates the VNCM with the management techniques considering vehicular efficiency. The performance metrics are computed to determine the most promising algorithm and to validate the consistency of the model. The results of the decision model facilitate the choice-making of the VNCM and also serve as the foundation for the construction of a predictive model.

Research and development in Next generation networks is a growing field, including subfields like Vehicle to Infrastructure (V2I), Vehicle to Vehicle (V2V) and Vehicle to everything (V2X) that helps application development and participating towards advanced smart city projects in India. Smart City projects in India demands fast communication channel, secure channel as well as data exchange policies to control infrastructure components. This article proposed a real-time simulation of a next generation network improving existing infrastructure of cities through the involvement of Quantum Computing. Quantum key based protocol BB84 has been simulated and tested in this article work for comparing Traditional and modern V2I based software defined network. Comparative results of proposed simulation in PoliQI and NS3 found effective in terms of security, performance, and scalability for different numbers of nodes. This article delivers a computational technique that may be used creates secure channels for key exchange between cloud service providers and customers, ensuring encryption keys

The efficacy of the vehicular networks is highly dependent on the choice of communication models. The vehicular network communication models outline the nature of interactions between the entities of vehicles, infrastructure, and the environment. Some of the most common communication models in existence are Vehicle-to-Vehicle, Vehicle-to-Infrastructure, Vehicle-to-Grid, and Vehicle-to-Network. These models are categorized based on the entities involved, direction, and purpose of communication. The choice-making of these communication models are hurdled by the primary challenges of heterogeneity of applications and dynamic network conditions. This chapter proposes solutions to these challenges by constructing a fuzzy-based multi-criteria decision-making method of LBWA (Level Based Weighting Assessment). This fuzzy model considers the factors to be taken into account in selecting the vehicular network communication models and also determines the weightage to each of the factors considered. This model developed in this chapter facilitates identifying the most significant factors in selecting the communication models. The optimal ranking of the factors is validated with other decision methods to determine the consistency of the results. The chapter concludes with future research directions.

The estimation of vehicular efficiency is highly significant in enhancing the sustainability of transportation and optimizing the allocation of resources. Machine learning algorithms are extensively applied in the domain of vehicular networks with a special focus on enriching vehicular potency. In this research work, a machine learning-based estimation model is developed for customer relationship management. The association between vehicular efficiency and customer acquisition cum retention is studied exclusively in this chapter. This work is indeed an intersection of machine learning algorithms, vehicular efficiency calculation, and customer relationship management. The interventions of the customer-oriented strategies with vehicular efficiency are explored in this study by determining the influence of customer-oriented metrics over vehicular efficiency. The results and the insights acquired from this work will certainly facilitate the policymakers and decision-makers in comprehending a comprehensive model integrating customer-centered metrics.

Communication protocols are guidelines governing the transmission of data between the entities of the vehicular networks. These protocols play a crucial role in enabling optimal connectivity, ensuring safety and traffic management. The effective functioning of the V2V network depends on the right choice of communication protocols, as the mismatch in selection results in incompatibility, performance degradation, congestion, and other security issues. This chapter focuses primarily on the intervention of a multi-criteria approach in making an optimal selection of the communication protocols used in the V2V network. The decision-making

Effective traffic management and congestion control are pivotal for sustainable urban development. This study proposes a Multi-Criteria Decision-Making (MCDM) approach, employing the Weighted Aggregated Sum Product Assessment (WASPAS) method, to address these obstacles. Integrating 30 diverse sub-criteria, including infrastructure capacity, environmental impact, and socio-economic elements, the model offers a comprehensive framework for decision-makers. By categorizing criteria through WASPAS, informed decisions can be made to optimize traffic flow and mitigate congestion. Here we highlight the innovative application of MCDM in transportation planning, emphasizing its potential to enhance urban mobility and promote sustainable development in increasingly congested urban environments.

Urban traffic congestion is a growing concern worldwide. Vehicular Adhoc Networks (VANETs) offer a glimpse into a future with smoother traffic flow and reduced congestion. These networks enable real-time communication between vehicles and infrastructure, creating a dynamic traffic management system. Imagine traffic signals that adjust based on real-time data, congestion being predicted and alleviated before it builds, and emergency services receiving faster response times. This is the potential of VANETs. Ensuring reliable communication and data integrity among constantly moving vehicles is crucial. Researchers are developing protocols and algorithms to address this, focusing on efficient routing, data dissemination, and network stability. The integration of emerging technologies like 5G, edge computing, and artificial intelligence holds promise for further enhancing network performance and robustness. While significant progress has been made, widespread adoption of

Foreword

In an era where intelligent transportation systems are rapidly evolving, vehicular networks have emerged as a crucial domain, bridging the gap between communication technologies and next-generation mobility solutions. The integration of networking, transport, and quality of service (QoS) in vehicular networks is pivotal in ensuring seamless connectivity, efficient data exchange, and enhanced safety on the roads.

This book, Networking, Transport, and Quality of Service in Vehicular Networks, serves as a comprehensive guide to the fundamental and advanced aspects of vehicular communication. It delves into the intricate mechanisms that enable reliable data transmission, robust transport layer protocols, and QoS strategies essential for the dynamic and latency-sensitive nature of vehicular environments.

With the increasing deployment of connected and autonomous vehicles, understanding the underlying communication frameworks becomes imperative for researchers, engineers, and industry professionals. This book not only provides in-depth theoretical insights but also explores real-world applications, challenges, and future trends in vehicular networks. By addressing key topics such as network architecture, congestion control, and service optimization, it equips readers with the knowledge to contribute to the advancement of intelligent transportation systems.

The authors have meticulously crafted this work to offer a balanced perspective, combining academic rigor with practical relevance. Whether you are a scholar seeking to deepen your expertise or a practitioner aiming to implement effective networking solutions, this book will be an invaluable resource.

I congratulate the authors on their outstanding contribution to this vital field and hope that readers will find this book both insightful and inspiring as they navigate the evolving landscape of vehicular networks.

Devasis Pradhan

Acharya Institute of Technology, Bangalore, India

Preface

OVERVIEW OF THE SUBJECT MATTER

The rapid advancement of vehicular networks has revolutionized the landscape of transportation, enabling a more connected, intelligent, and efficient system. The integration of networking technologies, transport protocols, and Quality of Service (QoS) mechanisms has become a cornerstone for enhancing vehicular communication, traffic management, and overall road safety. This book, Networking, Transport, and Quality of Service in Vehicular Networks, provides a deep dive into the emerging methodologies, technologies, and strategies that optimize vehicular communication networks.

Vehicular networks play a critical role in modern Intelligent Transportation Systems (ITS), facilitating vehicle-to-vehicle (V2V), vehicle-to-infrastructure (V2I), and vehicle-to-everything (V2X) communications. As the complexity of vehicular networks increases with the rise of connected and autonomous vehicles, the need for advanced networking solutions, traffic control mechanisms, and efficient data transmission strategies has become more pronounced. This book addresses these challenges through innovative approaches such as Artificial Intelligence (AI), Machine Learning (ML), fuzzy decision models, and quantum security protocols, among others.

Vehicular networks, which serve as a fundamental pillar of modern transportation systems, have evolved significantly over the past decade. The integration of wireless communication technologies, cloud computing, and edge computing has further amplified their effectiveness. Intelligent vehicular networks are not only confined to individual cars but also extend to fleet management, public transport systems, and emergency response vehicles. This expansion necessitates a comprehensive understanding of the core elements that enhance networking, transport efficiency, and quality of service within vehicular ecosystems.

Relevance in the World Today

The global push toward smart cities and sustainable mobility has elevated the significance of vehicular networks. Traffic congestion, road safety, and real-time decision-making are major concerns in urban transportation. The deployment of AI-driven traffic management, IoT-enabled vehicular networks, and secure communication protocols are transforming the efficiency and reliability of modern transportation systems. The research and methodologies discussed in this book align with these technological advancements, offering pragmatic solutions to contemporary transportation challenges. Moreover, the economic and environmental implications of intelligent vehicular networks further emphasize the urgency of integrating these cutting-edge technologies into real-world applications.

In today's world, urban areas are experiencing unprecedented population growth, leading to increased vehicular density and traffic congestion. Governments and urban planners are seeking innovative solutions to mitigate these challenges and improve transportation infrastructure. Vehicular networks provide an opportunity to enhance transportation efficiency through real-time data exchange, predictive analytics, and autonomous decision-making mechanisms. The inclusion of emerging technologies, such as blockchain for secure data sharing and 5G for high-speed communication, is further enhancing the potential of vehicular networks.

The introduction of electric and autonomous vehicles adds another layer of complexity to vehicular networking. These vehicles rely on advanced communication protocols and decision-making algorithms to navigate safely and efficiently. The ability to integrate various transportation modes, including ride-sharing services, public transit, and private vehicles, into a cohesive network has the potential to revolutionize urban mobility.

Target Audience

This book is intended for a diverse audience, including:

Researchers and Academicians: Those exploring the domains of vehicular networks, intelligent transportation, and network security will find this book a valuable resource for understanding emerging technologies and methodologies.

Industry Professionals: Engineers, developers, and decision-makers involved in vehicular communication systems, traffic management, and smart city initiatives will benefit from the practical applications discussed.

Graduate and Postgraduate Students: Students pursuing degrees in Computer Science, Electrical Engineering, Transportation Engineering, and related fields will gain insights into advanced vehicular network strategies and real-world implementations.

Policy Makers and Urban Planners: Those involved in designing policies and frameworks for future smart cities and connected transportation infrastructures will find useful guidance in implementing sustainable and efficient solutions.

This book is particularly useful for professionals engaged in developing intelligent transportation policies and implementing vehicular networking solutions. As autonomous driving gains traction, professionals in the automotive industry will benefit from the insights provided into AI-driven vehicular communication models. Urban planners working toward sustainable city designs will also find valuable methodologies in managing congestion and optimizing traffic flow.

CHAPTER SUMMARIES

Chapter 1: Enhancing Vehicular Communication through Neutrosophic Queueing Theory and Network Analysis-Based Traffic Signal Optimization

This chapter presents a novel approach to optimizing vehicular communication and traffic signals using Neutrosophic Queueing Theory and network analysis. The chapter explores the role of probabilistic decision-making in reducing congestion and improving traffic efficiency. The mathematical models and simulations discussed in this chapter provide a robust framework for improving vehicular data transmission and minimizing delays at traffic junctions.

Chapter 2: Explainable Artificial Intelligence and Fuzzy MCDM Approach in Location Selection of Roadside Units in Vehicular Networks with Economic Efficiency

This chapter integrates Explainable AI with a Fuzzy Multi-Criteria Decision-Making (MCDM) approach to optimize the placement of roadside units (RSUs), balancing economic feasibility with enhanced communication coverage. The increasing complexity of vehicular networks necessitates an explainable and transparent approach to decision-making, making this chapter particularly valuable for researchers and urban planners.

Chapter 3: Driving Towards Safer Cities: Innovations in Traffic Control and Emergency Services

Focusing on road safety, this chapter highlights innovative solutions for traffic control and emergency response. The discussion includes AI-driven traffic monitoring, predictive accident prevention, and smart emergency services deployment. The implementation of AI in emergency response planning has the potential to save lives and significantly improve crisis management on roads.

Chapter 4: Machine Learning for Optimal Matching Management Techniques with Vehicular Network Communication Models

This chapter explores how machine learning algorithms improve matching techniques within vehicular communication models, optimizing network performance, resource allocation, and real-time data exchange. ML-driven solutions provide scalability and adaptability to dynamic traffic conditions.

Chapter 5: Experiment and Simulation of Proposed V2I System: An Approach Using Quantum Key Distribution for E-Mobility Systems

Addressing security concerns, this chapter introduces a quantum key distribution approach for V2I communications, enhancing the security of e-mobility infrastructures. With increasing cyber threats in vehicular networks, quantum security techniques provide a reliable framework for ensuring data integrity and confidentiality.

Chapter 6: Fuzzy Decision Model for Selecting Core Factors in Vehicular Network Communications

The chapter proposes a fuzzy decision model to identify and prioritize key factors affecting vehicular communication efficiency, providing a structured approach for optimizing network parameters.

Chapter 7: Machine Learning In Estimating Vehicular Efficiency With Special Focus On Customer Acquisition And Retention

This chapter explores the integration of machine learning algorithms in estimating vehicular efficiency, with a unique focus on customer acquisition and retention. It presents a model that intersects vehicular performance analytics with customer relationship metrics, aiming to provide insights into how customer-oriented strategies influence vehicular efficiency. The chapter is particularly valuable for policymakers and decision-makers looking to adopt a comprehensive, data-driven approach to optimizing both vehicular systems and customer engagement.

Chapter 8: Multi-Criteria Approaches In Selecting Optimal Vehicle-To-Vehicle Communication Protocols

Focusing on the crucial task of selecting suitable communication protocols for Vehicle-to-Vehicle (V2V) networks, this chapter introduces a multi-criteria decision-making approach. Various protocols are evaluated against parameters such as security, scalability, and data rate using techniques like AHP, Entropy, and FUCOM. The chapter compares methods such as COPRAS, SMART, and MAIRCA to derive rankings and validate results, offering a detailed framework for choosing the most compatible and efficient V2V protocol configurations.

Chapter 9: Optimizing Vehicular Network Architecture And Communication Models With Machine Learning Approach

This chapter proposes a machine learning-based framework to identify optimal combinations of vehicular network architecture and communication models. By analyzing traffic patterns, network topologies, and communication requirements, the chapter leverages deep learning networks to enhance vehicular communication efficacy. It also includes a comparative performance evaluation with other learning algorithms, positioning the model as both an analytical and predictive tool for planning and implementing vehicular networks.

Chapter 10: Strategic Integration Of WASPAS Method For Effective Traffic Management And Congestion Control

Employing the Weighted Aggregated Sum Product Assessment (WASPAS) method, this chapter introduces a robust MCDM model to tackle traffic management and congestion issues. With 30 sub-criteria spanning infrastructure, environmental, and socio-economic factors, the chapter presents a comprehensive decision-making framework. It emphasizes the strategic use of quantitative methods to inform policy and design choices for sustainable urban mobility solutions.

Chapter 11: Traffic Management And Congestion Control In Vehicular Adhoc Networks

This chapter delves into the potential of Vehicular Adhoc Networks (VANETs) to transform urban traffic systems. By enabling real-time vehicle-to-infrastructure and vehicle-to-vehicle communication, VANETs promise responsive traffic signals, reduced congestion, and faster emergency response. The discussion covers routing algorithms, data dissemination techniques, and the integration of 5G, edge computing, and AI, while also addressing the ongoing challenges of scalability, security, and infrastructure readiness.

Chapter 12: Smart Cities On The Move Leveraging IoT Communication In Connected Cars For Enhanced Traffic Management

Highlighting the synergy between connected cars and IoT infrastructure, this chapter discusses how real-time data exchange in smart cities can revolutionize traffic management. It explores the use of sensor-rich vehicles and communication protocols like V2V and V2I to collect and integrate data for better road safety and congestion control. Applications of augmented reality and data analytics are also considered for their roles in enhancing situational awareness and optimizing transportation systems.

Chapter 13: Smart Cities, Smart Mobility: How Mobile Intelligence Is Transforming Urban Transportation

This chapter examines the role of mobile intelligence in the evolution of smart cities, emphasizing technologies like V2X communication, intelligent transportation systems, and telematics. It discusses how real-time data processing and autonomous systems contribute to traffic optimization, safety, and urban mobility. The chapter offers insights into the integration of digital and physical infrastructure to create interconnected, adaptive transportation environments.

Chapter 14: Future Of Mobility – Potential Challenges And Ethical Considerations

Focusing on the implications of quantum computing on vehicular communication, this chapter addresses the security challenges posed by quantum-powered adversaries. It presents the vulnerabilities of current cryptographic systems and explores the development of quantum-safe cryptographic solutions integrated with AI. The chapter concludes by discussing the ethical and practical challenges involved in implementing these advanced security measures at scale.

CONCLUSION

This book represents a significant contribution to the field of vehicular networks by providing a holistic overview of the latest research, innovative methodologies, and practical applications. By combining AI-driven solutions, machine learning techniques, and multi-criteria decision models, this book not only addresses current challenges but also lays the groundwork for future advancements in intelligent transportation systems. We hope that readers, whether researchers, professionals, or policymakers, find this work insightful and instrumental in shaping the next generation of vehicular communication and transport networks.

The advancements discussed in this book aim to redefine the future of smart mobility by integrating various cutting-edge technologies into an interconnected and optimized vehicular ecosystem. We believe that this work will spark further research and development in the domain, pushing the boundaries of vehicular networking toward a safer, more efficient, and sustainable future.

Acknowledgment

We extend our heartfelt gratitude to all those who have contributed to the successful completion of this book, Networking, Transport, and Quality of Service in Vehicular Networks. This work would not have been possible without the support, guidance, and encouragement of numerous individuals and institutions.

First and foremost, we express our deep appreciation to the contributing authors and researchers whose invaluable insights, innovative methodologies, and scholarly contributions have enriched the content of this book. Their dedication to advancing the field of vehicular networking and intelligent transportation has been instrumental in shaping the chapters contained herein.

We are immensely grateful to our editorial and peer review team for their meticulous assessment, constructive feedback, and unwavering commitment to maintaining the highest academic standards. Their efforts have significantly enhanced the quality and coherence of this publication.

Our sincere thanks go to our respective institutions and research organizations for providing the necessary resources, infrastructure, and intellectual environment that facilitated the development of this book. Their continued support has been a driving force behind our endeavors.

We would also like to acknowledge the support of our colleagues, mentors, and collaborators who have provided valuable discussions, insights, and encouragement throughout the writing and editing process. Their expertise and constructive critiques have greatly contributed to refining our work.

A special word of gratitude to our families and loved ones for their patience, understanding, and encouragement during this journey. Their unwavering support and belief in our vision have been a constant source of motivation.

Finally, we extend our appreciation to the publishing team for their professionalism and dedication in bringing this book to fruition. Their efforts in refining, formatting, and publishing this work have been instrumental in making it accessible to a global audience.

This book is a collective effort, and we sincerely hope it serves as a valuable resource for researchers, professionals, and students in the field of vehicular networking and intelligent transportation. We look forward to continued collaboration and further advancements in this ever-evolving domain.

Chapter 1
Enhancing Vehicular Communication Through Neutrosophic Queueing Theory and Network Analysis-Based Traffic Signal Optimization

M. Clement Joe Anand
https://orcid.org/0000-0002-1959-7631
Mount Carmel College, Bengaluru, India

V. A. Anjana
https://orcid.org/0009-0002-7573-2356
Mount Carmel College, Bengaluru, India

Manokamana Agarwal
https://orcid.org/0000-0001-5083-7384
Silver Oak College of Engineering and Technology, Silver Oak University, India

Utpal Saikia
https://orcid.org/0000-0003-0235-6871
Silapathar College, India

B. Ranjitha
Mohan Babu University, India

Chitra Ramaprakash
https://orcid.org/0000-0002-0421-2479
Dayananda Sagar College of Engineering, India

ABSTRACT

This chapter explores the integration of Neutrosophic Queuing Theory (NQT) and Neutrosophic Network Analysis (NNA) into the realm of vehicular communication at traffic signals to address the uncertainties and complexities of traffic flow, including

DOI: 10.4018/979-8-3693-6422-2.ch001

variable arrival rates, unpredictable service times, and dynamic network conditions. Using real-world traffic data, a neutrosophic queueing model is developed to optimize traffic signal timings specifically for enhancing vehicular communication by determining key traffic parameters such as traffic intensity, the probability of crossing the signal, and various performance measures emphasizing the expected waiting time and queue length. Also proposes a novel framework that integrates NNA into the design and optimization of communication protocols for vehicles at traffic signals with the help of the score function and modified Kruskal's algorithm. The results demonstrate significant improvements in communication reliability, reduced latency, and enhanced system throughput, highlighting the effectiveness of the proposed approach in enhancing vehicular communication efficiency.

1. INTRODUCTION

Queuing theory is a branch of mathematics and operations research developed by Erlang (1918), that focuses on the study of queues, which are waiting lines or systems where entities such as customers, jobs, or data packets arrive, wait in line for service, and then depart after being served. It's widely used in various fields such as telecommunications, computer networks, transportation, healthcare, and manufacturing to analyze and optimize the performance of systems involving waiting and service. Queuing theory finds extensive application in traffic engineering and transportation systems (Manoj et al., 2019) to analyze and optimize traffic flow, manage congestion, and improve overall system efficiency. In the context of traffic, queuing theory deals with the analysis of queues formed by vehicles at intersections, toll booths, traffic signals, parking lots, and other points where vehicles may wait in line for movement or service. Queuing theory was integrated with fuzzy set theory (Chen & Eur, 2009) in order to handle the uncertainties in the real-world. In fuzzy queuing models, the arrival rates and service times are expressed as fuzzy numbers characterized by membership functions. This approach allowed for a more realistic representation of systems with uncertain parameters. Recently neutrosophic set theory which was introduced by Florentin Smarandache (1999; 2002; 2005) in the 1990s, was also incorporated with queuing theory. This is extremely useful in environments with highly unpredictable and contradictory information. Neutrosophic queuing model (Mohamed, 2020a; 2020b) is a relatively recent development in the field of queuing theory and uncertainty modeling. Now it is a branch of mathematics that deals with the analysis and modeling of queues, particularly in traffic systems, using neutrosophic logic. Neutrosophic logic is an extension of fuzzy logic that allows for the representation of indeterminacy, uncertainty, and imprecise information. In the context of traffic systems, neutrosophic queuing theory helps in understanding and optimizing the flow of vehicles, pedestrians, or any other entities in a queuing system. It considers factors such as arrival rates, service times, queue lengths, and the behavior of drivers or individuals in the system, taking into account the uncertainty and vagueness that are often inherent in real-world traffic situations. Thus, it provides a more comprehensive and flexible framework for analyzing and managing traffic queues, leading to more accurate performance evaluation and optimization of vehicular communication protocols.

Network theory (Alhabib et al., 2018) deals systems of interconnected nodes that are linked by arcs with specific weights. The connected nodes enable to transfer of data from one node to another. Network analysis involves examining the structure and behavior of interconnected systems, providing insights into connectivity, flow, and communication patterns (Hussein et al., 2022). Concerning vehicular communication systems (Zhang et al., 2018), network analysis is applied to optimize the exchange of information between vehicles, enhancing traffic management, safety, and efficiency. In Vehicle-to-Vehicle (V2V) communication, network analysis helps ensure real-time information dissemination about traffic conditions, hazards, and optimal routes. Initially, network analysis in fuzzy logic was applied to handle uncertainty and partial truths within networks. Techniques like fuzzy cognitive maps (FCMs) were developed, where the relationships between nodes were represented with degrees of influence ranging between 0 and 1. This allowed for the modeling of complex systems where relationships were not binary but had varying strengths of connection. These fuzzy logic applications helped in fields such as decision making, pattern recognition, and data mining, where precise boundaries and clear-cut relationships were often impractical. As the need to handle even more nuanced uncertainties grew, neutrosophic logic emerged as an extension of fuzzy logic. Neutrosophy introduced three components: truth (T), indeterminacy (I), and falsity (F), each ranging independently between 0 and 1. Neutrosophic network analysis offers significant advantages over network analysis by effectively handling uncertainties and indeterminacies such as construction zones, network issues, and weather conditions inherent in complex systems, thus providing a more reliable and efficient understanding of network behavior. In computer networks, neutrosophic analysis enhances the reliability and security of data transmission by modeling uncertain and ambiguous connections, thereby improving fault tolerance and resilience against cyber-attacks. In highway systems, it aids in optimizing infrastructure maintenance and traffic management by accurately representing fluctuating road conditions and traffic flow uncertainties, leading to more effective planning and response strategies. Additionally, in transport network flow, neutrosophic network analysis facilitates the management of logistics and transportation systems by addressing the unpredictability of routes and transport schedules, ensuring more robust and efficient network operations. By capturing the nuanced variations in network parameters, neutrosophic network analysis provides a more adaptable and realistic approach to modeling and solving real-world problems in various interconnected systems.

2. REVIEW OF LITERATURE

This section explores the existing body of knowledge related to our research topic, providing a comprehensive review of literature. Mohamed Bisher Zeina (2020a; 2020b) have introduced a new approach to queuing theory by incorporating neutrosophic logic in it. This includes defining the neutrosophic performance measures and have also presented an event-based table for the neutrosophic queuing system. He has derived Neutrosophic Little's formula and has compared his results with the crisp solution. Patro and Smarandache (2016) focus on providing solutions to various statistical problems using neutrosophic approaches, which incorporate degrees of truth, falsity, and indeterminacy. The paper outlines how neutrosophic probability, which is broader than classical probability, can address real-world problems with inherent uncertainty and hence introduces neutrosophic statistics which is a generalization of classical and interval statistics. Bohlender and Kulisch (2010) have contributed to the operations and comparisons that can be used in interval arithmetic which deals with ranges of values instead of precise numbers. These definitions are necessary for creating a standardized interval arithmetic system, ensuring consistent and reliable results across various applications. Lartey's (2014) research suggests understanding the probabilistic behavior of traffic congestion. This study focuses on predicting traffic congestion using a stochastic process and initial value problem framework to describe and analyze vehicular flow as a queuing system. He have also used Markov chains to represent the state of traffic at discrete time points which allows for the estimation of steady-state probabilities.

Rashad & Mohamed (2021) have presented the three queuing models to neutrosophic environments to deal with the imprecise data. With the help of case studies, they have illustrated the accuracy of their results with a comparative study with the crisp queuing models. Hevey (2018) introduces the fundamental concepts of network analysis, including nodes, ties, and networks, and discusses different types of networks. He also delves into the methods and techniques used in network analysis, such as centrality measures, clustering algorithms, and visualization tools. Tamura et al. (2011) provides a comprehensive exploration of how graph and network theory can be applied to solve various issues in communication systems. In this the node colouring problem in graph theory is applied to the channel assignment problem in cellular mobile communication systems and thus analyzing how network theory results can address routing and network optimization issues. Zhang et al. (2018) surveys a wide array of studies focusing on the pivotal role of communication networks in advancing automated driving technologies. They highlight the challenges and opportunities posed by vehicular communication networks, such as ensuring low-latency and high-reliability communication for real-time decision-making in autonomous vehicles. The review also delves into various communication protocols, including DSRC and C-V2X, assessing their effectiveness in V2V and V2I communications essential for enhancing safety, traffic efficiency, and overall driving experience in the automated era. Hui and Dey (2020) explore a novel approach to modeling social networks using directed neutrosophic sets, aiming to capture the inherent uncertainties, indeterminacies, and inconsistencies present in social interactions. They have introduced some centrality measures for directed neutrosophic social network, because the arcs were associated with directed neutrosophic relations which would consist many information.

Meenakshi (2023a) have done their study on neutrosophic graphs which provide more flexibility and compatibility than Intuitionistic fuzzy graphs. They have developed neutrosophic graphs into a fully connected neutrosophic network using the product of graph, thus finding the efficient productivity in the flow of information on a single source network of a single-valued neutrosophic network. Kaviyarasu et al. (2024) explores the innovative application of neutrosophic graph theory to various network systems by providing a theoretical framework for the neutrosophic connectivity indices, detailing their mathematical formulations and properties. This is particularly relevant in complex systems like computer networks, highway systems, and transport networks. Also, through a series of case studies, they have demonstrated the practical application of neutrosophic connectivity indices. Nasir and Mamun (2013) delve into the intersection of graph theory and error-correcting codes within the context of VANETs. Since graph theory provides a powerful framework for modeling the network topology of VANETs they have used concepts such as connectivity (Hussein et al., 2022; Phull et al., 2022; Meenakshi & Mythreyi, 2023b), shortest path algorithms, and network flows, emphasizing their relevance to maintaining robust and efficient communication in the highly dynamic environment of VANETs, the various types of fuzzy and neutrosophic parameters were discussed (Varalakshmi et al., 2022; Anand & Bharatraj, 2017, 2018, 2019; Anand et al., 2014, 2017, 2018, 2019, 2023, 2024; Bharatraj & Anand, 2019; Devadoss, Anand, & Felix, 2014; Justin Raj et al., 2023; Kungumaraj et al., 2024; Manshath et al., 2024; Miriam & Martin, 2023; Prabha et al., 2024; Raj et al., 2023; Sudha et al., 2023). They have also discussed how the codes can detect and correct errors that occur during data transmission using hamming code, which is essential given the high-speed movement and frequent disconnections typical of vehicular networks.

3. PRELIMINARIES

In this section, we discuss the required basic definitions, which will be dealing with the upcoming sections.

Definition 2.1 Interval Arithmetic

Let $[a_1 , b_1]$ and $[a_2 , b_2]$ be two intervals where $a_1, a_2 , b_1, b_2 \in \mathbb{R}$ and for practical cases where $a_1, a_2 , b_1, b_2 > 0$, then:

a) $[a_1 , b_1] + [a_2 , b_2] = [a_1 + a_2 , b_1 + b_2]$

b) $[a_1 , b_1] - [a_2 , b_2] = [a_1 - b_2 , b_1 - a_2]$

c) $[a_1 , b_1] \times [a_2 , b_2] = [a_1 \times a_2 , b_1 \times b_2]$

d) $[a_1 , b_1] \div [a_2 , b_2] = [a_1 , b_1] \times \frac{1}{[a_2 , b_2]} = \left[\frac{a_1}{b_2}, \frac{b_1}{a_2}\right]$

Definition 2.2 Neutrosophic Density

A queuing system is said to be a neutrosophic queue if its parameters are neutrosophic numbers, i.e. average rate of customers entering the queuing system λ and the average rate of customers being served μ are neutrosophic numbers. Therefore, the neutrosophic interarrival rate is given by: $\lambda_N = [\lambda^L , \lambda^U]$. Similarly, the neutrosophic service rate is given by: $\mu_N = [\mu^L , \mu^U]$. Thus, if we have c servers, then the neutrosophic density is given as: $p_N = \frac{\lambda_N}{c\mu_N} = \frac{[\lambda^L , \lambda^U]}{c[\mu^L , \mu^U]}$

Definition 2.3 Queuing System

A queuing system is given in the form (A/B/c/b):(SP/CS/b) where A is the interarrival times distribution, B is servicing or processing times distribution, c is number of parallel servers, SP is the service policy, CS is the calling size and b is the system capacity. Also, we can use the short notation of the queuing system as (A/B/c/b).

In this paper, NP stands for neutrosophic probability, NM stands for the neutrosophic exponential distribution, and NL_s, NL_q, NW_s, NW_q are the performance measures that denote the neutrosophic expected number of customers in the system, the neutrosophic expected number of customers in the queue, the neutrosophic mean waiting time in the system, and the neutrosophic mean waiting time in the queue respectively.

Definition 2.4 Neutrosophic Probability

The neutrosophic probability that arriving customers will find 'k' customers in the queue is:

$$NP(k) = (1 - p_N)\, p_N^{\,k} ;\ k = 0, 1, 2, \ldots$$

where NP stands for neutrosophic probability for 'k' customers and 'p_N' denotes neutrosophic density.

Definition 2.5 Neutrosophic Expected Number of Customers in System

The neutrosophic expected number of customers in the system is denoted NL_s as and is given by:

$$NL_s = \frac{p_N}{1 - p_N} = \frac{\lambda_N}{\mu_N - \lambda_N}$$

where 'p_N' denotes neutrosophic density, 'λ_N' is neutrosophic interarrival and 'μ_N' is the neutrosophic service rate.

Definition 2.6 Neutrosophic Expected Number of Customers in Queue

The neutrosophic expected number of customers in the queue is denoted NL_q as and is given by:

$$NL_q = \frac{p_N^2}{1 - p_N} = \frac{\lambda_N^2}{\mu_N\,(\mu_N - \lambda_N)}$$

where 'p_N' denotes neutrosophic density, 'λ_N' is neutrosophic interarrival and 'μ_N' is the neutrosophic service rate.

Definition 2.7 Neutrosophic Expected Waiting Time in the System

The neutrosophic expected waiting time in the system is denoted NW_s as and is given by:

$$NW_s = \frac{1}{\mu_N - \lambda_N}$$

where 'λ_N' is neutrosophic interarrival and 'μ_N' is the neutrosophic service rate.

Definition 2.8 Neutrosophic Expected Waiting Time in the Queue

The neutrosophic expected waiting time in the queue is denoted NW_q as and is given by:

$$NW_q = \frac{p_N}{\mu_N - \lambda_N}$$

where 'λ_N' is neutrosophic interarrival and 'μ_N' is the neutrosophic service rate.

Definition 2.9 Score Function

Let N = (T, I, F) be a Single-Valued Neutrosophic Network (SVNN), then the score function S is defined as follows:

$$S(N) = \frac{2 + T - I - F}{3}$$

4. APPLICATION OF NEUTROSOPHIC QUEUING THEORY (NWT) TO TRAFFIC SYSTEM

In a vehicular system, vehicles are being considered as the customers and the traffic signal is being considered as the server. Therefore, the interarrival rate of vehicles is random and forms a queue to wait for their chance get served, that is to cross during green signal. Also, the interarrival times and servicing times have neutrosophic exponential distribution with parameters λ_N and μ_N respectively. Hence, let us consider the traffic flow at a junction which follows NM/NM/1 Queueing Model with a single traffic signal acting as a server. Since the number of vehicles passing daily through this traffic signal is large, we consider it as an unbounded system with infinite population. So, we can say that the calling source and system size are infinite. It's also clear that, this queue follows FCFS (First Come First Serve) discipline. Therefore the traffic queue model considered in this situation is (NM/NM/1):(FCFS/∞/∞).

The number of vehicles arriving at the junction during time interval of traffic light showing red signal forms the required arrival data and the number of vehicles that cross the junction during the time span of green signal forms the service data (or departure data). Here we have observed vehicles for complete one week and vehicles were counted during the six different time slots of the day.

Table 1. Count of vehicles arriving towards a junction at different time intervals for a complete one week

Time Slot	Day 1	Day 2	Day 3	Day 4	Day 5	Day 6	Day 7	Total
09:00 – 09:03	77 (25.667)	70 (23.333)	68 (22.667)	71 (23.667)	82 (27.333)	54 (18)	48 (16)	470 **0.373**
11:00 – 11:03	115 (38.333)	108 (36)	121 (40.333)	112 (37.333)	104 (34.667)	96 (32)	75 (25)	731 **0.580**
01:00 – 01:03	88 (29.333)	72 (24)	85 (28.333)	89 (29.667)	92 (30.667)	74 (24.667)	84 (28)	584 **0.464**
03:00 – 03:03	59 (19.667)	61 (20.333)	63 (21)	57 (19)	63 (21)	80 (26.667)	89 (29.667)	472 **0.375**
05:00 – 05:03	82 (27.333)	87 (29)	86 (28.667)	76 (25.333)	95 (31.667)	99 (33)	113 (37.667)	638 **0.506**
07:00 – 07:03	109 (36.333)	83 (27.667)	97 (32.333)	101 (33.667)	122 (40.667)	118 (39.333)	128 (42.667)	758 **0.602**
Column Total	530	481	520	506	558	521	537	3653
No. Veh /min	29.444	26.722	28.889	28.111	31.000	28.945	29.834	
No. Veh /sec	0.491	0.445	0.481	0.469	0.517	0.482	0.497	**3.382**

The above table represents the data of the number of vehicles arriving at the junction at a particular time slot of 3 minutes each day for a week. Above the count of vehicles is the number of vehicles arriving at the junction per minute, which is written inside the bracket.

Arrival rate for the volume of traffic for each day:

$$\lambda = \frac{3.382}{7} \approx 0.483$$

Arrival rate for the volume of traffic for each time slot:

$$\lambda = \frac{2.9}{6} \approx 0.483$$

Table 2. Count of vehicles served on green signal towards a junction at different time slots for complete one week

Time Slot	Day 1	Day 2	Day 3	Day 4	Day 5	Day 6	Day 7	Total
09:03 – 09:05	50 (25)	44 (22)	46 (23)	49 (24.5)	48 (24)	51 (25.5)	49 (24.5)	337 **0.401**
11:03 – 11:05	69 (34.5)	65 (32.5)	68 (34)	66 (33)	64 (32)	53 (26.5)	60 (30)	445 **0.530**
01:03 – 01:05	51 (25.5)	53 (26.5)	55 (27.5)	56 (28)	58 (29)	62 (31)	65 (32.5)	400 **0.476**
03:03 – 03:05	48 (24)	54 (27)	49 (24.5)	52 (26)	58 (29)	60 (30)	57 (28.5)	378 **0.45**
05:03 – 5:05	61 (30.5)	63 (31.5)	59 (29.5)	67 (33.5)	65 (32.5)	58 (29)	64 (32)	437 **0.520**
07:03 – 07:05	70 (35)	72 (36)	71 (35.5)	72 (36)_	78 (39)	79 (39.5)	69 (34.5)	511 **0.608**
Column Total	349	351	348	362	371	363	364	2508
No. Veh /min	29.083	29.25	29.000	30.167	30.917	30.25	30.333	
No. Veh /sec	0.485	0.488	0.483	0.503	0.515	0.504	0.506	**3.484**

The table below represents the data of the number of vehicles crossing the junction at a particular time slot of 2 minutes each day for a week. Below the count of vehicles is the number of vehicles crossing the junction per minute, which is written inside the bracket.

Arrival rate for the volume of traffic for each day:

$$\lambda = \frac{3.484}{7} \approx 0.498$$

Arrival rate for the volume of traffic for each time slot:

$$\lambda = \frac{2.985}{6} \approx 0.498$$

Vehicles arrive according to exponential distribution with a mean arrival rate between 22 to 28 vehicles per minute. Therefore the neutrosophic interarrival rate is the average of vehicles arriving at the junction to get served per unit time is denoted by λ_N.

Neutrosophic Interarrival Rate:

$$\lambda_N = [22 , 28] \text{ vehicles per minute}$$

Vehicle serving time is exponentially distributed with a service rate between 30 to 36 vehicles per minute. That is, the neutrosophic service rate is determined by the average number of vehicles served by unit time and is denoted by μ_N.

Neutrosophic Departure Rate:

$$\mu_N = [30 , 36] \text{ vehicles per minute}$$

Number of servers = 1 = c

Neutrosophic Traffic Intensity:

Therefore the neutrosophic traffic density becomes,

$$p_N = \frac{\lambda_N}{c\mu_N} = \frac{\lambda_N}{\mu_N}$$

$$p_N = [22 , 28] \div [30 , 36] = \left[\frac{22}{36}, \frac{28}{30}\right]$$

$$p_N = [0.611 , 0.933]$$

Neutrosophic Probability:

The neutrosophic probability that arriving vehicle will find 'k' vehicles in the traffic is:

$$NP(k) = (1 - p_N)\, p_N^{\,k} ;\ k = 0, 1, 2, \ldots$$

So the neutrosophic probability that the arriving vehicle will find the road empty is

$$NP(0) = (1 - p_N)p_N^{\,0} = 1 - [0.611, 0.933] = [1 , 1] - [0.611 , 0.933]$$

$$= [0.067\ ,\ 0.389]$$

Thus the probability that an arriving vehicle will find the road empty ranges between 0.067 and 0.389

The neutrosophic probability that the arriving vehicle will find at least one vehicle in the traffic is:

$$1 - NP(0) = 1 - [0.067, 0.389] = [1 - 0.389\ ,\ 1 - 0.067] = [0.611\ ,\ 0.933]$$

Therefore the neutrosophic probability that the arriving vehicle will find at least one vehicle in the traffic is the same as the neutrosophic traffic intensity.

Neutrosophic Performance Measures

The neutrosophic performance measures will be as follows:

1) Neutrosophic Expected Number of Vehicles in the System

The neutrosophic expected number of vehicles in the system can be given by:

$$NL_s = \frac{p_N}{1 - p_N} = \frac{\lambda_N}{\mu_N - \lambda_N}$$

$$NL_s = \frac{[0.611\ ,\ 0.933]}{1 - [0.611\ ,\ 0.933]}$$

$$NL_s = \frac{[0.611\ ,\ 0.933]}{[0.067\ ,\ 0.389]} = \left[\frac{0.611}{0.389}, \frac{0.933}{0.067}\right]$$

$$NL_s = [1.571\ ,\ 13.925]$$

Hence, we can say that the expected number of vehicles in the system ranges between 1 to 14 vehicles.

2) Neutrosophic Expected Number of Vehicles in the Queue

The neutrosophic expected number of vehicles in the queue can be given by:

$$NL_q = \frac{p_N^{\ 2}}{1 - p_N} = \frac{\lambda_N^{\ 2}}{\mu_N\,(\mu_N - \lambda_N)}$$

$$NL_q = \frac{[0.611\ ,\ 0.933]^2}{1 - [0.611\ ,\ 0.389]} = \frac{[0.373\ ,\ 0.870]}{[0.067\ ,\ 0.389]}$$

$$NL_q = \left[\frac{0.373}{0.389}, \frac{0.870}{0.067}\right]$$

$$NL_q = [0.959 , 12.985]$$

Thus we can say that expected number of vehicles in the queue ranges between 0 to 13 vehicles.

3) Neutrosophic Expected Waiting Time in the System

The neutrosophic expected waiting time in the system can be given by:

$$NW_s = \frac{1}{\mu_N - \lambda_N}$$

$$NL_s = \frac{1}{[30 , 36] - [22 , 28]}$$

$$NL_s = \frac{1}{[2 , 14]} = \left[\frac{1}{14}, \frac{1}{2}\right]$$

$$NL_s = [0.071 , 0.5]$$

Therefore the expected waiting time in the system ranges between 4.29 seconds to 30 seconds.

4) Neutrosophic Expected Waiting Time in the Queue

The neutrosophic expected waiting time in the queue can be given by:

$$NW_q = \frac{p_N}{\mu_N - \lambda_N}$$

$$NW_q = \frac{[0.611 , 0.933]}{[30 , 36] - [22 , 28]}$$

$$NW_q = \frac{[0.611 , 0.933]}{[2 , 14]} = \left[\frac{0.611}{14}, \frac{0.933}{2}\right]$$

$$NW_q = [0.004 , 0.467]$$

Hence the expected waiting time in the queue ranges between 2.62 seconds to 27.99 seconds.

Little's Formula

$$NL_s = [1.571\,,\,13.925]$$

$$\lambda_N \times NW_s = [22\,,28] \times \left[\frac{1}{14}, \frac{1}{2}\right] = \left[\frac{22}{14}, \frac{28}{2}\right]$$

$$\lambda_N \times NW_s = [1.571\,,\,14]$$

Therefore we can say that $\mathbf{NL_s = \lambda_N \times NW_s}$ is satified.

$$NL_q = [0.959\,,\,12.985]$$

$$\lambda_N \times NW_q = [22\,,28] \times \left[\frac{0.611}{14}, \frac{0.933}{2}\right]$$

$$\lambda_N \times NW_q = [0.960\,,\,13.062]$$

Therefore we can say that $\mathbf{NL_q = \lambda_N \times NW_q}$ is also satisfied.

5. APPLICATION OF NEUTROSOPHIC NETWORK ANALYSIS (NNA) TO TRAFFIC SYSTEM

Vehicle-to-Vehicle (V2V) communication in a traffic system can be represented with the help of networks. But, since vehicular communication in traffic systems has various uncertainties like weather conditions, signal interference, network latency, etc., that can significantly impact performance and reliability. Thus, we consider a neutrosophic network system, to get an efficient traffic management solution to handle the unpredictable transportation systems. Now let us consider the vehicular communication network system at a particular signal at an instance of time as given in Fig.1. Here each vehicle becomes each node and is named v1, v2, v3, and v4. The yellow line demonstrates that their wireless link exists between the respective nodes with certain weights.

Figure 1. Vehicular Communication Network System at a signal

This figure 1 shows four vehicles v1, v2, v3 and v4 at a traffic signal considered. The yellow links depicts whether the two vehicles are connected or not. If two vehicles are connected then those two vehcles form a network and thus they can communicate with one another.

Figure 2. Neutrosophic Network System

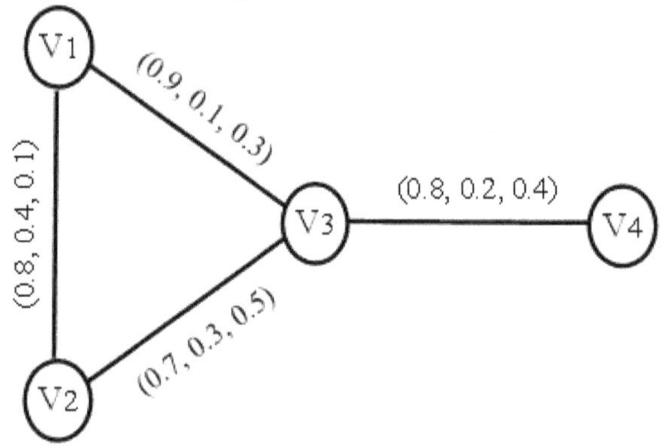

Figure 2 depicts the network system of vehicular communication at the traffic signal considered. Here each node represents each vehicle and the arc shows the link between those nodes. The arc has certain weights, represented as neutrosophic numbers. This practical situation can be depicted in the neutrosophic network and can be given as in Fig.2. Since the indeterminacy value (I) is uncertain and is not dependent upon both truth and falsity we assume it by taking 0.5 as both chances of truth and falsity which makes our Edge Score Function more efficient. Thus, the Modified Edge Score Function (MESF) is given by:

$$\text{MESF} = \frac{2 + T - (0.5 \times I) - F}{3}$$

Using the above-mentioned modified edge score function we convert the neutrosophic network with single-valued weights. Thus, we get,

Figure 3. Neutrosophic Network System with single-valued weights

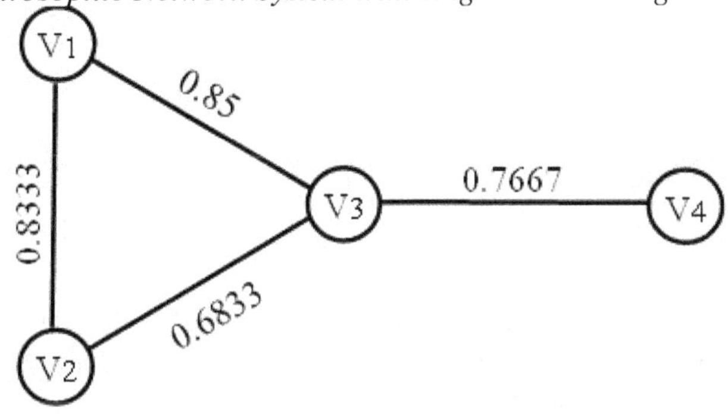

Figure 3 depicts the network system of vehicular communication at the traffic signal considered after converting the weighted arcs into crisp values with the help of the score function. Here each node represents each vehicle and the arc shows the link between those nodes. Now we use modified Kruskal's algorithm i.e. instead of considering the least weights at the initial stage we consider the highest weights and then continue with the steps of Kruskal's algorithm.

Table 3. Edge weights in descending order

Weight	Source	Destination
0.8500	V1	V3
0.8333	V1	V2
0.7667	V3	V4
0.6833	V2	V3

The table below represents the weights of each arc in decreasing order along with source nodes and destination nodes. The arc that forms a cycle have to elimated and thus is highlighted. Since the weights represent the strength of the flow of information from one node to another, we sort the edges in their descending order and thus pick up the links with no cycle formed from the sorted list. Hence the optimal network system for the vehicular communication system in the signal considered is:

Figure 4. Optimal Neutrosophic Network System

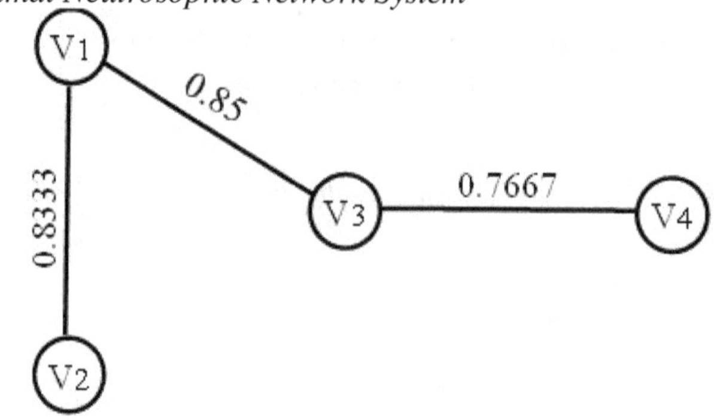

This figure 4 depicts the optimal network system of vehicular communication at the traffic signal considered after eliminating the weighted arc that forms a cyle according to Modified Kruskal's Algorithm. Here each node represents each vehicle and arc shows their link between those nodes.

The weighted value = 0.85 + 0.8333 + 0.7667

Weighted Value = 2.45

Similarly, let's consider the V2V communication system at a junction where four roads intersect as given in Fig.5. In this scenario, we have considered the junction with 14 vehicles which becomes the nodes named as v1, v2, v3, …, v13, v14 and the yellow arcs which links the nodes if and only if those vehicles are connected for data transmission. This practical situation can be represented as a neutrosophic network system as given in Fig.6.

Figure 5. Vehicular Communication System at a junction

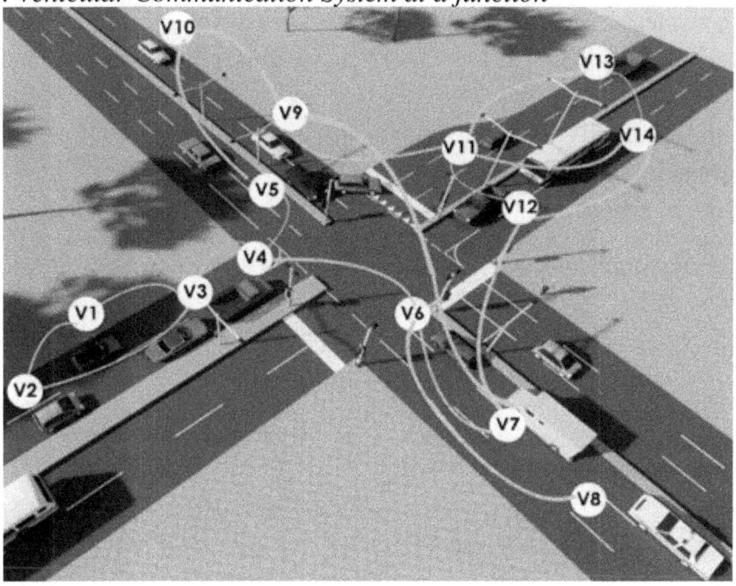

This figure 5 shows fourteen vehicles v1, v2, v3, …,v14 at a junction where four roads intersect is been considered. The yellow links depict whether the two vehicles are connected or not. If two vehicles are connected then those two vehcles form a network and thus they can communicate with one another.

Figure 6. Neutrosophic Network System

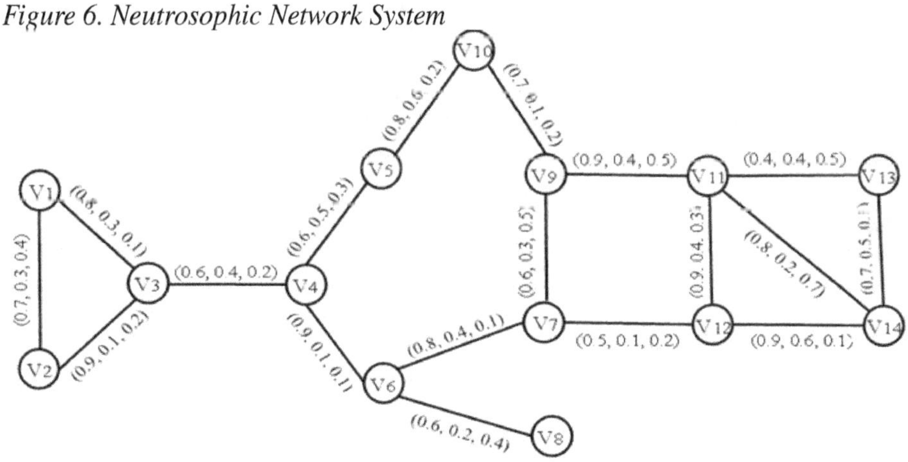

This figure 6 depicts the network system of the vehicular communication at the junction considered. Here each node represents each vehicle and arc shows their link between those nodes. The arc has certain weights, represenred as neutrosophic numbers.

Using the modified edge score function to convert the weights into single-valued weights, we get:

Figure 7. Neutrosophic Network System with single-valued weights

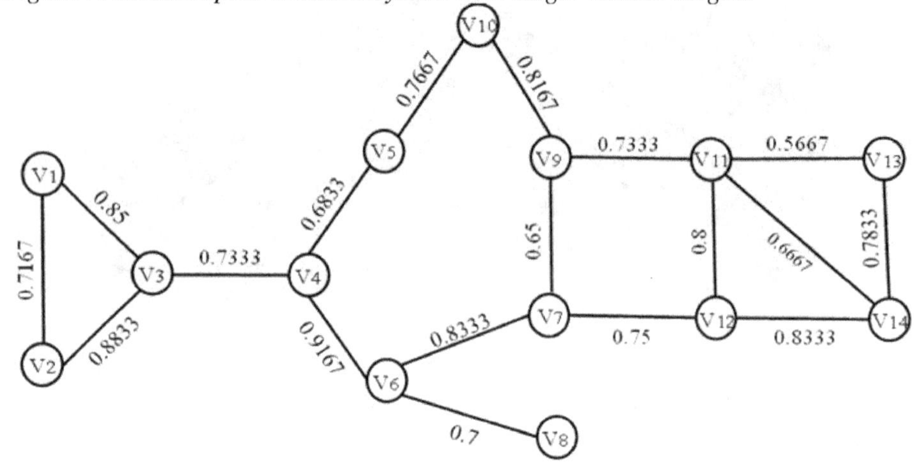

This figure 7 depicts the network system of vehicular communication at the junction considered after converting the weighted arcs into crisp values with the help of the score function. Here each node represents each vehicle and arc shows their link between those nodes.

Proceeding with the steps of modified Kruskal's algorithm we get:

Table 4. Edge weights in descending order

Weight	Source	Destination
0.9167	V4	V6
0.8833	V2	V3
0.8500	V1	V3
0.8333	V6	V7
0.8333	V12	V14
0.8167	V9	V10

continued on following page

Table 4. Continued

Weight	Source	Destination
0.8000	V11	V12
0.7833	V13	V14
0.7667	V5	V10
0.7500	V7	V12
0.7333	V9	V11
0.7333	V3	V4
0.7167	V1	V2
0.7000	V6	V8
0.6833	V4	V5
0.6667	V11	V14
0.6500	V7	V9
0.5667	V11	V13

The table below represents the weights of each arc in the decresing order along with source nodes and destination nodes. The arc which forms a cycle have to elimated and thus is highlighted. After picking up the edges from the sorted list with no cycle gives the optimal network. That is,

Figure 8. Optimal Neutrosophic Network System

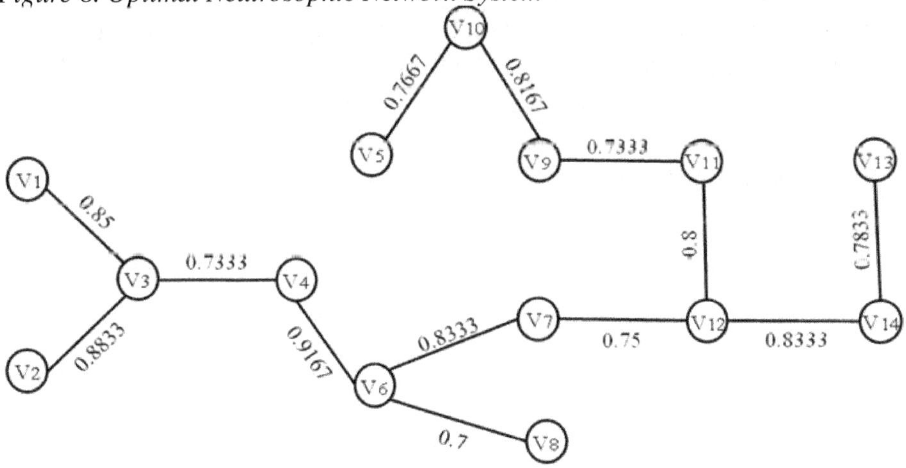

This figure 8 depicts the optimal network system of the vehicular communication at the junction considered after eliminating the weighted arc that forms a cycle according to Modified Kruskal's Algorithm. Here each node represents each vehicle

and arc shows their link between those nodes. The weighted value = 10.3999, this optimal neutrosophic network system gives more effective communication with better data communication strength.

6. CONCLUSION

By integrating neutrosophic sets into existing queueing models, we addressed the inherent uncertainties and indeterminacies present in urban traffic systems. This provides a comprehensive framework for determining critical traffic parameters such as the intensity of the traffic, average waiting time, and expected number of vehicles in the traffic system. Additionally, Little's formula is also effectively satisfied, which enhances the efficiency of the model considered and underscores the robustness of our approach. These results can lead to reduced congestion, shorter waiting times, and overall better traffic management in urban environments. In the future, this approach can also be extended to other aspects of traffic management, such as emergency vehicle prioritization.

We have also introduced the concept of neutrosophic into network analysis for handling the uncertainties and indeterminacy in the network parameters. A significant contribution of this work is the development of a modified edge score function, which effectively converts neutrosophic sets into single-valued weights. By prioritizing higher-weighted edges, we have modified Kruskal's algorithm to improve the signal strength and an efficient way to connect all nodes. Further, the modified Kruskal's algorithm optimizes the network system by identifying the most efficient connections between all the nodes and thus enhances the overall strength and reliability of the network systems. Thus, this paper offers a substantial advancement in network analysis and queuing methodologies to enhance the vehicular communication system by predicting the traffic flow and ensuring a resilient communication system.

REFERENCES

Alhabib, R., Ranna, M. M., Farah, H., & Salama, A. A. (2018). Some neutrosophic probability distributions. Infinite Study.

Anand, M. C. J., Kalaiarasi, K., Martin, N., Ranjitha, B., Priyadharshini, S. S., & Tiwari, M. (2023, September). Fuzzy C-Means Clustering with MAIRCA-MCDM Method in Classifying Feasible Logistic Suppliers of Electrical Products. In *2023 First International Conference on Cyber Physical Systems, Power Electronics and Electric Vehicles (ICPEEV)* (pp. 1-7). IEEE.

Anand, M. C. J., Priyadharshini, S. S., Martin, N., Anand, M. E., Poorani, B., & Mohiddin, S. K. (2023, September). Decision Trees in the Selection of Electric Vehicles Based on Various Parameters. In *2023 First International Conference on Cyber Physical Systems, Power Electronics and Electric Vehicles (ICPEEV)* (pp. 1-6). IEEE.

Bharatraj, J., & Clement Joe Anand, M. (2019). Power harmonic weighted aggregation operator on single-valued trapezoidal neutrosophic numbers and interval-valued neutrosophic sets. In *Fuzzy Multi-criteria Decision-Making Using Neutrosophic Sets* (pp. 45–62). Springer International Publishing. DOI: 10.1007/978-3-030-00045-5_3

Bohlender, G., & Kulisch, U. (2010). *Definition of the Arithmetic Operations and Comparison Relations for an Interval Arithmetic Standard*. Institut fur Angewandte und Numerische Mathematik, Universitat Karlsurhe.

Chen, S. P., & Eur, J. (2009). Parametric Nonlinear Programming for Analyzing Fuzzy Queues with Infinite Capacity, 157.

Clement Joe Anand, M., & Bharatraj, J. (2017). Theory of Triangular Fuzzy Number. *Proceedings of NCATM, 2017*, 80–83.

Clement Joe Anand, M., & Bharatraj, J. (2018). Gaussian qualitative trigonometric functions in a fuzzy circle. *Advances in Fuzzy Systems, 2018*, 1–9. DOI: 10.1155/2018/8623465

Clement Joe Anand, M., & Bharatraj, J. (2019). Interval-valued neutrosophic numbers with WASPAS. In *Fuzzy Multi-criteria Decision-Making Using Neutrosophic Sets* (pp. 435–453). Springer International Publishing. DOI: 10.1007/978-3-030-00045-5_17

Clement Joe Anand, M., & Bharatraj, J. (2019). Interval-valued neutrosophic numbers with WASPAS. In *Fuzzy Multi-criteria Decision-Making Using Neutrosophic Sets* (pp. 435–453). Springer International Publishing. DOI: 10.1007/978-3-030-00045-5_17

Clement Joe Anand, M., Martin, N., Clementking, A., Rani, S., Priyadharshini, S. S., & Siva, S. "Decision making on optimal selection of advertising agencies using machine learning," In 2023 International Conference on Information Management (ICIM), IEEE, 2023. https://ieeexplore.ieee.org/document/10145172

Clement Joe Anand, M., Moorthy, C. B., Sivamani, S., Indrakumar, S., Kalaiarasi, K., & Barhoi, A. (2023). Fuzzy intelligence inventory decision optimization model of sustainability and green technologies for mixed uncertainties of carbon emission. In *In: 2023 International Conference on Information Management (ICIM)*. IEEE., https://ieeexplore.ieee.org/document/10145085

Devadoss, A. V., Anand, M. C. J., & Felix, A. (2014, March). A CETD matrix approach to analyze the dimensions of personality of person. In *2014 International Conference on Computational Science and Computational Intelligence* (Vol. 2, pp. 40-45). IEEE.

Erlang, A. K. (1917). Solution of some problems in the theory of probabilities of significance in automatic telephone exchanges. *Post Office Electrical Engineer's Journal, 10*, 189–197.

Hevey, D. (2018). Network analysis: A brief overview and tutorial. *Health Psychology and Behavioral Medicine, 6*(1), 301–328.

Hui, Y., & Dey, A. (2020). A study on directed neutrosophic social networks. Infinite Study.

Hussein, N. H., Yaw, C. T., Koh, S. P., Tiong, S. K., & Chong, K. H. (2022). A Comprehensive Survey on Vehicular Networking [IEEE Access.]. *Communications, Applications, Challenges, and Upcoming Research Directions, 10*, 86127–86180.

Jojo, L. (2014). Predicting Traffic Congestion: A Queuing Perspective. [Open Journal of Modelling and Simulation.]. *Open Journal of Modelling and Simulation, 2*(2), 57–66. DOI: 10.4236/ojmsi.2014.22008

Kaviyarasu, M., Aslam, M., Afzal, F., Mohammed Saeed, M., Mehmood, A., & Gul, S. (2024). The connectivity indices concept of neutrosophic graph and their application of computer network, highway system and transport network flow. *Scientific Reports, 14*(1), 4891.

Kungumaraj, E., Lathanayagam, E., Saikia, U., Anand, M., Khanna, S. T., Martin, N., & Edalatpanah, S. A. (2024). Neutrosophic Topological Vector Spaces and its Properties. [IJNS]. *International Journal of Neutrosophic Science, 23*(2).

Kungumaraj, E., Lathanayagam, E., Saikia, U., Clement Joe Anand, M., Khanna, S. T., Martin, N., Tiwari, M., & Edalatpanah, S. A. (2024). Neutrosophic Topological Vector Spaces and its Properties. *International Journal of Neutrosophic Science, 23*(2), 63–76. DOI: 10.54216/IJNS.230206

Manshath, A., Kungumaraj, E., Lathanayagam, E., Anand, M. C., Martin, N., Muniyandy, E., & Indrakumar, S. (2024). Neutrosophic Integrals by Reduction Formula and Partial Fraction Methods for Indefinite Integrals. [IJNS]. *International Journal of Neutrosophic Science, 23*(1).

Meenakshi, A., & Mythreyi, O. (2023). RETRACTED: Applications of Neutrosophic social network using max product networks. *Journal of Intelligent & Fuzzy Systems, 45*(1), 407–420.

Meenakshi, A., Mythreyi, O., Bramila, M., Kannan, A., & Senbagamalar, J. (2023). Application of neutrosophic optimal network using operations. *Journal of Intelligent & Fuzzy Systems, 45*(1), 421–433.

Miriam, M. R., Martin, N., & Clement Joe Anand, M. (2023). Inventory model promoting smart production system with zero defects. *International Journal of Applied and Computational Mathematics, 9*(4), 47. DOI: 10.1007/s40819-023-01524-4

Modi, M., Agarwal, G., Patil, V., Khare, A., Shukla, S., & Sankhala, A. (2019). Minimization of traffic congestion by using queuing theory. *Int. J. Sci. Technol. Res, 8*(10), 532–541.

Nagaraj, C. T., Anand, M. C. J., Priyadharshini, S. S., & Aparna, P. (2023, December). GCNXG: Detecting Fraudulent Activities in Financial Networks: A Graph Analytics and Machine Learning Fusion. In International Conference on Renewable Energy, Green Computing, and Sustainable Development (pp. 17-32). Cham: Springer Nature Switzerland.

Nasir, M. K., & Al Mamun, S. Application of Graph theory and Error Correcting Code in VANETs.

Patro, S. K., & Smarandache, F. (2016). The neutrosophic statistical distribution, more problems, more solutions. Infinite Study.

Prabha, S. K., Anand, M., Vidhya, V., Nagarajan, G., Saikia, U., Martin, N., & Tiwari, M. (2024). Sorting Out Interval Valued Neutrosophic Fuzzy Shortest Cycle Route Problem by Reduced Matrix Method. [IJNS]. *International Journal of Neutrosophic Science, 23*(2).

Prabha, S. K., Clement Joe Anand, M., Vidhya, V., Nagarajan, G., Saikia, U., Martin, N., Kumari, M. S., & Tiwari, M. (2024). Sorting Out Interval Valued Neutrosophic Fuzzy Shortest Cycle Route Problem by Reduced Matrix Method. *International Journal of Neutrosophic Science*, *23*(2), 91–103. DOI: 10.54216/IJNS.230208

Raj, J., Prabhu, V. V., Krishnakumar, V., & Anand, M. C. J. (2023). Solar Powered Charging -of Fuzzy Logic Controller (FLC) Strategy with Battery Management System (BMS) Method Used for Electric Vehicle (EV). *International Journal of Fuzzy Systems*, *25*(7), 2876–2888. DOI: 10.1007/s40815-023-01537-7

Rajesh, K., Rathod, S., Kundale, J., Rathod, N., Anand, M., Saikia, U., & Martin, N. (2024). A Study on Interval Valued Temporal Neutrosophic Fuzzy Sets. [IJNS]. *International Journal of Neutrosophic Science*, *23*(1).

Rajesh, K., Rathod, S., Kundale, J., Rathod, N., Clement Joe Anand, M., Saikia, U., Tiwari, M., & Martin, N. (2024). A Study on Interval Valued Temporal Neutrosophic Fuzzy Sets. *International Journal of Neutrosophic Science*, *23*(1), 341–349. DOI: 10.54216/IJNS.230129

Rashad, H., & Mohamed, M. (2021). Neutrosophic theory and its application in various queueing models: Case studies. *Neutrosophic Sets and Systems*, *42*(1), 8.

Smarandache, F. (1999). *A Unifying Field in Logics: Neutrosophic Logic. Neutrosophy, Neutrosophic Set, Neutrosophic Probability*. American Research Press.

Smarandache, F. (2002). Neutrosophy and Neutrosophic Logic. First International Conference on Neutrosophy, Neutrosophic Logic, Set, Probability, and Statistics, University of New Mexico, Gallup, NM 87301, USA.

Smarandache, F. (2005). Neutrosophic set is a generalization of the intuitionistic fuzzy sets, 24, 287-297. Inter. J. Pure Appl. Math.

Sudha, S., Nivetha, M., Clement Joe Anand, M., Palanimani, P. G., Thirunamakkani, T., & Ranjitha, B. (2023). MACBETH-MAIRCA Plithogenic Decision-Making on Feasible Strategies of Extended Producer's Responsibility towards Environmental Sustainability. *International Journal of Neutrosophic Science*, *22*(2), 114–130. DOI: 10.54216/IJNS.220210

Tamura, H., Nakano, K., Sengoku, M., & Shinoda, S. (2011). On applications of graph/network theory to problems in communication systems. ECTI Transactions on Computer and Information Technology (ECTI-CIT), 5(1), 15-21.

Varalakshmi, A., Kumar, S. S., Shanmugapriya, M. M., Mohanapriya, G., & Clement Joe Anand, M. (2022). Markers location monitoring on images from an infrared camera using optimal fuzzy inference system. *International Journal of Fuzzy Systems*.

Zeina, M. B. (2020). Erlang service queueing model with neutrosophic parameters. *International Journal of Neutrosophic Science*, *6*(2), 106–112.

Zeina, M. B. (2020). Neutrosophic event-based queueing model. *International Journal of Neutrosophic Science*, *6*(1), 48–55.

Zhang, S., Chen, J., Lyu, F., Cheng, N., Shi, W., & Shen, X. (2018). Vehicular communication networks in the automated driving era. *IEEE Communications Magazine*, *56*(9), 26–32.

Chapter 2

Explainable Artificial Intelligence and Fuzzy MCDM Approach in Location Selection of Roadside Units in Vehicular Networks With Economic Efficiency

A. P. Pushpalatha
https://orcid.org/0000-0002-0799-7961
Velammal College of Engineering and Technology, India

Sakshi Taaresh Khanna
https://orcid.org/0009-0006-2936-8708
Ram Lal Anand College, New Delhi, India

Utpal Saikia
https://orcid.org/0000-0003-0235-6871
Silapathar College, India

Sulbha Raorane
St. Francis Institute of Management and Research, India

Y. Monikarchana
Mohan Babu University, India

Chitra Ramaprakash
Dayananda Sagar College of Engineering, India

ABSTRACT

Decision-making on the selection-based problem of the location of Roadside units (RSU) in vehicular networks is complicated. This intricate problem shall be resolved with the intervention of Explainable Artificial Intelligence (XAI) and fuzzy-based

DOI: 10.4018/979-8-3693-6422-2.ch002

multi-criteria decision-making methods (MCDM). This research proposes the integrated decision-making approach in making optimal location selection considering different criteria and economic aspects. The decision-making problem discussed in this chapter considers location selection of roadside units as the choice-making of the location is very significant in enhancing vehicular efficiency. This decision problem considers the criteria of coverage area, traffic density, accident frequency, infrastructure availability, environmental impact, energy efficiency, and cost components such as cost effectiveness, revenue competence, and investment returns. The integrated approach is leveraged to make an optimal choice of the RSU locations.

1. INTRODUCTION

The field of smart transportation systems is gaining intense advancements with the integration of robust technologies to foster and regulate traffic congestion and ensure safety and efficacy on the roadways. The vehicular efficiency is influenced by the location of roadside units (RSU). These components play a critical role in bridging the vehicular elements in a communication network. The vehicular network performance is also persuaded by the RSU locations and at the same time it facilitates economic efficiency. The decision-making on the location selection is a complicated kind of problem and it demands integrated and competent decision methods. In general, a decision-making process is more challenging as it is characterized by multiple alternatives and criteria. In addition, the existence of uncertainty also intensifies the complexity of the problem. To resolve such intricated problems fuzzy-based multi-criteria decision-making methods are the optimal choices. On the other hand, machine learning algorithms are applied in making optimal decisions, especially regarding vehicular efficiency. Machine learning and Deep learning are algorithm-based approaches and they are the subsets of artificial intelligence. These are characterized as the traditional approaches. However, these days these conventional trends are replaced by the techniques of Explainable Artificial Intelligence (XAI).

XAI refers to the advanced methods and techniques in artificial intelligence that are potent in enabling the decision-making process more comprehensive. These contemporary tools provide more insights into the modalities of making decisions and they aim in dealing with the 'black box' of the problem. These techniques of XAI are preferred as they possess the attributes of transparency, interpretability and reliability. There are several algorithms of XAI such as LIME (Local Interpretable Model-agnostic Explanations), SHAP (Shapley Additive explanations), Anchors, Saliency Maps, Partial Dependence Plots (PDP), Individual Conditional Expectation (ICE) Plots, Counterfactual Explanations, Gradient-weighted Class Activation Mapping (Grad-CAM), Explainable Boosting Machines (EBM), Rule-based Explanations. These algorithms are extensively applied in several decision-making circumstances based on the requirements.

This research work proposes an integrated approach of XAI with fuzzy based MCDM methods. The leveraging of these combined methods will certainly facilitate in making optimal decisions. The hybrid method obtained by combining these two methods shall be employed for different purposes. However, in general a decision-making problem comprises several criteria and this highly constraints the decision-makers. The selection of the optimal alternative has to fulfill the criteria to a maximum extent. Hence it is very evident that criteria selection is a crucial one. To resolve this chaos, the XAI algorithms shall be used in criteria reduction and as a result of it, the criteria number is optimized. By doing so the optimal ranking of the alternatives suiting the chosen criteria shall be accomplished by fuzzy-based multi-criteria decision-making methods. This is one of the advantages of hybrid decision models.

In the context of location selection of the RSU, this hybrid method shall be employed to determine the efficiency of the proposed combined method. As the criteria associated with location selection are categorized based on several aspects, cost efficiency is considered with special focus. The choosing of the criteria for location selection shall be made with the intervention of XAI algorithm and the ranking of the location sites shall be made with the application of fuzzy-based MCDM. The employing of this method is very instrumental in making optimal decisions. As there are several fuzzy decision methods, competent and efficient methods are to be chosen to rank the alternatives. This research work tests the versatility of the integrated approach in making optimal decisions concerning the choice-making of the RSU site locations. Also, the hybrid method discussed in this work is an intersection of XAI, MCDM, and smart transport systems. The remaining contents of the work are presented in the following sections. Section 2 comprises a detailed review of the literature of the related works. Section 3 consists of the procedure of the integrated method. Section 4 applies the proposed method in ranking the location sites of the RSU. Section 5 discusses the results and the last section concludes the work.

2. LITERATURE REVIEW

This section outlines the research works related to the methods applied in the placement of roadside units. This decision problem is highly significant in enhancing vehicular efficiency and the state of the art of the research is sketched out. The identification of the research gaps and the novel contributions made in this chapter are also well articulated at the end of this section.

(Aslam & Zou, 2011) applied the Balloon optimization techniques in making decisions on the placement framework of roadside units. (Aslam, Amjad, & Zou, 2012) used Binary Integer Programming and Balloon Expansion Heuristic methods. (Liang, Liu, & Rajan, 2012), (Liya, Chuanhe, Peng, & Junyu, 2013) and (Li, Liu, Huang, Wang, & Jia, 2015) applied the method of simulation in designing optimal solutions to the placement problem. The aspects of coverage, connectivity, and costs are considered in simulation. (Brahim, Drira, & Filali, 2014) applied the algorithms of Knapsack and PageRank. (Liu, Ding, Yang, & Yang, 2014) employed Expansion and Coloration Algorithm. (Mehar, Senouci, Kies, & Zoulikha, 2015) utilized the method of ODEL which comprises an algorithmic framework of Genetic and Dijkstra algorithm. (Massobrio, Toutouh, & Nesmachnow, 2020) used multi-objective and evolutionary algorithms. (Li, Huang, & Liu, 2014), (Jiang et al., 2018) (Jiang et al., 2019) demonstrated the Greedy algorithm to make constructive solutions. Researchers have applied algorithmic approaches in devising optimal solutions to the decision problems of RSU placement.

On the other hand, the methods of multi-criteria decision-making are also applied. (Patra, Mishra, & Murthy, 2014), applied the method of Analytical Hierarchy Process (AHP). (Malik, 2021) (Malik et al., 2022) discussed the applications of AHP-Entropy and VIKOR methods in making decisions. (Ma, Zheng, Wang, Wong, & Easa, 2024) discussed the implications of the method of probabilistic occupancy maps based cross entropy. (Jain, Jain, & Mishra, 2024) extended AHP to fuzzy AHP. In addition to it, the machine learning approach is also applied in making decisions. (Ackels, 2023) applied the techniques of Deep learning in resolving the RSU placement problem. The above-mentioned research works are tabulated in Table 1.

From the aforementioned literature, the following shortcomings are identified. The applications of fuzzy MCDM in RSU placement problems are very limited and the integrated methods of machine learning and MCDM do not exist in the literature. This has motivated the authors to develop an integrated hybrid decision model by combining the methods of XAI and Fuzzy MCDM. Explainable Artificial Intelligence (XAI) is a contemporary set of robust techniques applied in several fields and its implication is evolving in the domain of VANET at recent times. Henceforth this research work proposes a novel integrated method of XAI-Fuzzy MCDM in computing optimal solutions to RSU placement problems. The proposed method works in two phases. In the first phase, the criteria reduction is made with XAI techniques, in this case, the LIME tool is used. Then in the second phase, the methods of Fuzzy FUCOM-SMART are applied in computing criterion weights and ranking of the alternatives.

Table 1. State of Art of Research Works

Authors & Year	Methods Applied
Aslam & Zou., (2011)	Balloon optimization method
Liang et al., (2012)	Simulation
Liya et al., (2013)	Simulation
Brahim et al., (2014)	Knapsack and PageRank algorithms
Liu et al., (2014)	Expansion and Coloration Algorithm
Li et al., (2014)	Simulation
Patra et al., (2014)	AHP
Mehar et al., (2015)	ODEL, Genetic algorithm and Dijkstra algorithm
Li et al., (2015)	Greedy algorithm
Jiang et al., (2018)	Greedy algorithm
Jiang et al., (2019)	Greedy algorithm
Massobrio et al., (2020)	Multi-objective Evolutionary algorithms
Aslam et al., (2021)	Binary Integer Programming and Balloon Expansion Heuristic methods
Malik et al., (2021)	Analytical Hierarchy Process (AHP)-entropy and group-VIKOR
Malik et al., (2022)	AHP—Entropy & BORDA—VIKOR
Mohaghar et al., (2023)	Greedy algorithm
Ackels., (2023)	Deep Learning
Ma et al., (2024)	probabilistic occupancy maps based cross entropy
Jain et al., (2024)	Fuzzy AHP
Proposed Method	LIME & Fuzzy FUCOM-SMART

3. METHODOLOGY OF HYBRID METHOD

This section presents the procedure of the hybrid method of XAI & Fuzzy FUCOM-SMART. This proposed method works in two phases. The first phase plays a pivotal role in criteria reduction and the second phase performs the functions of computing of weights and ranking of alternatives. This section outlines the general procedures of XAI, Fuzzy FUCOM and Fuzzy SMART methods.

3.1 General Framework of LIME (Lundberg et al., 2022)(Arrieta et al., 2020) (Das & Rad, 2020)

LIME (Local Interpretable Model-agnostic Explanations) is a robust technique of XAI which is employed to make optimal interpretations. This tool shall be effectively employed in determining the most significant features of the model. The computations of coefficients facilitate in choosing the critical features and in dimensionality reduction. The algorithmic framework of the LIME technique of LIME is presented in Figure 1.

Figure 1. LIME Framework

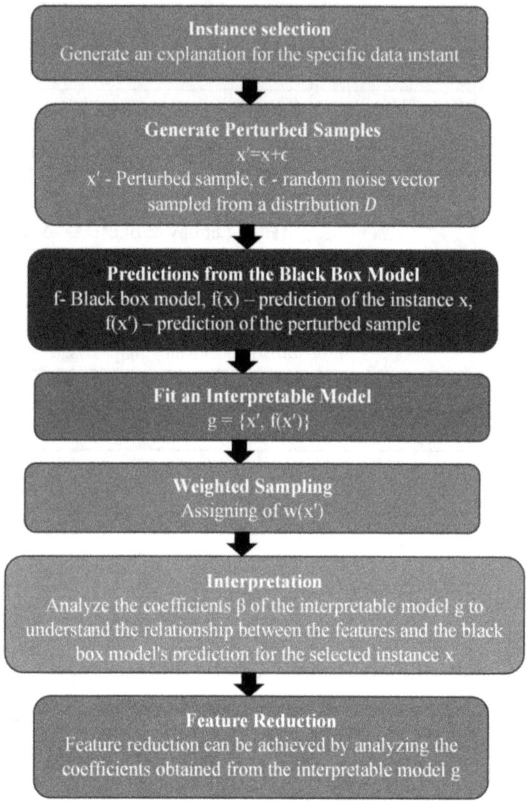

3.2 Fuzzy FUCOM (Pamucar & Ecer, 2020)

The method of FUCOM (Full Consistency Method) is employed to determine the weights of the criteria and the notion of fuzzy is applied to handle imprecise data.

Step 1: Defining of criteria and decision makers

The criteria set C and the decision makers set D is defined. Here C = {C1, C2, …, Cn} and D = {D1, D2, …, Dm}.

Step 2: Pairwise comparison with Fuzzy numbers

The decision-makers compare each criterion and construct a fuzzy matrix with triangular fuzzy numbers of the form = (l,m,u). The comparison matrix is constructed for each of the decision-makers D_k

Step 3: Determining the Combined Pairwise Comparisons

The fuzzy pairwise comparisons subjected to each of the decision makers is determined using the Geometric method.

Step 4: Computation of Fuzzy weights

The Eigenvector method is used to compute the fuzzy weights
The obtained fuzzy weights are defuzzified using centroid method

Step 5: Consistency check

The consistency ratio is calculated to determine the consistencies of the pairwise comparison matrices.

3.3 Fuzzy SMART

(Patel et al., 2017) Simple Multi-Attribute Rating Technique (SMART) is the method employed in ranking the alternatives. The integration of fuzzy logic with this method facilitates in handling imprecise and uncertain data.

Step 1: Defining of criteria and alternatives

The criteria set C and the decision makers set D is defined. Here C = { C1,C2,.. Cn} and A = {D1,D2,..Dm}.

Step 2: Conjoining Fuzzy Weights to Criteria

The fuzzy weights $= (l_i,m_i,u_i)$ is assigned to each of the criterion c_i

Step 3: Rating of the Alternatives using Fuzzy numbers

The alternatives (A_j) are ranked using a fuzzy rating with respect to each of the criterion C_i

Step 4: Computation of Weighted Fuzzy Rating

The weighted fuzzy rating is calculated using and hence as a result

Step 5: Determination of Aggregated weighted fuzzy ratings

The aggregated weighting scores S_j for each alternative A_j are calculated as follows where and the aggregate score

,

Step 6: Ranking of the alternatives

The final score values are defuzzified using the centroid method and the alternatives are ranked based on the scores obtained.

4. APPLICATION OF THE PROPOSED HYBRID METHOD IN RANKING LOCATION SITES OF RSU

This section applies the proposed hybrid method in ranking the location sites of Roadside units. The criteria of selection is based on the following factors as presented in Figure 2.

Figure 2. Criteria Classification

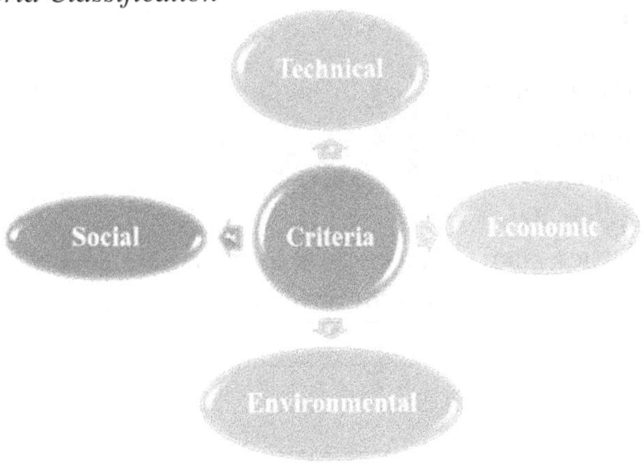

The sub-factors subjected to each of the core factors are as follows as in Figure 3

Figure 3. Description of the Criteria

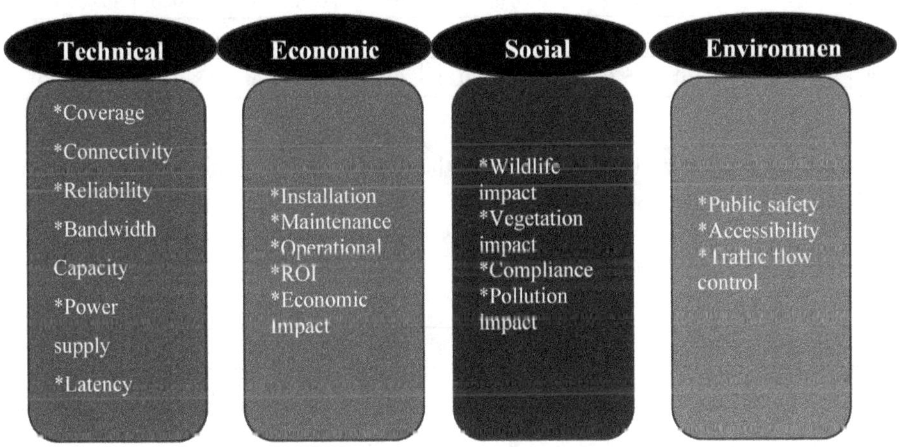

By using the package LIME in the R environment, the following criteria are determined to be more significant. The performance metric values are presented in Table 2.

Table 2. Performance Metric Measures

Accuracy	Precision	Recall	F1 Score	Specificity
0.85	0.836	0.915	0.8915	0.828

The significant criteria are presented as follows

C1 - Location (L)

C2 – Signal Coverage (SC)

C3 - Installation Costs (IC)

C4 – Maintenance Costs (MC)

C5 – Compliance (C)

C6 – Traffic Flow (TF)

The initial decision matrix comprises ten location sites namely L1,L2,..L10 and six significant criteria. The matrix with triangular fuzzy values is presented as follows

	L	SC	IC	MC	C	TF
L1	(0.7, 0.8, 0.9)	(0.4, 0.5, 0.6)	(0.5, 0.6, 0.7)	(0.6, 0.7, 0.8)	(0.8, 0.9, 1.0)	(0.6, 0.7, 0.8)
L2	(0.6, 0.7, 0.8)	(0.5, 0.6, 0.7)	(0.4, 0.5, 0.6)	(0.7, 0.8, 0.9)	(0.7, 0.8, 0.9)	(0.5, 0.6, 0.7)
L3	(0.8, 0.9, 1.0)	(0.6, 0.7, 0.8)	(0.6, 0.7, 0.8)	(0.5, 0.6, 0.7)	(0.9, 1.0, 1.0)	(0.7, 0.8, 0.9)
L4	(0.5, 0.6, 0.7)	(0.3, 0.4, 0.5)	(0.6, 0.7, 0.8)	(0.6, 0.7, 0.8)	(0.6, 0.7, 0.8)	(0.5, 0.6, 0.7)
L5	(0.7, 0.8, 0.9)	(0.5, 0.6, 0.7)	(0.7, 0.8, 0.9)	(0.7, 0.8, 0.9)	(0.8, 0.9, 1.0)	(0.6, 0.7, 0.8)
L6	(0.6, 0.7, 0.8)	(0.4, 0.5, 0.6)	(0.5, 0.6, 0.7)	(0.6, 0.7, 0.8)	(0.7, 0.8, 0.9)	(0.7, 0.8, 0.9)
L7	(0.8, 0.9, 1.0)	(0.6, 0.7, 0.8)	(0.6, 0.7, 0.8)	(0.5, 0.6, 0.7)	(0.8, 0.9, 1.0)	(0.8, 0.9, 1.0)
L8	(0.5, 0.6, 0.7)	(0.3, 0.4, 0.5)	(0.4, 0.5, 0.6)	(0.7, 0.8, 0.9)	(0.6, 0.7, 0.8)	(0.6, 0.7, 0.8)
L9	(0.7, 0.8, 0.9)	(0.4, 0.5, 0.6)	(0.7, 0.8, 0.9)	(0.6, 0.7, 0.8)	(0.8, 0.9, 1.0)	(0.7, 0.8, 0.9)
L10	(0.6, 0.7, 0.8)	(0.5, 0.6, 0.7)	(0.6, 0.7, 0.8)	(0.7, 0.8, 0.9)	(0.7, 0.8, 0.9)	(0.5, 0.6, 0.7)

By applying Fuzzy FUCOM method, the weights of the criteria are determined and presented in the Table 3.

Table 3. Criterion Weights

L	SC	IC	MC	C	TF
0.18	0.16	0.15	0.17	0.18	0.16

By using the criterion weights the alternatives are ranked based on the score values presented in Table 4.

Table 4. Final Defuzzified score values of the Alternatives

L1	L2	L3	L4	L5	L6	L7	L8	L9	L10
0.74	0.68	0.76	0.61	0.78	0.71	0.80	0.65	0.75	0.70

The above values are pictorially depicted in Fig. 4

Figure 4. Score values of the Alternatives

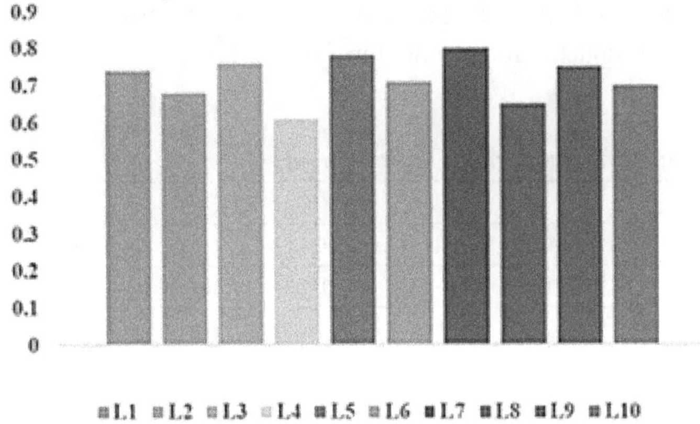

5. RESULTS AND DISCUSSIONS

From the score values presented in Table 4, the alternatives are ranked as follows: L7 > L5 > L3 > L9 >L1 >L6 >L10 >L2 >L8 >L4. The pictorial representation of the ranking I presented in Fig.5.

Figure 5. Ranking of the Alternatives

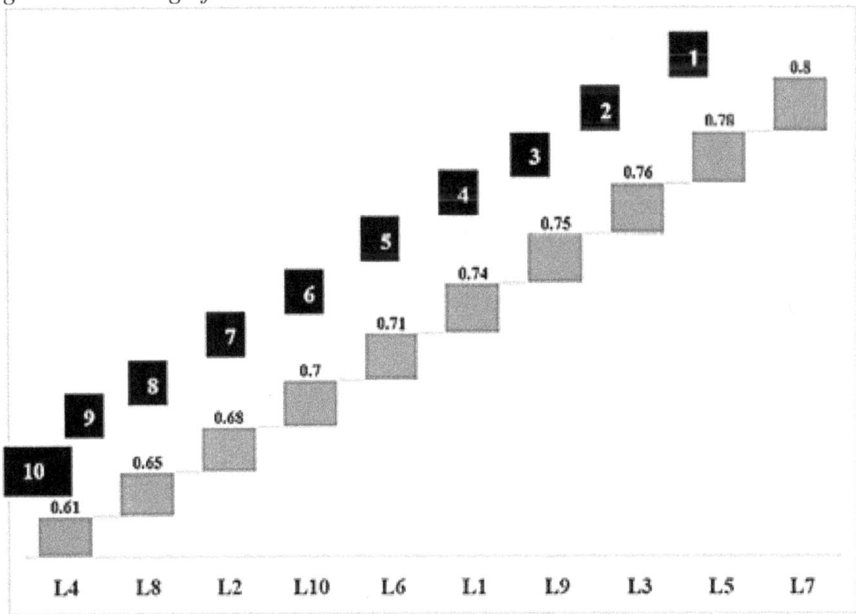

The consistency of the ranking results is determined with other weights computed using the standard methods of Fuzzy AHP and Fuzzy Entropy.

The weights computed are presented in Table 5

Table 5. Criterion Weights Using Other Methods

Methods	L	SC	IC	MC	C	TF
Fuzzy AHP	0.20	0.15	0.20	0.15	0.15	0.15
Fuzzy Entropy	0.102	0.1	0.107	0.096	0.139	0.457

Based on the criterion weights in Table 5, the score values are computed and presented in Table 6

Table 6. Score values with other combinations

Methods	L1	L2	L3	L4	L5	L6	L7	L8	L9	L10
Fuzzy AHP& SMART	0.64	0.65	0.78	0.59	0.77	0.70	0.78	0.58	0.74	0.71
Fuzzy Entropy & SMART	0.702	0.811	0.822	0.548	0.880	0.763	0.853	0.650	0.710	0.716
Equal Weights & Fuzzy SMART	0.634	0.699	0.867	0.7	0.882	0.767	0.833	0.6	0.850	0.870

The graphical representation of the score values is presented in Fig.6

Figure 6. Score values of Different Combinations of Methods

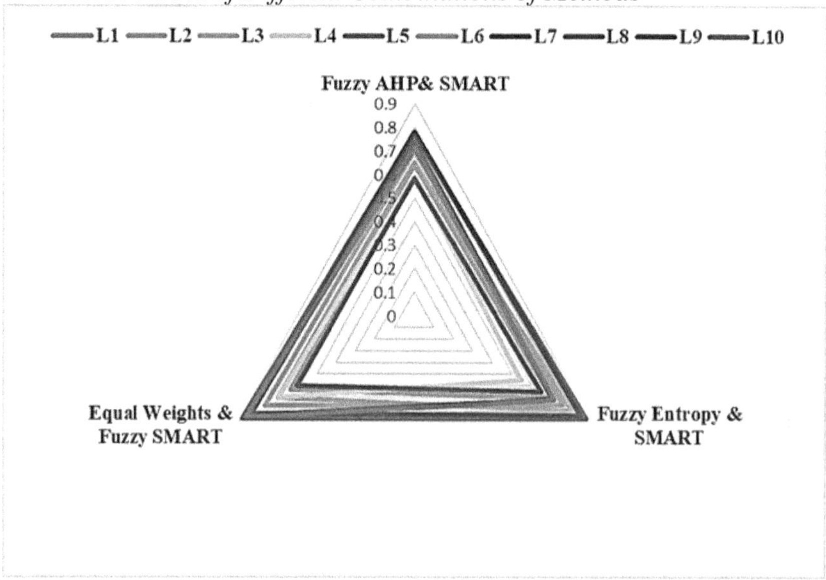

The rankings of the alternatives are presented in Table 7.

Table 7. Alternative Rankings

Methods	L1	L2	L3	L4	L5	L6	L7	L8	L9	L10
Fuzzy AHP& SMART	8	7	1	9	3	6	1	10	4	5
Fuzzy Entropy & SMART	8	4	3	10	1	5	2	9	7	6
Equal Weights & Fuzzy SMART	9	8	3	7	2	6	5	10	4	1

The consistency of the ranking results is tested with the values of correlation as presented in Table 8

Table 8. Correlation Coefficients

	Fuzzy FUCOM & SMART	Fuzzy AHP & SMART	Fuzzy Entropy & SMART	Equal Weights & SMART
Fuzzy FUCOM & SMART	1	0.881	0.770	0.527
Fuzzy AHP & SMART		1	0.827	0.772
Fuzzy Entropy & SMART			1	0.564
Equal Weights & SMART				1

From the correlation coefficients, it is highly evident that Fuzzy FUCOM & SMART and Fuzzy AHP & SMART are highly associated. Also, Fuzzy Entropy & SMART exhibits moderate similarity with Fuzzy FUCOM & SMART and Fuzzy AHP & SMART. The proposed hybrid method yields better results and its consistency is tested with other combinations.

6. CONCLUSION

This chapter presents a hybrid method of making optimal decisions on the location sites of the roadside units. The intervention of the XAI algorithm especially LIME is crucial in criterion reduction. The alternatives are ranked using the combination of fuzzy MCDM methods. This research work presents a novel approach to integrating XAI and Fuzzy MCDM. The efficacy of the method is tested with the application of ranking the location sites of the roadside units. The proposed model is more competent and its potency is validated with other MCDM methods. This method shall be applied to other domains of decision-making.

REFERENCES

Ackels, S. (2023). Deep Learning Techniques for Optimization of Roadside Unit Placement (Doctoral dissertation, The University of Texas at San Antonio).

Anand, M. C. J., & Bharatraj, J. (2017). Theory of Triangular Fuzzy Number. *Proceedings of NCATM, 2017*, 80–83.

Anand, M. C. J., & Bharatraj, J. (2018). Gaussian qualitative trigonometric functions in a fuzzy circle. *Advances in Fuzzy Systems, 2018*, 1–9. DOI: 10.1155/2018/8623465

Anand, M. C. J., & Bharatraj, J. (2019). Interval-valued neutrosophic numbers with WASPAS. In *Fuzzy Multi-criteria Decision-Making Using Neutrosophic Sets* (pp. 435–453). Springer International Publishing., https://link.springer.com/chapter/10.1007/978-3-030-00045-5_17 DOI: 10.1007/978-3-030-00045-5_17

Anand, M. C. J., Martin, N., Clementking, A., Rani, S., Priyadharshini, S., & Siva, S. (2023, March). Decision making on optimal selection of advertising agencies using machine learning. In *2023 International Conference on Information Management (ICIM)* (pp. 121-125). IEEE.

Anand, M. C. J., Moorthy, C. B., Sivamani, S., Indrakumar, S., Kalaiarasi, K., & Barhoi, A. (2023). *Fuzzy intelligence inventory decision optimization model of sustainability and green technologies for mixed uncertainties of carbon emission. In: 2023 International Conference on Information Management (ICIM)*. IEEE., https://ieeexplore.ieee.org/document/10145085 DOI: 10.1109/ICIM58774.2023.00020

Arrieta, A. B., Díaz-Rodríguez, N., Del Ser, J., Bennetot, A., Tabik, S., Barbado, A., & Herrera, F. (2020). Explainable Artificial Intelligence (XAI): Concepts, taxonomies, opportunities and challenges toward responsible AI. *Information Fusion, 58*, 82–115. DOI: 10.1016/j.inffus.2019.12.012

Aslam, B., Amjad, F., & Zou, C. C. (2012, July). Optimal roadside units placement in urban areas for vehicular networks. In 2012 IEEE symposium on computers and communications (ISCC) (pp. 000423-000429). IEEE.

Aslam, B., & Zou, C. C. (2011, January). Optimal roadside units placement along highways. In *2011 IEEE Consumer Communications and Networking Conference (CCNC)* (pp. 814-815). IEEE.

Bharatraj, J., & Anand, M. C. J. (2019). Power harmonic weighted aggregation operator on single-valued trapezoidal neutrosophic numbers and interval-valued neutrosophic sets. In *Fuzzy Multi-criteria Decision-Making Using Neutrosophic Sets* (pp. 45–62). Springer International Publishing., https://link.springer.com/chapter/10.1007/978-3-030-00045-5_3 DOI: 10.1007/978-3-030-00045-5_3

Brahim, M. B., Drira, W., & Filali, F. (2014, November). Roadside units placement within city-scaled area in vehicular ad-hoc networks. In *2014 International Conference on Connected Vehicles and Expo (ICCVE)* (pp. 1010-1016). IEEE.

Das, A., & Rad, P. (2020). Opportunities and challenges in explainable artificial intelligence (xai): A survey. arXiv preprint arXiv:2006.11371.

Devadoss, A. V., & Anand, M. C. J. (2014). *Felix., A.: A CETD matrix approach to analyze the dimensions of the personality of a person. In: 2014 International Conference on Computational Science and Computational Intelligence*. IEEE., https://ieeexplore.ieee.org/abstract/document/6822301

Jain, S., Jain, V. K., & Mishra, S. (2024). Fuzzy-AHP-based optimal RSU deployment (Fuzzy-AHP-ORD) approach using road and traffic analysis in VANET. *Ad Hoc Networks*, •••, 103529.

Jiang, P., Li, P., Zhang, T., Huang, W., He, H., Nie, L., & Liu, Q. (2019, August). Budget constraint roadside units placement for traffic flows monitoring system with reliability in vehicular networks. In 2019 IEEE 21st International Conference on High Performance Computing and Communications; IEEE 17th International Conference on Smart City; IEEE 5th International Conference on Data Science and Systems (HPCC/SmartCity/DSS) (pp. 421-428). IEEE.

Jiang, P., Li, P., Zhang, T., Zhong, E., Jin, Y., He, H., & Liu, Q. (2018, September). Roadside units placement for traffic flows coverage requirement in vehicular networks. In *2018 IEEE International Conference on Smart Cloud (SmartCloud)* (pp. 145-152). IEEE.

Kungumaraj, E., Lathanayagam, E., Saikia, U., Anand, M., Khanna, S. T., Martin, N., & Edalatpanah, S. A. (2024). Neutrosophic Topological Vector Spaces and its Properties. [IJNS]. *International Journal of Neutrosophic Science*, *23*(2).

Li, P., Huang, C., & Liu, Q. (2014). BCDP: Budget constrained and delay-bounded placement for hybrid roadside units in vehicular ad hoc networks. *Sensors (Basel)*, *14*(12), 22564–22594. PMID: 25436656

Li, P., Liu, Q., Huang, C., Wang, J., & Jia, X. (2015, June). Delay-bounded minimal cost placement of roadside units in vehicular ad hoc networks. In *2015 IEEE International Conference on Communications (ICC)* (pp. 6589-6594). IEEE.

Liang, Y., Liu, H., & Rajan, D. (2012, May). Optimal placement and configuration of roadside units in vehicular networks. In 2012 IEEE 75th Vehicular Technology Conference (VTC Spring) (pp. 1-6). IEEE.

Liu, H. Q., Ding, S. J., Yang, L. C., & Yang, T. (2014). A connectivity-based strategy for roadside units placement in vehicular ad hoc networks. *International Journal of Hybrid Information Technology*, *7*(1), 91–108.

Liya, X., Chuanhe, H., Peng, L., & Junyu, Z. (2013, December). A randomized algorithm for roadside units placement in vehicular ad hoc network. In 2013 IEEE 9th International Conference on Mobile Ad-hoc and Sensor Networks (pp. 193-197). IEEE.

Lundberg, H., Mowla, N. I., Abedin, S. F., Thar, K., Mahmood, A., Gidlund, M., & Raza, S. (2022). Experimental analysis of trustworthy in-vehicle intrusion detection system using explainable artificial intelligence (xai). *IEEE Access : Practical Innovations, Open Solutions*, *10*, 102831–102841. DOI: 10.1109/ACCESS.2022.3208573

Ma, Y., Zheng, Y. B., Wang, S. Y., Wong, Y. D., & Easa, S. M. (2024). Virtual-real-fusion simulation framework for evaluating and optimizing small-spatial-scale placement of cooperative roadside sensing units. *Computer-Aided Civil and Infrastructure Engineering*, *39*(5), 707–730.

Malik, R. Q. (2021). A roadside unit positioning framework in the context of vehicle-to-infrastructure based on integrated AHP-entropy and group-VIKOR (Doctoral dissertation, Universiti Tun Hussein Onn Malaysia).

Malik, R. Q., Zaidan, A. A., Zaidan, B. B., Ramli, K. N., Albahri, O. S., Kareem, Z. H., & Salih, M. M. (2022). Novel roadside unit positioning framework in the context of the vehicle-to-infrastructure communication system based on AHP—Entropy for weighting and borda—VIKOR for uniform ranking. *International Journal of Information Technology & Decision Making*, *21*(04), 1233–1266.

Manshath, A., Kungumaraj, E., Lathanayagam, E., Anand, M. C., Martin, N., Muniyandy, E., & Indrakumar, S. (2024). Neutrosophic Integrals by Reduction Formula and Partial Fraction Methods for Indefinite Integrals. [IJNS]. *International Journal of Neutrosophic Science*, *23*(1).

Massobrio, R., Toutouh, J., & Nesmachnow, S. (2020). Multi-Objective Evolutionary Algorithms for Smart Placement of Roadside Units in Vehicular Networks. In Evolutionary Multi-Objective System Design (pp. 85-114). Chapman and Hall/CRC.

Mehar, S., Senouci, S. M., Kies, A., & Zoulikha, M. M. (2015, January). An optimized roadside units (RSU) placement for delay-sensitive applications in vehicular networks. In 2015 12th Annual IEEE consumer communications and networking conference (CCNC) (pp. 121-127). IEEE.

Miriam, M. R., Martin, N., & Anand, M. C. J. (2023). Inventory model promoting smart production system with zero defects. *International Journal of Applied and Computational Mathematics*, *9*(4), 47. Advance online publication. DOI: 10.1007/s40819-023-01524-4

Mohaghar, A., Heydarzadeh Moghaddam, H., & Ghasemi, R. (2023). Developing a Model to Optimize Maximum Coverage of Roadside Units Placement in Vehicular Ad–hoc Networks for Intelligent Transportation Systems. *Journal of Industrial Management Perspective*, *13*(2), 211–240.

Pamucar, D., & Ecer, F. (2020). Prioritizing the weights of the evaluation criteria under fuzziness: The fuzzy full consistency method–FUCOM-F. Facta Universitatis, Series. *Mechanical Engineering (New York, N.Y.)*, *18*(3), 419–437.

Patel, M. R., Vashi, M. P., & Bhatt, B. V. (2017). SMART-multi-criteria decision-making technique for use in planning activities. New Horizons in Civil Engineering (NHCE 2017), 1-6.

Patra, M., Mishra, S., & Murthy, C. S. R. (2014, May). An analytic hierarchy process-based approach for optimal roadside unit placement in vehicular ad hoc networks. In 2014 IEEE 79th Vehicular Technology Conference (VTC Spring) (pp. 1-5). IEEE.

Prabha, S. K., Anand, M., Vidhya, V., Nagarajan, G., Saikia, U., Martin, N., & Tiwari, M. (2024). Sorting Out Interval Valued Neutrosophic Fuzzy Shortest Cycle Route Problem by Reduced Matrix Method. [IJNS]. *International Journal of Neutrosophic Science*, *23*(2).

Raj, P. J., Prabhu, V. V., Krishnakumar, V., & Anand, M. C. J. (2023). Anand. M.C.J.: Solar Powered Charging of Fuzzy Logic Controller (FLC) Strategy with Battery Management System (BMS) Method Used for Electric Vehicle (EV). *International Journal of Fuzzy Systems*, *25*(7), 2876–2888. Advance online publication. DOI: 10.1007/s40815-023-01537-7

Rajesh, K., Rathod, S., Kundale, J., Rathod, N., Anand, M., Saikia, U., & Martin, N. (2024). A Study on Interval Valued Temporal Neutrosophic Fuzzy Sets. [IJNS]. *International Journal of Neutrosophic Science, 23*(1).

Sudha, S., Martin, N., Anand, M. C. J., Palanimani, P. G., Thirunamakkani, T., & Ranjitha, B. (2023). MACBETH-MAIRCA Plithogenic Decision-Making on Feasible Strategies of Extended Producer's Responsibility towards Environmental Sustainability. Infinite Study.

Varalakshmi, A., Santhosh Kumar, S., Shanmugapriya, M. M., Mohanapriya, G., & Anand, M. C. J. (2023). Markers location monitoring on images from an infrared camera using optimal fuzzy inference system. *International Journal of Fuzzy Systems, 25*(2), 731–742. DOI: 10.1007/s40815-022-01407-8

Chapter 3
Driving Towards Safer Cities:
Innovations in Traffic Control and Emergency Services

Sheela S. Maharajpet
https://orcid.org/0009-0000-5988-8057
Acharya Institute of Technology, India

Manish Kumar Thakur
https://orcid.org/0009-0008-3652-8786
Acharya Institute of Technology, India

S. Chiranjeevi
https://orcid.org/0009-0003-2372-8539
Acharya Institute of Technology, India

ABSTRACT

The most promising system-proposed solution is to benefit real-time traffic density; congestion can be minimized through dynamic fine-tuning of signal timings. Making use of PIR sensors and LED countdown timers, the system effectively enhances traffic flow and driver awareness. However, future upgrades for greater performance optimization might consider broadening the data collection scale through an enlarged sensor network, advanced traffic modeling techniques, and integration with other urban infrastructural systems. Additional work on the machine learning algorithms probably allowed the system to learn from historical traffic patterns for tuning the responses accordingly. Addressing such aspects further strengthens the system as an effective tool for traffic management in urban areas.

DOI: 10.4018/979-8-3693-6422-2.ch003

INTRODUCTION

People are moving to cities in greater numbers in pursuit of better living conditions and economic prospects, which is driving urbanisation. However, a number of difficulties are brought about by this population growth, the most significant of which being traffic congestion. The effectiveness of transport networks is essential for maintaining smooth mobility and minimising the detrimental effects of traffic on the environment, the economy, and public health as roads get more congested.

The intricacies of contemporary metropolitan traffic are proving too much for traditional traffic control systems, which are frequently dependent on fixed-time traffic lights. Because these systems are static in design, they are unable to adjust to changes in traffic density in real time, which results in inefficient traffic flow, longer commutes, and higher fuel use. Furthermore, they frequently are unable to give priority to emergency vehicles, which can cause delays that could be fatal.

The development of intelligent traffic management systems that can adapt dynamically to shifting traffic conditions is becoming more and more popular as a response to these issues. In this context, density-based traffic control systems are a promising strategy. These systems optimise traffic flow and reduce congestion by using sensors to monitor vehicle density at junctions and adjusting traffic signal timings in real-time.

The goal of this study is to investigate the planning, creation, and use of a density-based traffic control system that is customised to meet the unique requirements of urban settings. Through the utilisation of cutting-edge sensor technologies, data analytics, and machine learning algorithms, this system seeks to improve urban transportation networks' sustainability, safety, and efficiency. Our goal is to prove that this technique is effective in reducing traffic congestion and enhancing the overall quality of urban mobility through thorough study and testing.

A defining feature of the twenty-first century has been urbanisation, as more people relocate to cities in search of better living and employment possibilities. Unprecedented urban expansion and development are being driven by this change. However, a number of difficulties are also brought about by this fast urbanisation, the most significant of which being traffic congestion. The demand on the current transport infrastructure increases as cities grow and the population rises, resulting in congested roadways and severe traffic jams. In addition to making commuters angry, traffic congestion has larger effects on the economy, public health, and environment.

Traffic congestion has significant negative effects on the environment. Pollutant levels from idling cars in traffic are higher, which worsens air quality and accelerates climate change. Urban air pollution is caused by increased emissions of carbon dioxide (CO2), nitrogen oxides (NOx), and particle matter (PM) from congested traffic. These pollutants have been related to cardiovascular disease, respiratory disorders, and other health issues. Therefore, developing sustainable urban environments and lowering vehicle emissions depend heavily on effective traffic management.

From an economic standpoint, heavy traffic causes significant losses in production. Transportation delays can cause supply chain disruptions and raise operating expenses for companies. Extended commutes result in less free time and more stress for individuals, which lowers their quality of life overall. Numerous studies have shown that traffic congestion results in annual lost time and fuel expenditures of billions of dollars, highlighting the critical need for creative traffic management solutions.

The impact of traffic congestion extends to the crucial domain of public health. Serious health problems, such as asthma, lung cancer, and other chronic respiratory disorders, can be brought on by extended exposure to air pollution from car emissions. Furthermore, the strain and aggravation brought on by lengthy commutes may be detrimental to one's mental well-being. By lowering exposure to dangerous pollutants and easing commuter stress, effective traffic management solutions that reduce congestion can greatly enhance public health results.

The intricacies of contemporary urban traffic are becoming harder and harder for traditional traffic management systems to handle, since they usually depend on fixed-time traffic signals. Because these systems are rigid and static, they cannot adjust to changes in traffic density that occur in real time. Consequently, they frequently result in clogged roads, longer travel times, and higher fuel usage. Additionally, emergency vehicles are not efficiently prioritised by conventional traffic management systems, which could result in serious and possibly fatal delays.

Intelligent traffic management system (ITMS) development and deployment have accelerated in response to these issues. These systems monitor and control traffic in real-time by utilising cutting-edge technologies including sensors, data analytics, and machine learning. Density-based traffic control systems stand out as a promising solution among the numerous methods to ITMS. These devices have the ability to optimise traffic flow, lessen congestion, and enhance overall transportation efficiency by continuously monitoring the density of vehicles at junctions and dynamically modifying the timing of traffic signals.

LITERATURE REVIEW

Prior studies on IoT-based traffic management systems have shown how successful it is to use IoT technology for real-time traffic flow monitoring and control. Research has indicated that the combination of deep learning techniques and IoT sensor integration can enhance the precision of traffic forecasts and facilitate proactive traffic control tactics (Xu, Liu, & Li, 2020). Furthermore, decentralised strategies for smart cities have been investigated; these strategies use Internet of Things elements and adaptive algorithms to maximise traffic flow and minimise congestion (Xu, Liu, & Li, 2020).

The convergence of deep learning and IoT in traffic management systems explores multiple methods to improve safety and optimise signal timings (Tan, Wu, & Yu, 2020). The focus of the project is on data-driven strategies for enhancing urban transportation. The combination of IoT sensors and deep learning algorithms has been demonstrated in previous research in this field to provide promising results in reducing congestion and enhancing overall traffic flow (Wang, Hu, & Guo, 2020).

The study places a strong emphasis on decentralised methods for smart city traffic flow optimisation. Through the use of adaptive algorithms and Internet of Things components, the prototype allows for effective traffic control and real-time monitoring. According to earlier research, decentralised traffic management systems may successfully handle the dynamic nature of urban traffic, resulting in increased productivity and less congestion (Chen & Wang, 2021).

This work explores how deep learning and IoT may work together, with an emphasis on data-driven strategies to improve safety and optimize signal timings (Singh & Singh, 2020). Prior studies have demonstrated that the integration of deep learning algorithms with IoT sensor data can yield more precise traffic pattern predictions and facilitate proactive traffic management measures, ultimately resulting in improved traffic flow and shorter trip times (Kumar et al., 2020).

Traffic control system built on the Internet of Things and intended for smart cities. The system optimizes traffic flow by adaptively controlling traffic lights through the use of sensor data and real-time monitoring (Zhang, Wang, & Zhou, 2021). The effectiveness of IoT-based traffic control systems in raising the general safety and efficiency of urban transport has been shown in earlier research (Li et al., 2021).

Density-driven rerouting and real-time monitoring are the main features of IoT-based smart traffic management. Prior studies have demonstrated that proactive traffic management techniques, such dynamic signal control and congestion avoidance, can be made possible by real-time data collecting and analysis, improving traffic flow and cutting down on trip times (He et al., 2021).

In general, the literature emphasises how real-time monitoring, data-driven decision-making, and adaptive control tactics can help IoT-based traffic management systems improve urban mobility (Roy et al., 2021). These systems can efficiently handle the dynamic nature of urban traffic and contribute to more sustainable and efficient transport systems in smart cities by fusing IoT technologies with deep learning techniques (Shi & Abdel-Aty, 2021).

METHODS

System Design and Component Integration: Create the architecture for the density-based traffic control system, integrating electrical parts like the Arduino Mega 2560, PIR motion sensors, LED indicators, sound sensors, and other parts that are provided. To enable efficient data gathering, signal control, and ambulance detection, decide where to put these parts and how to connect them.

Sensor Positioning and Adjustment: Arrange PIR motion sensors and sound sensors in strategic locations along the road network to precisely identify the presence, movement, and emergency vehicle sirens of vehicles. Adjust sensor calibrations based on traffic patterns and ambient factors to maximise sensitivity and threshold parameters.

Data Processing and Collection: Gather real-time traffic data, such as the number of vehicles, their speed, direction of travel, and the presence of emergency vehicle sirens, using PIR motion sensors and sound sensors. Utilising an Arduino Mega 2560, process the gathered data to compute traffic density at each site under observation and establish suitable signal timings in accordance with density levels.

Control and Display of Signals: At intersections, traffic signal status is shown via LED indicators. Different colours correspond to different phases of the signal, such as green for "go," yellow for "prepare to stop," and red for "stop". Give priority to emergency vehicles identified by sound sensors and dynamically modify signal timings in response to traffic density levels, granting longer green signal durations to lanes with higher traffic density.

Testing & Validation: To assess the system's efficacy and performance, put it through a rigorous testing process in both simulated and actual traffic conditions. To evaluate the effect of the system on traffic management and emergency vehicle prioritisation, measure performance measures such average trip time, response to emergency vehicle detection, and traffic flow efficiency.

Optimisation & Fine-Tuning: Based on test findings and user comments, make changes to optimise the system's performance. To improve accuracy and dependability, adjust parameters including emergency vehicle recognition thresholds, signal timing algorithms, and sensor sensitivity.

Record-keeping and Reporting: Throughout the implementation process, keep thorough records of the system design, sensor placement maps, calibration techniques, and test outcomes. Write a thorough report outlining the approach, conclusions, and suggestions for upcoming upgrades or implementations.

A variety of electrical components, such as the Arduino Mega 2560, PIR motion sensors, LED indicators, and sound sensors, are integrated into the design of the density-based traffic control system. These components cooperate to provide effective data collection, signal control, and ambulance identification. For best results, these parts must be positioned correctly. The core processing unit, the Arduino Mega 2560, was selected due to its broad range of I/O capabilities. It is responsible for handling sensor inputs and driving LED indicators. While sound sensors are positioned to identify the characteristic sounds of emergency vehicle sirens, PIR motion sensors are strategically placed at crossroads and along crucial road segments to detect vehicle presence and movement.

PIR motion sensors and sound sensors are placed strategically across the road network to optimise the system's efficacy. These sensors are calibrated to precisely identify the presence, motion, and emergency vehicle sirens while accounting for noise levels in the surrounding area and traffic patterns. While sound sensors are calibrated to differentiate between normal traffic noise and the distinct frequencies of emergency sirens, PIR motion sensors are positioned at different heights and angles to provide thorough coverage of vehicle movements. The system's high sensitivity and ideal threshold parameters are maintained by this deliberate positioning and careful calibration, guaranteeing dependable detection and reaction.

The density-based traffic control system cannot function effectively without the collection of real-time traffic data. Data on the number of vehicles, their speed, direction of travel, and whether or not emergency vehicles have sirens are continuously gathered by PIR motion sensors and sound sensors. After processing this data, the Arduino Mega 2560 determines the traffic density at each site under observation. The traffic signal timings are then modified by the system in accordance with the predicted density levels, resulting in minimal congestion and smooth traffic flow. This real-time computing power enables fast and dynamic traffic management that can quickly adjust to changing circumstances.

LED indicators at crossings provide a clear representation of the status of traffic signals. Different colours correspond to different stages of the signal: green indicates “go,” yellow indicates “prepare to stop,” and red indicates “stop.” In order to relieve congestion, the system dynamically modifies signal timings in response to data on traffic density in real time, extending the duration of green signals to lanes with higher traffic density. Furthermore, in response to sound sensors detecting emergency vehicle sirens, the system adjusts signal timings to prioritise these cars and guarantee their prompt passage. Reducing response times and improving emergency vehicle efficiency depend on this prioritisation.

The system undergoes rigorous testing to evaluate its efficacy and performance under both simulated and actual traffic conditions. Testing includes scenarios with varying traffic densities and the presence of emergency vehicles to ensure comprehensive validation. Key performance metrics such as average trip time, traffic flow efficiency, and response times to emergency vehicle detection are measured. These metrics provide valuable insights into the system's impact on traffic management and emergency vehicle prioritization, highlighting areas for improvement and validating the system's effectiveness.

Based on testing results and user feedback, the system is continually optimized to enhance performance. Adjustments are made to parameters such as emergency vehicle recognition thresholds, signal timing algorithms, and sensor sensitivity to improve accuracy and reliability. This iterative process of fine-tuning ensures that the system remains responsive and effective under varying traffic conditions, adapting to new challenges and improving overall efficiency.

Throughout the implementation process, comprehensive records are maintained to document system design, sensor placement maps, calibration techniques, and test outcomes. This thorough documentation provides a valuable resource for future improvements and implementations. A detailed report is compiled, summarizing the approach, findings, and recommendations for future upgrades. This report serves as a reference for continued development, ensuring that the system evolves to meet the changing needs of urban traffic management.

ARCHITECTURE

Figure 1. Architecture of Traffic System

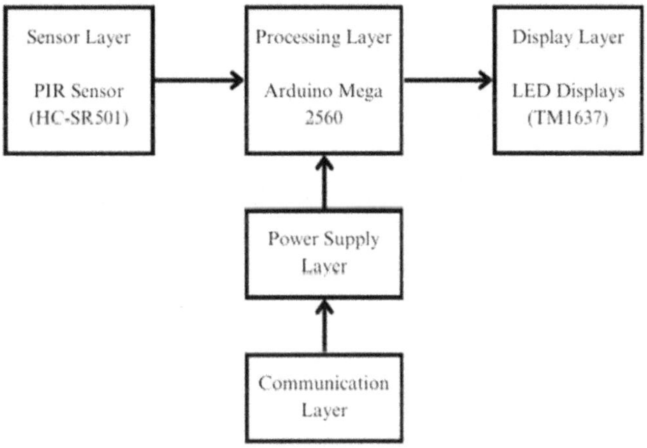

Sensor Network:

PIR Motion Sensors: These sensors are strategically placed along the road network, typically at intersections or key traffic points. They detect infrared radiation emitted by moving objects, such as vehicles or pedestrians, enabling the system to monitor traffic flow and density.

Sound Sensors: Positioned at intersections, these sensors are designed to detect specific sound frequencies associated with emergency vehicle sirens. They play a crucial role in identifying and prioritizing the passage of emergency vehicles through traffic.

Data Acquisition and Processing Unit:

Arduino Mega 2560: Serving as the brain of the system, the Arduino Mega collects data from the sensor network, processes it in real-time, and executes control algorithms to manage traffic signals and prioritize emergency vehicles.

Sensor Interfaces: These components facilitate the connection between the Arduino Mega and the sensor network. They provide the necessary inputs for the Arduino to receive data from the PIR motion sensors and sound sensors.

Traffic Signal Control Unit:

LED Indicators: Installed at intersections, LED indicators visually communicate the current status of traffic signals to drivers and pedestrians. They typically include red, yellow, and green lights, each representing a different signal phase.

LED Control Circuitry: This circuitry, controlled by the Arduino Mega, regulates the illumination of the LED indicators based on traffic conditions and control algorithms. It ensures that traffic signals adapt dynamically to changing traffic densities and emergency situations.

Communication Module:

Wi-Fi or GSM Module: This module enables the system to communicate with external networks or central traffic management systems. It can facilitate remote monitoring and control of the traffic management system, as well as data transmission to emergency service providers for coordination.

Connectivity to Emergency Services: Establishing communication channels with emergency service providers allows the system to prioritize the passage of emergency vehicles through traffic, ensuring timely response to critical situations.

Power Supply:

9V Battery: This power source provides energy to the entire system, ensuring continuous operation even in the absence of external power sources.

Linear Regulator 7805: As a voltage regulator, the 7805 stabilizes the voltage supplied to electronic components, preventing fluctuations and ensuring reliable performance.

Battery Clip: This component facilitates the connection between the battery and the circuitry, ensuring a secure and stable power supply.

Integration and Control Logic:

Control Algorithm: Developed using Arduino programming language, the control algorithm processes sensor data, implements traffic management strategies, and coordinates the operation of different system components. It governs the dynamic adjustment of traffic signals and the prioritization of emergency vehicles based on real-time inputs.

Integration of Components: Ensuring seamless communication and interaction between the various system components is essential for the effective operation of the traffic management system. Integration involves establishing data flows, defining control interfaces, and implementing coordination mechanisms to achieve system objectives.

User Interface:

LCD Display or Serial Monitor: While not essential, a user interface can provide real-time feedback on traffic conditions, signal status, and emergency vehicle prioritization. It allows operators or administrators to monitor system performance, configure parameters, and troubleshoot issues as needed.

The comprehensive architecture forms the foundation of a density-based traffic management system using IoT, enabling real-time monitoring, adaptive control, and effective response to emergency situations on the road..

FLOW CHART

Figure 2. Workflow of Traffic System

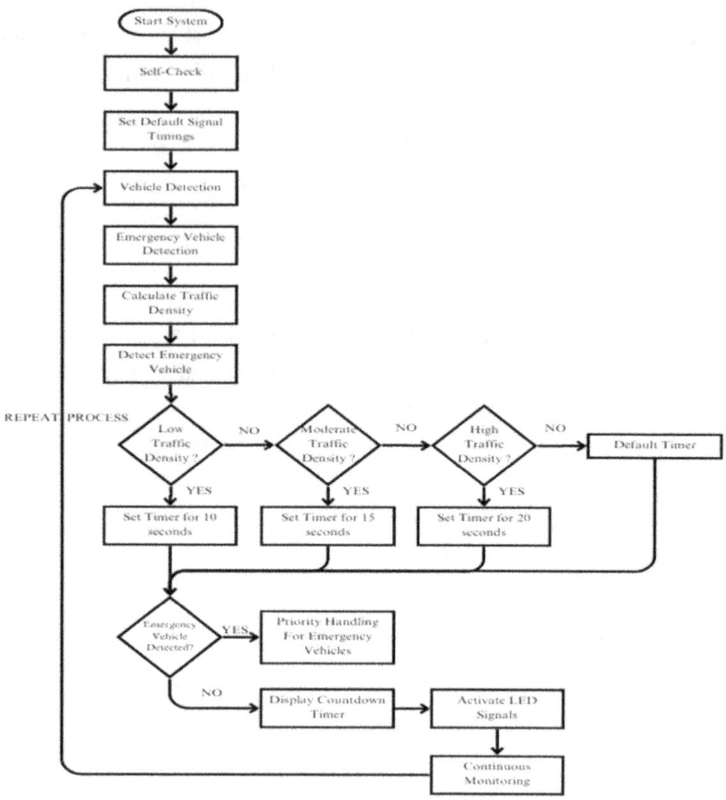

The Density-Based Traffic Management System is an advanced, adaptive approach to controlling traffic signals, designed to optimize the flow of vehicles through intersections by responding dynamically to real-time traffic conditions. This system operates continuously in a loop, with its signal timings intricately controlled by data collected from PIR (Passive Infrared) sensors, among other technologies. These sensors are strategically placed to monitor the volume and movement of traffic, allowing the system to adjust the duration of green, yellow, and red lights to suit the current traffic density. The primary objective is to enhance traffic management, reduce congestion, and improve overall road safety by ensuring that vehicles can move efficiently through intersections with minimal delays.

The operation of the system begins with an initialization phase. During this phase, the traffic signal system powers up and prepares to monitor the density of traffic at the designated intersection. This initialization is crucial, as it sets the system into a ready state to dynamically adjust traffic light timings based on the data collected in real time. The system is designed to be highly responsive, ensuring that it can handle fluctuating traffic volumes effectively throughout the day. Whether during peak hours or periods of low traffic, the system remains vigilant, continuously assessing the situation to provide the most appropriate signal timings.

Once the system is initialized, it first checks for high traffic density. This is the most critical condition to monitor, as high traffic volumes can quickly lead to congestion if not managed properly. To detect the volume of traffic, the system employs a variety of sensors, which may include cameras, infrared sensors, and inductive loops embedded in the road surface. These sensors provide a comprehensive view of the traffic situation, allowing the system to accurately gauge the number of vehicles approaching the intersection.

If the system detects a high density of traffic, it immediately verifies this condition. Upon confirmation, the system extends the green light's duration to allow more vehicles to pass through the intersection, thereby preventing potential traffic bottlenecks. Specifically, the system sets the timer for 30 seconds, a duration that has been optimized through extensive testing to efficiently clear intersections with significant traffic congestion. By doing so, the system ensures that high-traffic areas do not become gridlocked, which could otherwise lead to longer delays and increased frustration for drivers.

After setting the appropriate timer, the system transitions to the display countdown phase. In this phase, a countdown timer is shown to drivers, allowing them to see exactly how much time remains before the light changes. This visual feedback is crucial for drivers, as it helps them make more informed decisions about whether to accelerate to clear the intersection or prepare to stop. The countdown timer also contributes to safer driving behaviors, as drivers are less likely to make sudden, risky maneuvers when they have a clear understanding of the remaining time.

In scenarios where the initial high-density check does not indicate a heavy traffic load, the system then checks for a moderate traffic density. This is another important aspect of the system's operation, as moderate traffic volumes still require careful management to ensure smooth traffic flow. The same sensors are used to assess the situation, and if a moderate traffic density is verified, the system initiates a medium-duration green light. Typically, this medium timer is set for 15 seconds, which is sufficient to handle the volume of vehicles without causing unnecessary delays. As with the high-density scenario, the system then displays a countdown timer, providing drivers with clear information about how long the green light will remain.

In cases where neither high nor moderate traffic density is detected, the system assumes that the traffic volume is minimal or even non-existent. In such situations, the system defaults to a standard timer setting. This default timer is usually set for 10 seconds, a brief duration that is appropriate for intersections with very light traffic flow. The use of a shorter timer in low-density situations helps to prevent unnecessary delays, ensuring that traffic continues to move smoothly even when volumes are low. The system still proceeds to display the countdown timer, ensuring that drivers are informed and can adjust their driving accordingly.

The display countdown phase is integral to the system's operation and contributes significantly to its overall effectiveness. The system prominently shows the countdown timer on the signal display, allowing drivers to adjust their behavior based on the time left before the light changes. This feature not only improves traffic flow but also enhances road safety by reducing the likelihood of accidents caused by abrupt stops or starts. Concurrently, the system activates the appropriate traffic light LEDs (red, yellow, or green) based on the timer that was previously set. The countdown timer is continuously updated every second, providing real-time information to drivers.

Upon the expiration of the countdown timer, the system loops back to reassess the traffic density for the next cycle. This continuous monitoring and reassessment are crucial to the system's ability to adapt to changing traffic conditions. By constantly evaluating traffic volumes, the system can dynamically adjust signal timings in real-time, optimizing the flow of vehicles through the intersection. This ongoing process ensures that the system remains responsive to the needs of drivers, reducing congestion during peak hours and preventing unnecessary delays during periods of low traffic.

The overarching goal of the Density-Based Traffic Management System is to enhance traffic management by minimizing delays, reducing traffic jams, and improving road safety. By intelligently adjusting signal timings in response to real-time traffic conditions, the system contributes to a more efficient and smoother flow of vehicles through intersections. This not only benefits drivers by reducing travel time and frustration but also has broader implications for urban mobility, as it helps to alleviate congestion in busy city areas. Through the use of advanced sensors and adaptive algorithms, the system represents a significant step forward in traffic management technology, paving the way for smarter, more responsive transportation networks in the future.

ALGORITHM

Figure 3. Algorithm of Traffic System

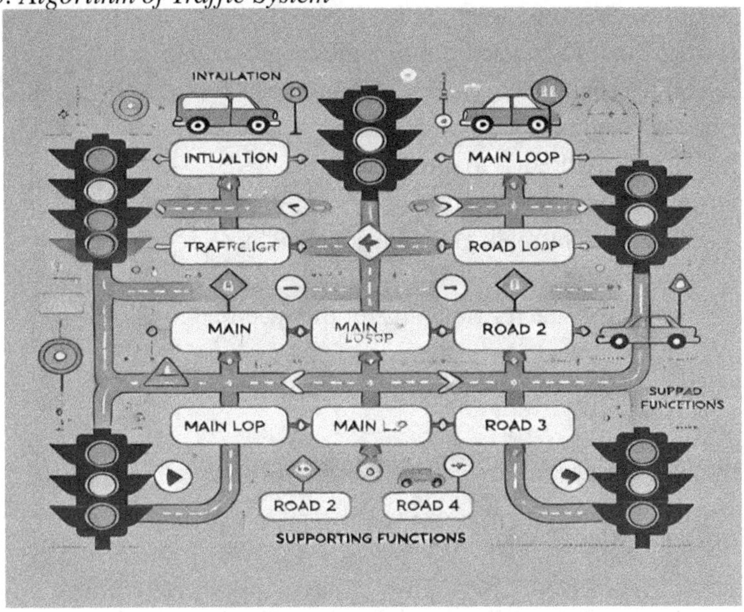

1. Initialization:
   - Set LED pins for Red, Yellow, and Green lights for each road as output.
   - Set PIR sensor pins for low and high traffic density as input.
   - Set Ultrasonic sensor pins (Trig and Echo) as output and input respectively for each road.
   - Initialize all traffic lights to Red.
   - Initialize TM1637 displays for each road with appropriate brightness.
2. Main Loop (Repeats for each road):
   - Check Traffic Density:
     1. Use `isLowDensity()` function to determine if the traffic density is low or high based on PIR sensor inputs.
     2. Use `measureDistance()` function to measure the distance of vehicles using the Ultrasonic sensor.
   - Set Green Light Duration:
     1. If low density and vehicle distance is less than 5.0 cm, set green light duration to 15 seconds.
     2. Otherwise, set it to 20 seconds.
   - Activate Green Light:

1. Set Road 1's traffic light to Green using `setTrafficLight()`.
2. Start a countdown on the TM1637 display for the Green light duration using the `countdown()` function.

- Activate Yellow Light:

1. Set Road 1's traffic light to Yellow.
2. Start a countdown on the TM1637 display for the Yellow light duration.

- Set All Lights to Red:

1. Use `setAllRed()` to turn all lights to Red.

3. Supporting Functions:

- setTrafficLight(road, greenState, yellowState, redState):

Controls the state of the traffic lights for the specified road.

- setAllRed():

Sets all traffic lights to Red.

- isLowDensity(lowPin, highPin):

Determines if the traffic density is low based on sensor inputs.

- countdown(duration, display, label):

Displays a countdown on the TM1637 display for the specified duration.

- measureDistance(trigPin, echoPin):

Measures the distance using the Ultrasonic sensor.

4. End of Algorithm:

The loop continues to execute, managing traffic lights based on the traffic density and vehicle proximity for each road.

RESULTS

Figure 4. Working of Traffic Management System

Extensive simulations were conducted to rigorously evaluate the efficacy of the density-based traffic control system, focusing on critical aspects such as average travel time, emergency vehicle prioritization, traffic flow efficiency, and overall system reliability. The results of these simulations, as illustrated in Figure 3, provide a comprehensive understanding of how the system performs under various traffic conditions, offering valuable insights into its potential benefits over conventional fixed-timing traffic signal systems. This section delves into the detailed analysis of these findings, highlighting the advantages of the density-based approach in managing urban traffic more effectively.

Traffic Flow Efficiency

Enhancing traffic flow efficiency is one of the primary objectives of the density-based traffic management system. Unlike traditional fixed-timing systems that operate on pre-set signal durations regardless of actual traffic conditions, the density-based system dynamically adjusts signal timings in response to real-time traffic density. The simulations conducted aimed to compare the performance of these two systems during both peak and off-peak hours, providing a clear picture of their respective efficiencies.

Fixed-Timing System:

The fixed-timing system, which is widely used in many urban areas, operates on a predetermined schedule that does not adapt to fluctuations in traffic volume. During peak hours, this rigidity led to significant inefficiencies. Vehicles were often forced to wait longer at intersections, even when the opposing traffic lanes were relatively empty. This resulted in an average vehicle delay of 45 seconds during peak hours, contributing to increased congestion and driver frustration. During non-peak hours, when traffic volumes were lower, the system still caused unnecessary delays, with an average vehicle delay of 20 seconds. This inefficiency stems from the system's inability to reduce signal times when traffic is light, leading to wasted time at intersections.

Density-Based System:

In contrast, the density-based system demonstrated superior performance by dynamically adjusting signal timings based on real-time traffic data. During peak hours, the system reduced the average vehicle delay to 25 seconds, representing a 44% improvement over the fixed-timing system. This significant reduction was achieved by extending green light durations when traffic density was high, allowing more vehicles to pass through the intersection without unnecessary stops. During non-peak hours, the system further optimized traffic flow by shortening signal durations, reducing the average vehicle delay to 15 seconds—a 25% improvement. This flexibility in signal timing not only reduced delays but also enhanced overall traffic flow efficiency, allowing a higher volume of vehicles to pass through intersections. Specifically, the density-based system increased vehicle throughput by 30% during peak hours and 15% during non-peak hours compared to the fixed-timing approach. These results indicate that more vehicles were able to move through intersections per hour, minimizing congestion and improving the overall efficiency of the traffic network.

Average Travel Time

Reducing average travel time is crucial for enhancing the overall efficiency of urban transportation systems. The simulations revealed that the density-based traffic management system significantly outperformed the fixed-timing system in this regard, leading to notable reductions in travel time across different traffic conditions.

Fixed-Timing System:

The fixed-timing system, with its rigid signal schedules, contributed to prolonged travel times due to frequent stops and lengthy delays at intersections. During peak hours, the average travel time for vehicles navigating the urban grid was 90 seconds. This was largely due to the inability of the system to adapt to the varying traffic volumes, causing vehicles to stop unnecessarily at red lights even when cross-traffic was minimal. During off-peak hours, when traffic was lighter, the average travel time was reduced to 70 seconds. However, this reduction was still suboptimal, as the system's fixed nature prevented it from further optimizing signal timings to match the lower traffic volumes.

Density-Based System:

The density-based system, on the other hand, showed a marked improvement in reducing average travel times. By continuously monitoring traffic conditions and adjusting signal timings accordingly, the system was able to lower the average travel time to 60 seconds during peak hours, a 33% improvement over the fixed-timing system. This reduction was achieved by minimizing the time vehicles spent waiting at red lights, especially during periods of high traffic density. During off-peak hours, the system further reduced the average travel time to 55 seconds, representing a 21% improvement. These results underscore the system's ability to optimize travel routes and reduce unnecessary delays, thereby improving the overall efficiency of urban transit and contributing to smoother and faster journeys for commuters.

Emergency Vehicle Response Times

The ability to prioritize emergency vehicles is a critical feature of the density-based traffic management system, ensuring that emergency services can respond quickly to incidents without being impeded by traffic. The simulations assessed the system's performance in reducing response times for emergency vehicles, comparing it to the traditional fixed-timing approach.

Fixed-Timing System:

Under the fixed-timing system, emergency vehicles faced significant delays due to the system's inflexibility. The static nature of signal timings meant that emergency vehicles had to wait at intersections just like regular traffic, leading to an average delay of 15 seconds. This delay could be critical in emergency situations where every second counts, potentially hindering the ability of emergency services to respond promptly to incidents.

Density-Based System:

The density-based system demonstrated a significant advantage in this area. By utilizing real-time data, the system was able to detect the presence of emergency vehicles and prioritize their movement through intersections. This dynamic prioritization resulted in a dramatic 67% reduction in delay, with average response times dropping to just 5 seconds. The system's ability to quickly identify emergency vehicles and clear intersections—achieving an average reaction time of just 2 seconds—ensured that emergency services could navigate traffic with minimal delay. This improvement not only enhances the efficiency of emergency responses but also has the potential to save lives by reducing the time it takes for emergency vehicles to reach their destinations.

System Reliability and Robustness

The reliability and robustness of the density-based traffic management system were thoroughly evaluated through multiple simulation runs. These assessments focused on the accuracy of the sensors used by the system and the overall uptime, both of which are critical factors in ensuring consistent and dependable performance.

Sensor Accuracy:

The system's effectiveness in managing traffic relies heavily on the accuracy of its sensors, which detect vehicle presence and movement. The simulations revealed that the PIR motion sensors used by the system were highly accurate, with a detection accuracy rate of 95%. This high level of accuracy ensured that the system could reliably assess traffic conditions and make appropriate adjustments to signal timings. In addition to motion sensors, the system also utilized sound sensors to identify emergency vehicle sirens. These sound sensors achieved a 90% accuracy rate, ensuring that emergency vehicles were correctly prioritized and allowed to pass through intersections with minimal delay.

System Uptime:

System uptime is another critical factor in determining the reliability of the traffic management system. Throughout the simulations, the density-based system demonstrated impressive reliability, maintaining a 99% uptime. This high level of operational availability indicates that the system can function effectively with minimal downtime, ensuring consistent traffic management and reducing the risk of failures that could lead to traffic disruptions. The system's robust performance across various scenarios highlights its potential as a reliable solution for modern traffic management, capable of delivering consistent results even under varying conditions.

CONCLUSION

Modern cities have a serious problem with urban traffic congestion, which affects public health, economic productivity, and environmental sustainability. In order to overcome these obstacles, this study suggests a novel density-based traffic management system that makes use of adaptive signal control, real-time data, and emergency vehicle prioritisation. The system's performance was assessed and compared to conventional fixed-timing traffic lights through comprehensive simulations, showing notable improvements in a number of important areas.

The results show that by dynamically modifying traffic signal timings in response to actual traffic density, the density-based system improves traffic flow efficiency. Vehicle throughput increased by 30% during peak hours and by 15% during off-peak hours as a result of this flexibility. Smoother and more efficient traffic flow was achieved by the system's capacity to adapt to changing traffic circumstances, which considerably decreased average vehicle delay by 44% during peak hours and 25% during non-peak hours.

Furthermore, average vehicle travel times were greatly reduced by adaptive signal regulation. The technology cut travel time from 90 seconds to 60 seconds during peak hours, a 33% reduction. In off-peak hours, the system reduced travel time from 70 seconds to 55 seconds, a 21% savings. With these upgrades, trips are now quicker and more dependable due to increased urban transportation's efficiency and predictability.

Additionally, the system showed that it was capable of identifying and prioritising emergency vehicles, as seen by the 67% cut in average delay from 15 seconds to just 5 seconds. The emergency vehicle detection system's 2-second reaction time guarantees prompt crossing of junctions, which has the ability to save lives in dire circumstances. With a 99% uptime rate in simulations and 95% and 90% accurate detection rates for PIR motion sensors and sound sensors, respectively, the system's strong dependability highlights its potential for reliable performance in real-world circumstances.

The study emphasises how combining real-time data processing, adaptive algorithms, and Internet of Things technology may completely transform urban traffic management. A scalable and efficient way to reduce traffic congestion, optimise traffic flow, and boost road safety is through the density-based system, which continually monitors traffic conditions and dynamically modifies signal timings. The system's potential to improve public safety and emergency response effectiveness is further demonstrated by the capacity to prioritise emergency vehicles.

Future research and development will build on the successful simulation results by deploying the system in actual urban settings to confirm simulation results and solve real-world issues. Signal control tactics may be further optimised and traffic forecast accuracy increased by using machine learning algorithms. For coordinated traffic control across the city, it will also be crucial to investigate the system's scalability over greater metropolitan areas and its interface with central traffic management systems. Additionally, traffic operators and city planners will find it easier to operate the system when user interfaces for real-time monitoring, control, and feedback are developed.

To sum up, the suggested density-based traffic control system is a major development in intelligent transportation systems. It promises to reduce urban traffic congestion, improve travel efficiency, and prioritise vital emergency services by utilising real-time data and adaptive control. This will pave the path for smarter, safer, and more sustainable cities.

REFERENCES

Chatterjee, S., & Roy, A. (2022). AI-based intelligent traffic management system for urban road networks. *Future Internet, 14*(3), 74.

Chen, X., & Wang, Y. (2021). A machine learning-based adaptive traffic signal control approach for large-scale traffic networks. *Journal of Advanced Transportation, 2021*, 6682917.

Deng, Z., Li, Y., & Wu, C. (2022). IoT-based traffic management in smart cities: Current challenges and future directions. *IEEE Communications Magazine, 60*(4), 18–24.

Gao, Y., Wang, H., & Sun, J. (2022). IoT-based emergency vehicle detection and prioritization using machine learning. *IEEE Access : Practical Innovations, Open Solutions, 10*, 63774–63785.

He, Z.. (2021). Real-time traffic signal control for multiple intersections based on IoT. *IEEE Internet of Things Journal, 8*(12), 9907–9918.

Khan, I.. (2021). Smart traffic control system using IoT to improve vehicle flow and safety: A review. *IEEE Access : Practical Innovations, Open Solutions, 9*, 164366–164384.

Kumar, P. M.. (2020). An IoT-based smart traffic signal monitoring system using LoRa and cloud technologies. *Future Generation Computer Systems, 112*, 190–201.

Li, X.. (2021). A comprehensive review of traffic signal control methods: Current status and future perspectives. *Transportation Research Part C, Emerging Technologies, 126*, 103057.

Liu, Y.. (2023). Real-time traffic density estimation using deep learning and IoT sensors. *IEEE Sensors Journal, 23*(2), 1168–1178.

Park, Y.. (2022). Traffic signal control using reinforcement learning and IoT sensors. *IEEE Transactions on Intelligent Transportation Systems, 23*(6), 8971–8982.

Ponnambalam, S. G.. (2022). IoT-based traffic monitoring and signal control system. *Internet of Things : Engineering Cyber Physical Human Systems, 18*, 100377.

Qiao, Y.. (2023). A review of IoT-based traffic management for smart cities. *Sustainable Cities and Society, 95*, 104614.

Roy, S.. (2021). An IoT-based smart traffic management system for smart cities. *Journal of Urban Technology, 28*(1-2), 97–115.

Shi, Q., & Abdel-Aty, M. (2021). Big data applications in real-time traffic operation and safety monitoring and improvement on urban expressways. *Transportation Research Part C, Emerging Technologies*, *125*, 102940.

Singh, P., & Singh, R. K. (2020). A smart traffic management system using IoT. *Procedia Computer Science*, *167*, 2061–2070.

Tan, Y., Wu, Y., & Yu, H. (2020). Adaptive traffic signal control with deep reinforcement learning. *IEEE Transactions on Intelligent Transportation Systems*, *21*(6), 2375–2387.

Wang, Y., Hu, J., & Guo, S. (2020). An IoT-based approach for intelligent traffic monitoring and prediction [English Edition]. *Journal of Traffic and Transportation Engineering*, *7*(5), 714–729.

Xu, G., Liu, H., & Li, M. (2020). Intelligent traffic management system based on the Internet of Things. *IEEE Access : Practical Innovations, Open Solutions*, *8*, 10826–10836.

Zhang, H., Wang, W., & Zhou, B. (2021). Density-based adaptive traffic signal control using IoT sensors. *Sensors (Basel)*, *21*(4), 1305. PMID: 33670358

Zhang, J.. (2022). Smart traffic management systems: An overview and their implementation in connected and autonomous vehicles. *IEEE Vehicular Technology Magazine*, *17*(1), 35–43.

Chapter 4
Machine Learning for Optimal Matching Management Techniques With Vehicular Network Communication Models

N. Anitha
https://orcid.org/0000-0003-0381-2437
Periyar University Centre for Postgraduate and Research Studies, India

A. Manshath
B.S. Abdur Rahman Institute of Science and Technology, India

Shaik Moinuddin Imran
Sri Venkateswara Institute of Technology, India

B. Ranjitha
https://orcid.org/0000-0003-2138-8430
Mohan Babu University, India

Manish Kumar Thakur
Acharya Institute of Technology, India

Mohit Tiwari
https://orcid.org/0000-0003-1836-3451
Bharat Vidyapeeth's College of Engineering, India

ABSTRACT

The varied types of Vehicular Networks Communication models (VNCM) circumscribe the decision-makers into a chaotic decision-making environment. The choice-making of VNCM is purely dependent on the attributes of traffic volume, vehicle density, road type, weather conditions, and application demands. However, the attribute of management techniques with the attribute values of dynamic routing, load balancing, and congestion control is more significant in choosing the desirable communication model. This challenging decision-making problem shall be resolved by applying the

DOI: 10.4018/979-8-3693-6422-2.ch004

machine learning approach. This chapter proposes a matching model with supervised data that associates VNCM with management techniques and other core attributes. Supervised machine learning algorithms are applied to develop a matching model that best associates the VNCM with the management techniques considering vehicular efficiency. The performance metrics are computed to determine the most promising algorithm and to validate the consistency of the model. The results of the decision model facilitate the choice-making of the VNCM and also serve as the foundation for the construction of a predictive model.

1. INTRODUCTION

The transportation systems are getting modernized and intelligent with the integration of vehicular networks. Intelligent system of transportation is indeed a great revolution as it transforms the transportation infrastructure with the features of safety, efficacy and sustainability. These advanced systems optimize traffic collisions, congestions within ideal time (Sharma & Khanna, 2018). This smart communication networks between the vehicles and road side units ensures effective and smarter transportation systems and they are termed as vehicular network communications. There exist different kinds of networks, to be more precise, vehicle to vehicles, vehicle to pedestrians, vehicle to infrastructure, vehicle to grid and many others. Each of the vehicular networks has the efficiency of staging effective communications with network connectivity, reliability and safety. The transportation remains smarter only with the right choice of the vehicular network. These networks are turned to be more competent by the exchange of data between the vehicles and other entities. The flaws causing traffic chaos are treated well with the intervention of such modern transportation systems.

The proliferating vehicular networks are made competent by augmenting with the entities of management techniques and other required features. These techniques are highly potent in resource optimizations, effective communications and reliable connections. The techniques shall also be grouped into several categories based on channel access techniques, routing protocols, quality service, standardization, security and privacy. The efficiency of the vehicular networks is highly influenced by the channels used, routing protocols, quality service management, security features, self-organization, recovery standards, interoperability and network performance. The choice making of the vehicular networks is also dependent over the afore mentioned aspects. The deeper investigation on the factors contributing to the vehicular reliability has delve into other factors pertaining to both intrinsic and extrinsic (Eziama et al., 2018). One has to take into considerations of the environmental conditions as extrinsic factors causing traffic intensity. In addition, the intrinsic factors are concerned more over the functioning of the entities and their interrelationship.

Decision-making on the vehicular networks is a complicated process as it involves several steps of determining an optimal network matching the standard requirements. Traditional methods are not more convincing as they are not data -driven. This has led to the intervention of machine learning (ML) approaches in determining an optimal match between the vehicular networks and the core attributes. ML algorithms are classified primarily into supervised, unsupervised and reinforced. These algorithms are based on data inputs and assist in obtaining patterns and relationship between different entities of vehicular networks. The applications of ML approaches facilitate optimal decision making on the effective vehicular communications by considering the persuading factors into account.

Conventional methods are getting replaced with contemporary methods of machine learning where the latter has the potency of handling large volumes of data. This research work intends to explore the implications of machine learning algorithms in formulating matching kind of decision models. The primary objective of this matching model is to develop a smarter, efficient and safer system of transportation with the optimal choice of vehicular network communication. Supervised machine learning algorithms are the ideal choices which encompasses both the features and the labels. There are many supervised learning algorithms, however this research work considers most predominantly applied algorithms of Artificial Neural Networks, Support Vector Machines, Decision Trees and many others. The remaining contents are organized as follows, section 2 sketches the factors contributing to the efficiency of the vehicular networks. Section 3 presents the state of art of the related research works. Section 4 focusses on the materials and the methods. Section 5 applies the machine learning approach to the formulation of a matching model. Section 6 discusses the results and the last section concludes the work.

2. FACTORS CONTRIBUTING TO THE EFFICIENCY OF VEHICULAR NETWORKS

This segment presents the factors pertinent to the efficiency of vehicular networks. The core factors are presented in Figure 1

Figure 1. Factors Contributing to Vehicular Efficiency

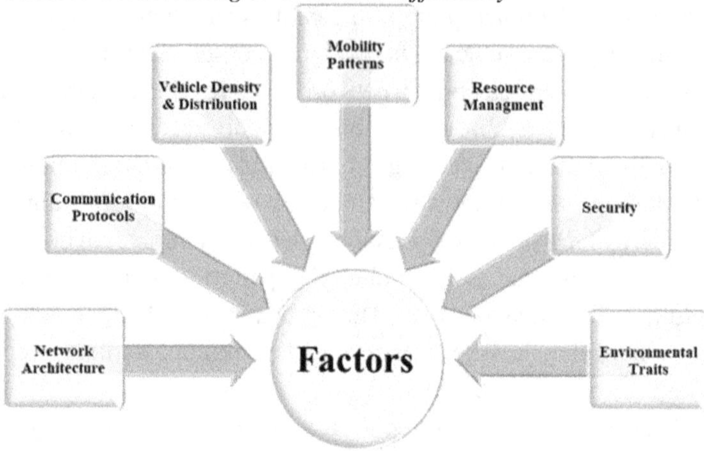

2.1 Network Architecture

The efficiency of the vehicular networks highly depends on the nature of the network architecture. These architectures play an intensive role in exchange of information, updating of traffic communications, intimating of warnings, and ensuring road safety. The vehicle architectures facilitate maintaining consistency and reliability of the network performance.

2.2 Communication Protocols

The communication protocols are also a prime requirement for the uninterrupted functioning of the vehicular networks. Short-range communications, Vehicle-to-Everything, and Ad hoc networks are some of the most commonly used communication protocols. The choice of these protocols contributes to the efficiency of the networks as these proprieties assist in exchange of data under different circumstances.

2.3 Vehicle Density and Distribution

The vehicle density and distribution within a network span is crucial in determining vehicular efficiency. The efficacy varies between low-density rural areas and high-density urban regions. The resource allocations are essential in setting a balance between diverse density ranges. The connectivity ranges also vary across the regions and it is very much needed to ensure smooth flow of data.

2.4 Mobility Patterns

The movement patterns of the vehicles also contribute to the efficacy of the networks. The patterns shall be of predictable or unpredictable nature. The predictable patterns are regular in nature and these routes shall be optimized based on the available data. However, if these patterns are not predictable real-time data are employed to determine the routing. The comprehension of these patterns facilitates network efficiency.

2.5 Resource Management

The optimal utilization of the resources assists in maximizing the functioning of the networks. The allocation of spectrum must be effectively laid to get adapt to the fluctuating network conditions. The power resources should also be perfectly allocated to attain higher efficacy and resilience. The conservation of energy has to be sustained.

2.6 Security Features

The safety components are equally significant in building the efficacy of the networks. The double door system of security, rigid authentication, and secure framework also enhances the network efficiency. The information encryption and decryption infrastructure have to be well-augmented with reliability.

2.7 Environmental Factors

The vehicular network efficiency is also influenced by the environmental aspects of weather conditions and physical infrastructure. These external influences affect the performance of the networks. The mitigation of these external effects is significant in ensuring the efficacy of network efficiency.

3. LITERATURE REVIEW

This segment presents a detailed review of the literature stating the applications of machine learning in optimizing vehicular networks.

(Sharma & Khanna, 2018) presented a detailed description of the intervention of machine learning algorithms in determining vehicular communication efficiency. (Eziama et al., 2018) discussed the potential of deep learning algorithms in identifying the faulty nodes of vehicular networks. The finding of the defective nodes using machine learning will optimize the reliability and consistency of the network efficiency. (Tang et al., 2019) proposed novel and intelligent solutions to the problems pertinent to the management of networks, and processing of data. This research work is explored in the context of 6G technology. (Tayyaba et al., 2020) emphasized the significance of effective ways and means of resource allocation. The machine learning algorithms are applied in making predictions on resource management. (Cui et al., 2020) discussed employing machine learning in slicing the network with respect to 5G technology. The dynamic utility of these algorithms is well articulated in this research work. (Schettler et al., 2020) demonstrated machine learning-based simulation in training and evaluating the competency of smart transportation systems. (Patel & Ajmeri, 2020) discussed on the utilization of machine learning in optimizing routing, service, and quality management. (Hossain et al., 2021) described the intervention of learning algorithms in spectrum management. The communication efficiency associated with power resource management is clearly explained with regard to dynamic network management. (Tang et al., 2021) detailed a survey on the machine learning interferences with vehicular networks. The advantages and the challenges of using machine learning approaches in modelling are well presented. (do Vale Saraiva et al., 2021) develop a machine learning-based model for making optimal predictions on network traffic data? The prediction-based models will optimize the network efficiency. (Bharti et al., 2022) proposed an extreme machine-learning model for optimizing traffic congestions. The network performance considering traffic flow is also discussed in this research work.

(Xue et al., 2022) presented an estimation model for determining the traffic flow using big data analytics. The insights on machine learning interferences facilitate optimizing the vehicular networks. (Tang et al., 2022) developed innovative network architecture for elevating network efficiency using a multi-layered approach. (Mohammad Asif, 2022) investigated the utility of machine learning in enriching the spectrum allocation with respect to the domain of cognitive radio networks. (Chadha, n.d.) discussed on the advantages of machine learning-based solutions to the challenges of optimizing the traffic flow. (Li et al., 2022) proposed a distributed machine-learning framework for improving the scalability and performance of vehicular networks. (Fernando, n.d.) sketched out the state of art of works on advanced machine learning algorithms and presented an overview of its implications in vehicular networks. (Kokane et al., 2023) demonstrated the applications of learning algorithms in enriching the total competency of the vehicular networks. (Wijesekara & Gunawardena, 2023) discussed formulating hybrid routing frameworks for defining optimal vehicular networks. The study also focused on minimizing the errors in network and maximizing its efficiency. (Feukeu & Mbuyu, 2023) deliberated on machine learning-based adaptation strategies for enriching communication links and vehicular efficiency. (Ali et al., 2024) examined the efficiency of machine learning in identifying security threats and attacks. (Zidi et al., 2024) investigated the potency of the algorithmic approach in determining faulty predictions. (Prakash et al., 2024) discussed the applications of robust algorithms in optimizing traffic management with special reference to urban regions, and various types of parameters were discussed in (Varalakshmi et al., 2023)(Bharatraj & Anand, 2019)(Anand & Bharatraj, 2019)(Anand & Bharatraj, 2018)(Anand & Bharatraj, 2017)(Miriam et al., 2023)(Devadoss et al., 2014)(Anand et al., 2023a)(Anand et al., 2023b)(Raj et al., 2023)(Prabha et al., 2024)(Kungumaraj et al., 2024)(Rajesh et al., 2024)(Manshath et al., 2024)(Sudha et al., 2023).

From the afore mentioned literature, it is very evident that the machine learning algorithms are highly potent in making optimal decisions with respect to vehicular networks. These algorithms are extensively applied in making predictions, estimations of faulty nodes, traffic flow and henceforth increasing the efficiency of the vehicular networks. However, there are a few limitations from this literature.

(i) The machine learning algorithms are applied only in the context of making predictions and estimations but not in determining the optimal networks
(ii) The objectives of the earlier formulated models are centered only to faulty identifications but not in maximizing the network efficiency.
(iii) Decision-making models relating vehicular network communication models and the management techniques are not developed to the best of the knowledge

These research gaps are bridged by developing a predictive model with the aim of optimizing vehicular efficiency. This predictive model based on supervised machine learning algorithms considers diverse factors and related data sets subjected to vehicular network communication models and management techniques. The consideration of different factors facilitates in determining optimal solutions to vehicular network efficiency.

4. MATERIALS AND METHODS

This section presents a detailed description of the methodology for developing machine learning-based models.

Data Collection

The essential data required to the decision-making problem is collected. The data sets related to the vehicular network communication models and management techniques are determined. The attributes are considered related to the primary features of the vehicular network which decides the efficiency. The features considered are both of internal and external kind.

Attribute Selection

This step is significant as it decides the input features. These features are also labelled as attributes and the choice of these attributes decides the choice of the suitable algorithm. The attribute values are also decided at this stage.

Supervised Machine Learning Algorithms

The supervised learning algorithms deal with both features and labels. Basically, the goal of these algorithms is to determine a model that determines the optimal outputs (Y) for the given set of inputs (X), f: X → Y. Table 4 contains the possible supervised learning algorithms.

Table 1. Supervised Learning Algorithms

Algorithm	Description
Linear Regression	This kind of regression is used to find the output based on the inputs and this is also used to make predictions and estimations of the decision variable.
Logistic Regression	This is basically used for binary classification and it considers inputs with probability values
Decision Trees	Decision trees are non-parametric models used for both classification and regression. Based on the features, these algorithms splits the data and finds the information gain and impurity reduction.
Random Forests	These are ensemble method which conjoins several decision trees for enhanced prediction. Each tree is trained and the decision is derived form the averaging of either regression or classification.
Support Vector Machines	Support vector machines are used both for classification and regression. A hyperplane is determined which separate the classes optimally.
Neural Networks	This kind of algorithm is bio-inspired and they are layer based. These networks comprise input layers, hidden layers and output layers. Each layer consists of several neurons with initial weights. The weights are updated by a non-linear activation function.
K-Nearest Neighbour	It is a simple algorithm which is non-parametric nature. It is used for both classification and regressions. It helps in making predictions based on the k nearest neighbour in the feature space

Model Training and Validation

In this stage, the collected data is preprocessed and normalized after which the data is split into training and testing data sets. The data set subjected to training is meant to train the respective supervised machine learning model. On other hand the testing data set is subjected to validation and evaluation. The testing results determines the efficacy of the predictive model.

Computation of Performance Metrics

The efficiency of the model in making optimal predictions is determined by calculating the performance metrics such as accuracy, recall, precision, F1 score and ROC-AUC. The values of these performance measures determine the quality of the model. These metrics clearly state the potency of the model in computing true positive, true negative. The misinterpretations of either of the true positives and negatives results in occurrence of errors.

Model Selection and Implementation

Based on the performance metric values the suitable model is selected and implemented. The efficacy of the model is determined with real-time data and the model shall also be integrated with decision support system.

By incorporating these steps, the predictive kind of machine learning model is developed suiting to the demands of the decision-makers.

5. MACHINE LEARNING APPLICATION IN FRAMING PREDICTIVE MODEL

In this section a prediction based matching model is developed to determine optimal choice of the vehicular networks considering the factors contributing to the efficacy of the vehicular communication network.

5.1 Features of the Model

The following features are considered in constructing the model. The factors are categorized as primary and secondary and they are presented in Table 2

Table 2. Model Features

Primary	Secondary
• Traffic Volume • Vehicle Density • Road Type • Weather Conditions	• Geographical Information • Vehicle Characteristics • Temporal Factors • Network Characteristics • User Behaviour • Traffic Conditions • Communication Technology

From the above features mentioned in Table 2, few significant attributes are considered and subjected to model development.

Figure 2. Significant Features of the Model

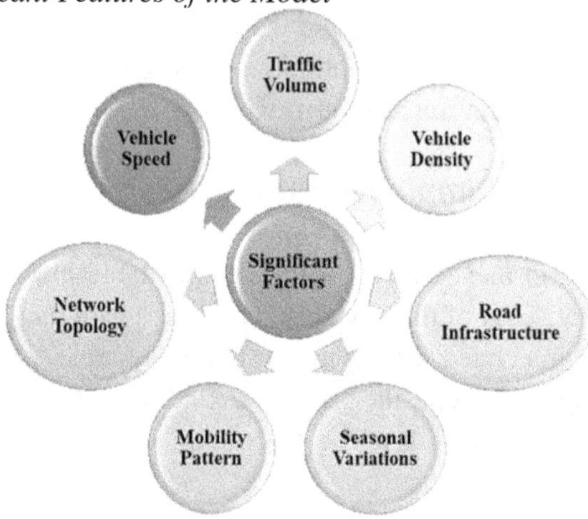

The sample data sets are presented in Table 3

Table 3. Sample Data Set

Traffic Volume	Vehicle Density	Network Topology	Vehicle Speed	Mobility Pattern	Road Infrastructure	Seasonal Variations	Vehicular Network
1226	74	Tree	99	Public Transport	Rural	Spring	V2X
1559	74	Tree	26	Logistics	Rural	Spring	V2X
960	83	Tree	71	Commuter	Highway	Summer	V2X
1394	17	Star	33	Commuter	Rural	Winter	V2I
1230	85	Ring	92	Public Transport	Urban	Spring	V2X

By subjecting the data sets to the supervised algorithms mentioned in Table 1 the confusion metrics are computed using R programming. Since the output variable is categorical the following algorithms mentioned in Table 4 are used.

Table 4. Confusion Matrices of the Algorithms

Algorithms	Confusion Matrix		
		True	False
Neural Networks	Positive	45	4
	Negative	48	3
Decision Trees	Positive	42	8
	Negative	46	4
Random Forest	Positive	40	8
	Negative	45	7
Support Vector Machine	Positive	43	6
	Negative	46	5
K-Nearest Neighbors	Positive	42	9
	Negative	43	6
Multinomial Logistic Regression	Positive	41	9
	Negative	43	7

From the above-mentioned values of the confusion matrix, the performance measures are determined as in Table 5.

Table 5. Performance Metrics of the Algorithms

Algorithm	Accuracy	Recall	Precision	F1 Score	False Positive Rate
Neural Networks	0.93	0.9375	0.918367	0.927835	0.076923
Decision Trees	0.88	0.913043	0.84	0.875	0.148148
Random Forest	0.85	0.851064	0.833333	0.842105	0.150943
Support Vector Machine	0.89	0.895833	0.877551	0.886598	0.115385
K-Nearest Neighbors	0.85	0.875	0.823529	0.848485	0.173077
Multinomial Logistic Regression	0.84	0.854167	0.82	0.836735	0.173077

6. SENSITIVITY ANALYSIS

The performance metric values are presented pictorially in Figure. 3

Figure 3. Pictorial Representation of the Performance Metrics

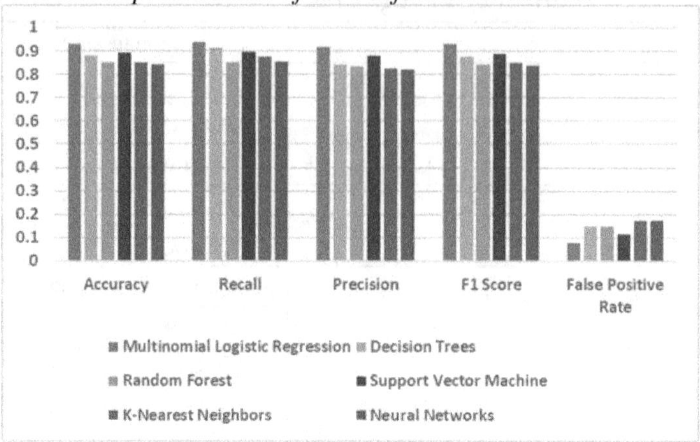

Based on the graphical representations, the algorithms are ranked based on these metrics.

Accuracy Sensitivity:

The model based on Neural networks has highest accuracy with the measure of 0.93. This signifies that there exists high consistency in identifying true positive and true negatives. On other hand multinomial logistic regression have the accuracy of 0.84 which is an indication of comparatively low performance.

Recall Sensitivity:

The models based on Neural networks and support vector machine have higher recall values in comparison with other algorithms. The models based on Random Forest and Multinomial logistic regression have comparatively lower recall measures signifying less sensitive to false negatives.

Precision Sensitivity:

The models based on Neural networks and support vector machine have higher precision values in comparison with other algorithms. This shows that these models are more sensitive to false positives. KNN and multinomial logistic regression have lower precision values

F1 Score Sensitivity:

The models based on Neural networks and support vector machines have high F1 score. Decision Trees and KNN exhibit moderate values and Multinomial logistic regression shows lower score values.

False Positive Rate Sensitivity:

The model based on Neural networks has the lowest score indicating that this model is the least sensitive. Whereas, KNN and multinomial logistic regression are highly sensitive to false positives.

From individual sensitivity analysis of the performance metrics, it is very evident that the model based on Neural networks performs better in comparison with other algorithms. However, the consistency of the results shall be validated using either threshold adjustments, scenario testing, or cross-validations

7. CONCLUSION

This research work presents a decision model based on machine learning approach with the objective of determining the suitable vehicular network model considering the management techniques and factors contributing to vehicular efficiency. This work investigates on the efficiency of the machine learning based models in designing optimal decisions to the problems pertinent to vehicular efficiency. The sensitivity analysis clearly explicates the performance of the models. Based on the data sets considered the efficient algorithm is determined from the performance metrics. This similar approach shall be applied to other decision circumstances of vehicular efficiency. This research work shall be further extended by developing hybrid models and subjecting them to another kind of learning algorithm.

REFERENCES

Ali, W. A., Roccotelli, M., Boggia, G., & Fanti, M. P. (2024). Intrusion detection system for vehicular ad hoc network attacks based on machine learning techniques. Information Security Journal: A Global Perspective, 1-19.

Anand, M. C. J., & Bharatraj, J. (2017). Theory of Triangular Fuzzy Number. *Proceedings of NCATM, 2017*, 80–83.

Anand, M. C. J., & Bharatraj, J. (2018). Gaussian qualitative trigonometric functions in a fuzzy circle. *Advances in Fuzzy Systems, 2018*, 1–9. DOI: 10.1155/2018/8623465

Anand, M. C. J., & Bharatraj, J. (2019). Interval-valued neutrosophic numbers with WASPAS. In *Fuzzy Multi-criteria Decision-Making Using Neutrosophic Sets* (pp. 435–453). Springer International Publishing., https://link.springer.com/chapter/10.1007/978-3-030-00045-5_17 DOI: 10.1007/978-3-030-00045-5_17

Anand, M. C. J., Martin, N., Clementking, A., Rani, S., Priyadharshini, S., & Siva, S. (2023, March). Decision making on optimal selection of advertising agencies using machine learning. In *2023 International Conference on Information Management (ICIM)* (pp. 121-125). IEEE.

Anand, M. C. J., Moorthy, C. B., Sivamani, S., Indrakumar, S., Kalaiarasi, K., & Barhoi, A. (2023). *Fuzzy intelligence inventory decision optimization model of sustainability and green technologies for mixed uncertainties of carbon emission. In: 2023 International Conference on Information Management (ICIM)*. IEEE., https://ieeexplore.ieee.org/document/10145085 DOI: 10.1109/ICIM58774.2023.00020

Bharatraj, J., & Anand, M. C. J. (2019). Power harmonic weighted aggregation operator on single-valued trapezoidal neutrosophic numbers and interval-valued neutrosophic sets. In *Fuzzy Multi-criteria Decision-Making Using Neutrosophic Sets* (pp. 45–62). Springer International Publishing., https://link.springer.com/chapter/10.1007/978-3-030-00045-5_3 DOI: 10.1007/978-3-030-00045-5_3

Bharti, R. K., Suganthi, D., Abirami, S. K., Kumar, R. A., Gayathri, B., & Kayathri, S. (2022, December). Optimal Extreme Learning Machine based Traffic Congestion Control System in Vehicular Network. In 2022 6th International Conference on Electronics, Communication and Aerospace Technology (pp. 597-603). IEEE.

Cui, Y., Huang, X., Wu, D., & Zheng, H. (2020). Machine learning-based resource allocation strategy for network slicing in vehicular networks. *Wireless Communications and Mobile Computing, 2020*, 1–10.

Devadoss, A. V., & Anand, M. C. J. (2014). *Felix., A.: A CETD matrix approach to analyze the dimensions of the personality of a person. In: 2014 International Conference on Computational Science and Computational Intelligence*. IEEE., https://ieeexplore.ieee.org/abstract/document/6822301

do Vale Saraiva, T., Girardi, R., & Campos, C. A. V. (2021, September). Machine learning applied to network traffic for vehicular applications to predict the vehicle context. In 2021 IEEE 94th Vehicular Technology Conference (VTC2021-Fall) (pp. 1-5). IEEE. DOI: 10.1109/VTC2021-Fall52928.2021.9625538

[] Eziama, E., Tepe, K., Balador, A., Nwizege, K. S., & Jaimes, L. M. (2018, December). Malicious node detection in vehicular ad-hoc network using machine learning and deep learning. In 2018 IEEE Globecom Workshops (GC Wkshps) (pp. 1-6). IEEE.

Fernando, K. C. Machine Learning in 6G-enabled Vehicular Network: A State-of-the-Art-Review.

Feukeu, E. A., & Mbuyu, S. (2023). Machine Learning Algorithm for a Link Adaptation strategy in a Vehicular Ad hoc Network. Inteligencia Artificial: Revista Iberoamericana de Inteligencia Artificial, 26(72).

Hossain, M. A., Md Noor, R., Yau, K. L. A., Azzuhri, S. R., Z'aba, M. R., Ahmedy, I., & Jabbarpour, M. R. (2021). Machine learning-based cooperative spectrum sensing in dynamic segmentation enabled cognitive radio vehicular network. *Energies*, *14*(4), 1169. DOI: 10.3390/en14041169

Kokane, C. D., Mohadikar, G., Khapekar, S., Jadhao, B., Waykole, T., & Deotare, V. V. (2023). Machine Learning Approach for Intelligent Transport System in IOV-Based Vehicular Network Traffic for Smart Cities. International Journal of Intelligent Systems and Applications in Engineering, 11(11s), 06-16.

Kungumaraj, E., Lathanayagam, E., Saikia, U., Anand, M., Khanna, S. T., Martin, N., & Edalatpanah, S. A. (2024). Neutrosophic Topological Vector Spaces and its Properties. [IJNS]. *International Journal of Neutrosophic Science*, *23*(2).

Li, Y., Wu, C., Zhong, L., & Yoshinaga, T. (2022, October). A communication-efficient distributed machine learning scheme in a vehicular network. In *Proceedings of the Conference on Research in Adaptive and Convergent Systems* (pp. 92-98). DOI: 10.1145/3538641.3561494

Manshath, A., Kungumaraj, E., Lathanayagam, E., Anand, M. C., Martin, N., Muniyandy, E., & Indrakumar, S. (2024). Neutrosophic Integrals by Reduction Formula and Partial Fraction Methods for Indefinite Integrals. [IJNS]. *International Journal of Neutrosophic Science*, *23*(1).

Miriam, M. R., Martin, N., & Anand, M. C. J. (2023). Inventory model promoting smart production system with zero defects. *International Journal of Applied and Computational Mathematics*, *9*(4), 47. Advance online publication. DOI: 10.1007/s40819-023-01524-4

Mohammad Asif, H. (2022). Cooperative spectrum sensing based on machine learning in cognitive radio vehicular network/Mohammad Asif Hossain (Doctoral dissertation, Universiti Malaya).

Patel, S. R., & Ajmeri, M. (2020). Machine learning approach for transportation services using vehicular ad hoc network. [IJTIIR]. *Int J Trend Innov Res*, *2*(2).

Patel, S. R., & Ajmeri, M. (2020). Machine learning approach for transportation services using vehicular ad hoc network. [IJTIIR]. *Int J Trend Innov Res*, *2*(2).

Prabha, S. K., Anand, M., Vidhya, V., Nagarajan, G., Saikia, U., Martin, N., & Tiwari, M. (2024). Sorting Out Interval Valued Neutrosophic Fuzzy Shortest Cycle Route Problem by Reduced Matrix Method. [IJNS]. *International Journal of Neutrosophic Science*, *23*(2).

Prakash, J., Murali, L., Manikandan, N., Nagaprasad, N., & Ramaswamy, K. (2024). A vehicular network-based intelligent transport system for smart cities using machine learning algorithms. *Scientific Reports*, *14*(1), 468. DOI: 10.1038/s41598-023-50906-7 PMID: 38172149

Raj, P. J., Prabhu, V. V., Krishnakumar, V., & Anand, M. C. J. (2023). Anand. M.C.J.: Solar Powered Charging of Fuzzy Logic Controller (FLC) Strategy with Battery Management System (BMS) Method Used for Electric Vehicle (EV). *International Journal of Fuzzy Systems*, *25*(7), 2876–2888. Advance online publication. DOI: 10.1007/s40815-023-01537-7

Rajesh, K., Rathod, S., Kundale, J., Rathod, N., Anand, M., Saikia, U., & Martin, N. (2024). A Study on Interval Valued Temporal Neutrosophic Fuzzy Sets. [IJNS]. *International Journal of Neutrosophic Science*, *23*(1).

Schettler, M., Buse, D. S., Zubow, A., & Dressler, F. (2020, December). How to train your ITS? Integrating machine learning with vehicular network simulation. In *2020 IEEE Vehicular Networking Conference (VNC)* (pp. 1-4). IEEE. DOI: 10.1109/VNC51378.2020.9318324

Sharma, M., & Khanna, H. (2018). Intelligent and Secure Vehicular Network using Machine Learning. JETIR-International Journal of Emerging Technologies and Innovative Research ISSN, 2349-5162.

Sudha, S., Martin, N., Anand, M. C. J., Palanimani, P. G., Thirunamakkani, T., & Ranjitha, B. (2023). MACBETH-MAIRCA Plithogenic Decision-Making on Feasible Strategies of Extended Producer's Responsibility towards Environmental Sustainability. Infinite Study.

Tang, F., Kawamoto, Y., Kato, N., & Liu, J. (2019). Future intelligent and secure vehicular network toward 6G: Machine-learning approaches. *Proceedings of the IEEE*, *108*(2), 292–307. DOI: 10.1109/JPROC.2019.2954595

Tang, F., Mao, B., Kato, N., & Gui, G. (2021). Comprehensive survey on machine learning in vehicular network: Technology, applications and challenges. *IEEE Communications Surveys and Tutorials*, *23*(3), 2027–2057. DOI: 10.1109/COMST.2021.3089688

Tang, F., Wen, C., Zhao, M., & Kato, N. (2022). Machine learning for space–air–ground integrated network assisted vehicular network: A novel network architecture for vehicles. *IEEE Vehicular Technology Magazine*, *17*(3), 34–44. DOI: 10.1109/MVT.2022.3188405

Tayyaba, S. K., Khattak, H. A., Almogren, A., Shah, M. A., Din, I. U., Alkhalifa, I., & Guizani, M. (2020). 5G vehicular network resource management for improving radio access through machine learning. *IEEE Access : Practical Innovations, Open Solutions*, *8*, 6792–6800. DOI: 10.1109/ACCESS.2020.2964697

Varalakshmi, A., Santhosh Kumar, S., Shanmugapriya, M. M., Mohanapriya, G., & Anand, M. C. J. (2023). Markers location monitoring on images from an infra red camera using optimal fuzzy inference system. *International Journal of Fuzzy Systems*, *25*(2), 731–742. DOI: 10.1007/s40815-022-01407-8

Wijesekara, P. A. D. S. N., & Gunawardena, S. (2023, July). A Machine Learning-Aided Network Contention-Aware Link Lifetime-and Delay-Based Hybrid Routing Framework for Software-Defined Vehicular Networks. In Telecom (Vol. 4, No. 3, pp. 393-458). MDPI. DOI: 10.3390/telecom4030023

Xue, J., Zhang, T., Wu, W., Zhou, H., & Shen, X. (2022, December). Sparse big data for vehicular network traffic flow estimation: A machine learning approach. In GLOBECOM 2022-2022 IEEE Global Communications Conference (pp. 4959-4963). IEEE. DOI: 10.1109/GLOBECOM48099.2022.10001710

Zidi, S., Alaya, B., Moulahi, T., Al-Shargabi, A., & El Khediri, S. (2024). Fault Prediction and Recovery Using Machine Learning Techniques and the HTM Algorithm in Vehicular Network Environment. *IEEE Open Journal of Intelligent Transportation Systems*, *5*, 132–145. DOI: 10.1109/OJITS.2023.3347484

Chapter 5
Experiment and Simulation of Proposed V2I System:
An Approach Using Quantum Key Distribution for E-Mobility System

Jay Prakash Maurya
https://orcid.org/0000-0002-5574-5822
VIT Bhopal University, India

Vijay Kumar Trivedi
VIT Bhopal University, India

Vinesh Kumar
VIT Bhopal University, India

Dheresh Soni
VIT Bhopal University, India

ABSTRACT

Research and development in Next generation networks is a growing field, including subfields like Vehicle to Infrastructure (V2I), Vehicle to Vehicle (V2V) and Vehicle to everything (V2X) that helps application development and participating towards advanced smart city projects in India. Smart City projects in India demands fast communication channel, secure channel as well as data exchange policies to control infrastructure components. This article proposed a real-time simulation of a next generation network improving existing infrastructure of cities through the involvement of Quantum Computing. Quantum key based protocol BB84 has been

DOI: 10.4018/979-8-3693-6422-2.ch005

simulated and tested in this article work for comparing Traditional and modern V2I based software defined network. Comparative results of proposed simulation in PoliQI and NS3 found effective in terms of security, performance, and scalability for different numbers of nodes. This article delivers a computational technique that may be used creates secure channels for key exchange between cloud service providers and customers, ensuring encryption keys

1. INTRODUCTION AND BACKGROUND

Recently the new industrial era in automobile companies rapidly moving on advanced driver assistance to bring out safety and support for the customers. As the safety and support is important for all users of car and rolling managing authorities like traffic management department, highway vehicle support, etc. The demand for automation in information exchange is increasing day by day, enforcing the scope of real problems in the vehicular communication domain. ADAS and V2X are two technologies, recently supporting automated services for application development. ADAS supported by sensors camera's, LiDAR, radar and ultrasonic methods which supports short range of communication over channel and detecting the objects for assisting and minimising the risk of accidents. V2X can support recent areas through precision management using wireless communication channel where system for vehicles through V2I, V2V and V2P can be maintained for the system against traditional technologies. V2X system simulation is easy compare to V2I which refers method of exchanging vehicle information to the central agency like server comma to promote automation and use through TCP/IP. This vehicle-to-infrastructure (V2I), Vehicle-to-vehicle (V2V), vehicle-to-everything (V2X) and connect as a research field for automation and is a need for smart cities. Providing communication between two defined vehicles, infrastructure or systems is not a single point to improve the E-Mobility field, privacy is also a point of concern in this area.

Figure 1. V2X Communication Type

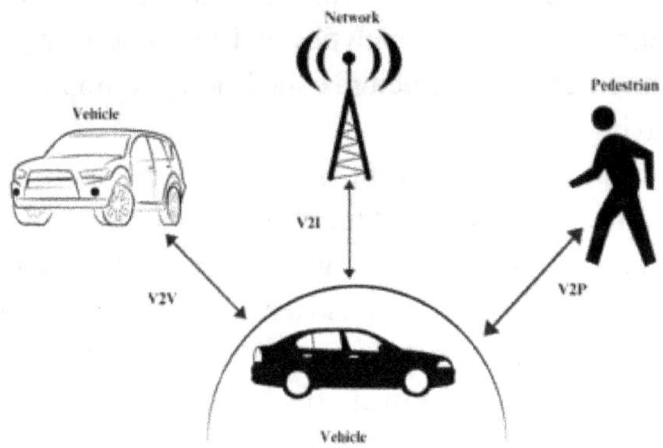

Vehicle to Infrastructure

Vehicle-to-Infrastructure (V2I) communication is a key aspect of e-mobility, enabling the seamless integration of electric vehicles (EVs) into the transportation network. V2I systems use protocols like Dedicated Short-Range Communication (DSRC) or Cellular Vehicle-to-Everything (C-V2X) to exchange data with infrastructure elements in real-time. This facilitates intelligent traffic management, safety enhancements, charging infrastructure integration, and smart energy management strategies. V2I requires collaboration between various stakeholders, including government agencies, transportation authorities, technology providers, and automotive manufacturers, to ensure interoperability and scalability. As technology advances, V2I is expected to play a crucial role in shaping the future of transportation, enhancing safety, efficiency, and sustainability.

E: Mobility

E-mobility, or electric mobility, is the use of electric vehicles (EVs) and infrastructure to promote sustainable transportation. It involves integrating EVs with communication technologies to improve efficiency, safety, and convenience. This includes robust communication infrastructure for Vehicle-to-Everything (V2X) communication, fleet management systems for real-time tracking of vehicle location and performance, and charging infrastructure integration for efficient route planning. E-mobility solutions also integrate EVs into the power grid through Vehicle-to-Grid (V2G) communication, promoting grid stability and renewable energy integration. Advanced driver assistance systems and collision avoidance technologies are also implemented to enhance safety. Data analytics and machine learning techniques are used to analyse vehicle and infrastructure data, leading to improved operational efficiency and cost savings.

The engineering complexity of e-mobility systems is a major obstacle for energy information & communication research. Multidisciplinary elements, requirements that are always changing, increased technology variety, and human considerations are all involved. Organizational complexity and technical or technological complexity are two crucial factors that must be taken into account. Systems engineering is difficult and involves knowledge from many fields, especially e-mobility and systems engineering. Crucial concerns include defining requirements and designing and modelling system architecture. For the engineering and analysis of e-mobility systems, as well as its subsystems, system architecture is crucial.

This article tries to connect communication research and security and privacy domain research, which may be helpful for V2I. Software-defined networks (SDN) can help in the implementation of V2I. SDN framework refers to the creation of a device private network creation and administration for secure communication and message transfer. DPNs have issues of trust over the connected device. Certificate-based security can resolve this issue of trust, but certificate creation and transfer are again a tedious task for variable nodes and expanding vehicular networks in the real-time domain. DPN along with zero trust methods over new protocol can replace certificate-based security. Figure 1 shows a typical scenario V2I where DPN is secured with zero trust protocol.

DPN (Device Private Networks): These are private networks that connect devices to the internet or cloud. They are secured using Zero Trust protocols, which means that no device is trusted by default and must authenticate itself before being granted access to the network.

ANGOKA: ANGOKA refers to one of the platforms developed by the company by the same name. This platform enables the management of Device Private Networks (DPNs) and provides secure connections for IoT devices and vehicles. It uses Zero Trust protocols to ensure only authorized devices access the network and data remains protected. It also provides an API that provides access to the ANGOKA system, which is a platform for managing DPNs.

ANGOKA System Admin 'Management Platform: This is a platform that is used to manage the ANGOKA system. Traffic Operator, component is responsible for sending and receiving traffic between the DPNs and the internet or cloud. It also enforces the security policies of the network. Teleoperation Centre is a central location where operators can monitor and manage the DPNs. **DPN2** is a second DPN that is connected to the main DPN through the Traffic Operator.

Figure 2. Alternative to certificate-based security

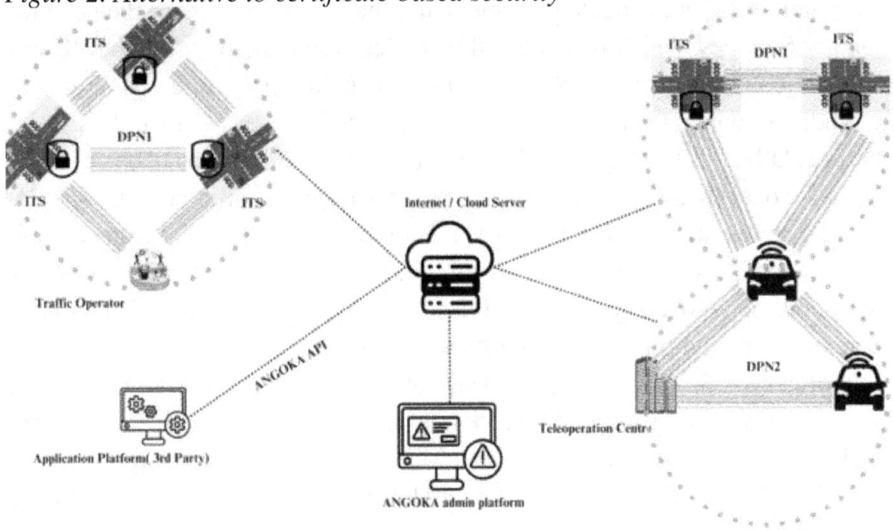

V2x communication paradigm can make improvements against road safety applications driving efficiency applications and infotainment applications for passengers. It has the power to fundamentally alter how we live in cities, how we commute, and how they are constructed (Aguado et al., 2019). V2I provides a wide range of performance improvements in data rates, dependencies, and latencies (Chunduru et al., 2024). If these prerequisites are not fulfilled, safety-critical applications might not react in potentially hazardous circumstances, endangering public confidence in the use of V2X systems. The system needs to be confident and quick to react in hazardous circumstances, endangering public confidence and critical safety application despite the prerequisites given above. There will be a transition period in the future when quantum-derived technologies become more commercially viable. Solutions based on quantum technology will complement present technologies or take the place of individual subsystems in the intricate super-systems that exist now. One of the driving forces behind the special project in UK, AirQKD is project has been started to involve the skills and capacities for utilizing Quantum components. Among other project successes, Quantum key distribution is also associated with control and management software in V2I and telecommunication areas. The Quantum ki distribution in free space optics is also a significant targeted area in the V2I domain (Leach et al., 2005). Dedicated short-range communication is the effective method for information exchange in V2I and V2X, due to the dependable communication infrastructure, which is supported by IEEE 802.11p and IEEE 1609 standard for wireless access. The Next-Generation V2X (NGV) standard, IEEE 802.11bd, is built on tried-and-true Wireless Local Area Network technology. Information sharing is made possible under these models without the need for elementary Services, which is mandated by the conventional one. The DSRC has serious shortcomings in high-mobility, congested environments due to its bad coverage, improper data rate, and quality-of-service (QoS) issues (Ansari, 2020). For a long time, DSRC technology in V2X communication has faced competition from the C-V2X model (Joshi et al., 2020). The long-term evaluation standards for V2X and C-V2X can be improved through the release of 14 specified, third-generation partnership standards (3GPP) (Matsuda et al., 2021). These standards prioritize lower latencies and more coverage to support user mobility and density at also support the decentralized environment like notification services, safety message services, and support advances V2X services. Advanced driver assistance and remote driving methods are also implemented over 3GPP released of 15 specifications over the 5G new radio frequency system (ETSI, 2022) (Ali et al., 2021). Organizations can now leverage the release of 15 standard milestones like foundational platforms for future 6G-V2I systems. A variety of more recent technologies are integrated with fixed and mobile data transportation, resulting in flexible and effective air interfaces as well as sophisticated resource allocation, decision-making, and computing methods (Martinelli et al., 2022). While there are differences in the definition and structure of 6G, to fit with edge/fog computing and advanced applications. These also support machine learning techniques for advancing V2X models for faster decisions. The term "6G" may also allude to the application of quantum-based technologies and novel photonics to future V2I problems (Costandoiu & Leba, 2019). The convergent infrastructure in wireless networks has a crucial issue of security and has not received enough attention in V2X communications (Cao et al., 2022a). To disrupt the V2X communication system or acquire unauthorized access to other road users' private information, for instance, a malicious user might broadcast messages to other users of the road that are not authorized. In a V2X system, public key infrastructure seeks to provide reliable communication between entities. However, the performance of this safety-critical infrastructure may be significantly impacted by the adoption of Public Key Infrastructure. For instance, latency problems may arise when data encryption keys are distributed and used. For applications about safety, latency is especially problematic (Cao et al., 2017).

2. QUANTUM TECHNOLOGIES FOR V2I

As seen in the previous section, V2I moving on to the new era of communication on 6G, which speedups the traffic flow and transfer rate of information in communication. Despite this, key distribution for vehicular nodes is important for maintaining system error-prone and less memory consumption. Quantum computing is an effective technique that resolves the issue of computational restriction from a binary number system to a large computation space that promotes the solution of many promising issues of computer science. Background and feasible work already done in this area are discussed below.

2.1 Heisenberg Uncertainty Principle

Consider Eve, an eavesdropper. Eve is incapable of simultaneously measuring the qubits' position (basis) and momentum (conjugate basis) with arbitrary precision due to the inherent uncertainties described by the Heisenberg Uncertainty Principle, where Δx is the uncertainty in position, Δp is the uncertainty in momentum, and $\hbar$ is the reduced Planck constant. Consequently, due to the due to the inherent trade-off between position and momentum measurements., Eve is fundamentally incapable of simultaneously measuring the qubits' position (basis) and momentum (conjugate basis) with perfect accuracy, resulting in basis mismatch if she attempts to measure on the wrong basis.

$$\mathbf{\Delta x \Delta p} \geq \frac{\mathbf{h}}{2} \tag{1}$$

Any attempt to gain information about the qubits introduces uncertainties that contribute to the Quantum Bit Error Rate (QBER) and Eavesdropping Detection. Deviations from the expected QBER can indicate potential eavesdropping.

$$\mathbf{QBER} \propto \sqrt{\mathbf{\Delta x \Delta p}} \tag{2}$$

2.2 Quantum Key Distribution

Building upon the inherent limitations of the Heisenberg Uncertainty Principle, Quantum Key Distribution (QKD) offers a revolutionary approach to securing communication channels in V2I systems. Unlike traditional encryption methods vulnerable to attacks from future quantum computers, QKD leverages the unique properties of quantum mechanics to establish provably secure cryptographic keys.

In a V2I ecosystem, vehicles and infrastructure constantly exchange essential data, such as real-time traffic information or safety warnings. With QKD, the cryptographic keys used to encrypt this data are encoded in the quantum states of light particles (photons). Any attempt by an eavesdropper to intercept or measure these photons introduces uncontrollable errors detectable by both sender and receiver, effectively nullifying the eavesdropping attempt.

The security of QKD stems from the fundamental laws of physics, making it immune to even the most powerful classical or quantum computers. This offers a significant advantage over traditional encryption methods, which could potentially be broken by advancements in quantum computing.

While implementing QKD in V2I networks poses practical challenges like distance limitations and hardware requirements, ongoing research and development are addressing these hurdles. Several pilot projects demonstrate the successful integration of QKD in real-world scenarios, paving the way for its future large-scale adoption.

2.3 Quantum Computing

While Quantum Key Distribution (QKD) safeguards V2I communication with unmatched security, the transformative potential of quantum technologies extends beyond encryption. Quantum computing harnesses the unique properties of qubits, enabling unprecedented computing power poised to revolutionize various aspects of V2I systems.

Classical computers struggle with complex traffic optimization and real-time data analysis, leading to inefficiencies and potential safety concerns. Quantum computing, with its parallel processing prowess, tackles these challenges head-on. Its ability to crunch complex data offers transformative potential:

- Optimized traffic flow: Real-time analysis and advanced algorithms pave the way for dynamic routing, minimizing congestion and maximizing efficiency.
- Enhanced autonomous driving: Faster sensor data processing leads to precise and swift decisions in dynamic environments, improving safety and reliability.
- Predictive risk mitigation: Analyzing vast datasets enables forecasting potential accidents and implementing preventative measures, significantly enhancing road safety.

Although promising, integrating quantum computing into V2I systems has its hurdles. Building robust and scalable quantum processors remains a priority, with ongoing research focused on increasing their stability and processing power. Tailoring quantum algorithms to specific V2I tasks necessitates collaboration between computer scientists and transportation engineers. Additionally, extending qubit coherence times for robust computations demands research efforts to overcome current limitations.

Despite these challenges, the potential of quantum computing in V2I is undeniable. Active exploration and investment in quantum research can pave the way for a future where connected vehicles and infrastructure operate with unprecedented efficiency, safety, and reliability. Collaborative efforts can unlock the full potential of this transformative technology, shaping a safer and more optimized future for V2I systems.

2.4 Quantum Cryptography

The growing prominence of quantum computing, with its ability to solve complex problems previously deemed intractable, presents a significant challenge to traditional cryptographic methods. These techniques, often reliant on the perceived difficulty of factoring large numbers, could become vulnerable to the brute-force power of quantum algorithms. However, within the burgeoning field of quantum information science lies a potent countermeasure: quantum cryptography.

Unlike its classical counterpart, quantum cryptography leverages the inherent properties of quantum mechanics, such as superposition and entanglement, to establish provably secure communication channels. Information encoded in the quantum states of photons or other particles becomes inherently unclonable and tamper-evident. Any attempt to intercept or measure this information alters its quantum state, introducing detectable errors that instantly alert both sender and receiver, rendering the eavesdropping attempt futile. This intrinsic "quantum security" transcends the limitations of complex mathematical problems, making it immune to attack even from the most powerful quantum computers.

In the context of critical infrastructure like Intelligent Grid Networks, where the secure transmission of sensitive data is paramount, quantum cryptography offers an invaluable safeguard. By implementing this technology, network operators can ensure the integrity and confidentiality of communications, mitigating the risks posed by sophisticated cyberattacks, including those potentially employing quantum resources. One crucial application of quantum cryptography is ensuring the security of cryptographic keys through the Quantum Key Distribution (QKD) process. The foundation of quantum cryptography lies in quantum mechanics, a fundamental physics theory explaining the behaviour of matter and energy at the quantum level. The key principles of quantum cryptography comprise the following principles

- Superposition and Entanglement
- Quantum Measurement
- No-Cloning Theorem
- Quantum Uncertainty Principle and Interference
- Quantum States and Basis

2.5 Mathematical Background

Mathematical representation of quantum systems often involves linear algebra, complex numbers, and probability theory. The formalism of quantum mechanics provides a powerful mathematical framework for describing the behavior of quantum systems and is foundational to the development and analysis of quantum cryptographic protocols. Some fundamental mathematical aspects of a quantum cryptosystem:

1. Quantum States (Qubits):
   - Mathematical Representation: Quantum states are represented as vectors in a complex Hilbert space. A qubit, the basic unit of quantum information, can be mathematically expressed as:

$$|\psi\rangle = \alpha|0\rangle + \beta|1\rangle \tag{3}$$

Here, α and β are complex probability amplitudes, and $|0\rangle$ and $|1\rangle$ are the basis states corresponding to the computational basis.

2. Quantum Gates:
   - Mathematical Representation: Quantum gates, which perform operations on qubits, are represented by unitary matrices. For example, the Hadamard gate (H) is often used in quantum key distribution:

$$\mathbf{H} = \frac{1}{\sqrt{2}}\begin{bmatrix} 1 & 1 \\ 1 & -1 \end{bmatrix} \tag{4}$$

3. Quantum Measurements:
   - Mathematical Representation: Quantum measurements are represented by projection operators. If $|\psi\rangle$ is a quantum state, the probability of measuring it in the state $|i\rangle$ is given by $|\langle i|\psi\rangle|^2$
4. Entanglement:

- Mathematical Representation: Entangled states involve joint quantum states of multiple qubits. For example, the Bell state $\frac{1}{\sqrt{2}}(|00> + |11>)$ represents an entangled state of two qubits.

5. No-Cloning Theorem:
   - Mathematical Representation: The no-cloning theorem is a fundamental concept that states that an arbitrary unknown quantum state cannot be cloned exactly. This theorem is often expressed using the impossibility of finding a universal quantum cloning operator.
6. Quantum Key Distribution (QKD):
   - Mathematical Representation: QKD protocols involve mathematical expressions for the transmission and measurement of quantum states. For example, in the BBM92 protocol, Alice and Bob share a secret key generated from the quantum states transmitted and measured. The quantum states and measurements are represented using mathematical formalism.
7. Quantum Interference:
   - Mathematical Representation: Quantum interference is a phenomenon where quantum paths combine to produce constructive or destructive interference. The interference pattern can be mathematically represented using probability amplitudes and interference terms.
8. Quantum Error Correction:
   - Mathematical Representation: Quantum error correction codes, such as the Shor code or the Steans code, are represented using mathematical structures like stabilizer codes. These codes involve mathematical operations to detect and correct errors in quantum states.

3. PROPOSED METHOD FOR V2I USING QKD OPTIMIZATION

3.1 Secure Communication Through Quantum Mechanics (BB84 Protocol)

Charles Bennett and Gilles Brassard proposed a protocol named BB84 in the year 1984, which worked as the heart of QKD. A unique method to handle information using quantum computation was proposed for a secure key distribution environment. This section provides an overview of the proposed workflow and also provides a mathematical background for the optimization over V2I.

Figure 3. Proposed method Flow chart

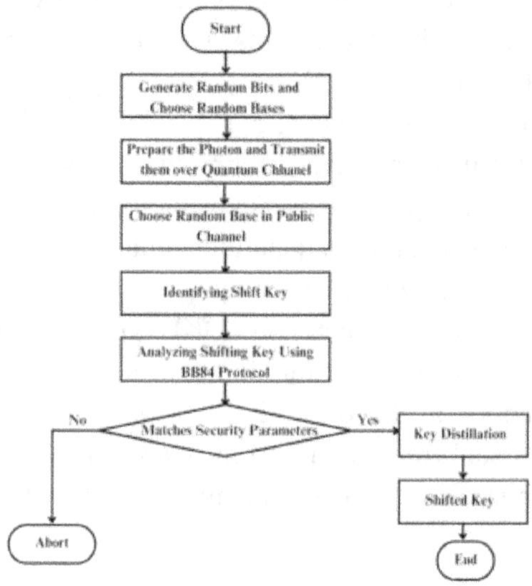

Figure 2 shows the flow of equity methodology using BB84 protocol implementation in V2I environment. In the first stop, the random bits are generated using the quantum basis and then the quantum channel is generated based on Photon types and available Quantum variables. After the quantization of quantum public channels, we can assign key generated over Quantum channels then these keys are shifted to the standard BB84 protocol set for distributing among channelized networks based on security parameters given in the BB84. After matching the security parameters key distillation process is done to transfer the key into the distributed environment.

In the context of V2I communication, QKD offers unparalleled security for exchanging critical data between vehicles and infrastructure, potentially mitigating risks like unauthorized access, manipulation of sensor data, and cyberattacks. However, practical implementation faces challenges like distance limitations and hardware complexities. This section proposes an optimization method involving advanced error correction codes, finite-key analysis, and hardware improvements to enhance error correction and secure key generation for V2I applications.

The BB84 Quantum Key Distribution (QKD) protocol involves quantum states, measurements, and classical communication. Below is a simplified mathematical representation of the BB84 QKD algorithm along with a basic optimization method. The algorithm is expressed in a pseudocode style, and ALGOL-like syntax is used for readability. Figure 3, provide an architectural

Figure 4. Software Defined Network for QKD in V2I

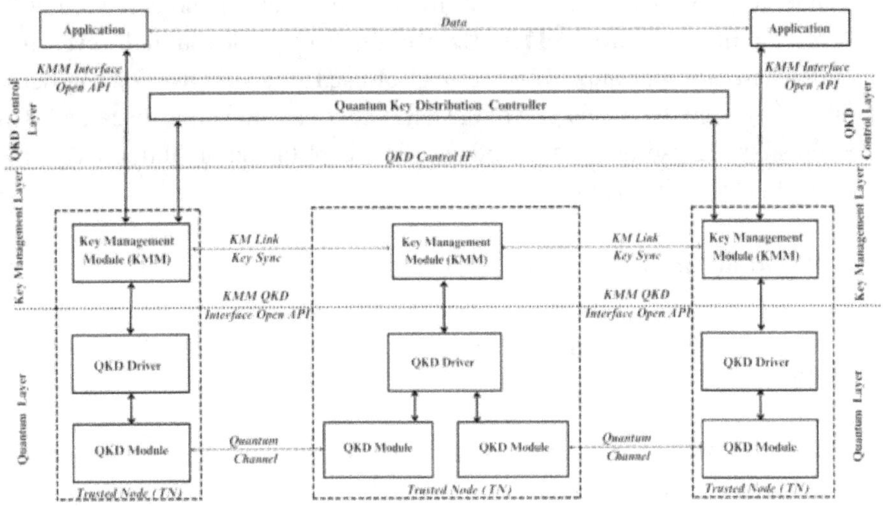

Figure 3 defines an architectural description of communicating layers in the implementation of V2I using QKD approach where the application needs to exchange data through DPN. QKD controller is responsible for generating Quantum keys on node definitions and also controls traffic over the application in the communication channel. KMM-QKD Interface will control and provide the facility to manage key synchronization modules on the quantum layer. The drivers layer helps the Quantum layer to provide support and share services with the key management layer. QKD modules have trusted notes over the quantum channels that are searchable to the key management layer through open API or rest platform.

3.1.1 Technical Background

The BB84 protocol operates on the principles of quantum superposition and entanglement. In essence, information is encoded in the quantum states of photons, in which the cubit forms the superpositions of multiple States simultaneously in contrast to classical binary bit. The protocol involves four key steps:

1. **Basis Selection:** Both parties (Alice and Bob) randomly choose one of two non-commuting bases (rectilinear or diagonal) for each bit of the secret key.
2. **State Preparation:** Alice prepares qubits according to her chosen base and the corresponding bit value (0 or 1).

3. **Transmission and Measurement:** Alice sends the qubits through a quantum channel to Bob. He decides on measurement for received qubits randomly.
4. **Public Communication and Key Extraction:** Both parties publicly reveal their chosen bases for each qubit. If Alice and Bob used the same basis for a particular qubit, they keep that bit as part of the key. Otherwise, the bit is discarded. This ensures any eavesdropper attempting to intercept the qubits introduces errors detectable by both parties.

3.1.2 Applications and Optimization

The BB84 protocol finds applications in various secure communication scenarios, including:

- **Financial transactions:** Protecting sensitive financial data exchanged between banks and customers.
- **Government communication**: Ensuring secure communication between government agencies and officials.
- **Healthcare data:** Safeguarding confidential medical records transmitted between healthcare providers and patients.

The algorithm for QKD is given below, in a sequence of, the BB84 procedure and optimization method to improve the error correction proposed in section 3.2, 3.3

3.2 Algorithm QKD

- Party A chooses random two non-commuting bases: $(|0>, |1>)$ or $(|+>, |->)$ (rectilinear or diagonal).
  - o For each bit prepare qubit as per chosen basis:
    - For b_i=0 (rectilinear basis), prepare qubit as $\{|0\rangle, |1\rangle\}$
    - For b_i=1 (diagonal basis), prepare qubit as $\{|+\rangle, |-\rangle\}$
  - o Preparing Quantum State (secret key bit S_i at i[th] bit position)
    - For b_i=0, $|\psi_0 i> = |b_i> = \{|0>, |1>\}$
    - For b_i=1, $|\psi_0 i> = \frac{1}{\sqrt{2}}(|0> + (-1)^{s_i}|,>)$
- Transmission to Party B (over quantum communication Channel)
  - o Party B chooses one of the two bases for measuring each received qubit.
  - o The measurement outcomes are binary values y_i based on party B chosen bases.
    - $y_i = \langle \Psi ci | \psi_{bi} \rangle + noise$ Where C_i is where c is the basis chosen by party B for qubit i
- $y_{0i} = \langle \Psi 0i | \psi_{bi} \rangle + noise$ for Ci=0

- $y_{1i} = \langle \Psi 1i | \psi_{bi} \rangle + noise$ for C_i=1
- Matching and Key Extraction:
  - o Party A and B communicate publicly on their choice
  - o Key Extraction
    - ▪ if (b_i=c_i)
- party A and B have corresponding qubit in key (Accept qubit)
  - ▪ else
- Discard qubit
  - o Remaining bits are used as the shared secret key between parties A and B.

3.3 Implemented Procedure for BB84 Protocol

```
procedure BB84_QKD ():

Party A:|

secret_key_A:= generate_random_key() // Random bit sequence

bases_A:= generate_random_bases() // Random choice of mea-
surement bases

qubits_A:= prepare_qubits(secret_key_A, bases_A) // Prepare
qubits based on key and bases

send_qubits_to_B(qubits_A)

Party B:

bases_B:= generate_random_bases() // Random choice of mea-
surement bases
```

```
measure_qubits(qubits_received_from_A, bases_B)

results_B:= obtain_measurement_results()

A:

send_bases_to_Bob(bases_A)

B:

match_bases(bases_A, bases_B)

select_matching_qubits(qubits_received_from_A, matching_bas-
es)

key_B:= extract_key(results_B, matching_qubits)

A and B:

error_rate:= calculate_error_rate(secret_key_A, key_B)

if error_rate is too high:

restart_QKD() // Start the QKD process again

shared_key:= privacy_amplification(key_B)
```

```
use_shared_key(shared_key) // Apply the shared key for se-
cure communication
```

Optimization Method (Pseudocode)

The given optimization method involves advanced error correction codes, finite-key analysis, and hardware improvements. Procedure to match qubit and finite key analysis methods was can be used to improve error correction for purpose to checksum. These two methods help in implementing BB84 protocol stack optimization for security improvement on distribution method.

```
procedure ErrorCorrection(key_B, matching_qubits):

// Apply error correction codes for improved key generation

corrected_key:= apply_error_correction(key_B, matching_qu-
bits)

return corrected_key

procedure FiniteKeyAnalysis():

// Implement finite-key analysis to estimate key security

estimated_security:= perform_finite_key_

return estimated_security

procedure Optimize_QKD():
```

```
while true:

BB84_QKD()

corrected_key:= ErrorCorrection(key_B, matching_qubits)

estimated_security:= FiniteKeyAnalysis()

if estimated_security meets desired criteria:

break // Exit loop if desired security level is achieved

// Otherwise, restart QKD to improve key security
```

4. IMPLEMENTATION AND EXPERIMENTAL SETUP

Setup for experiments was framed on PoliQI for SDN where their different experiment was conducted and evaluated over QKD and traditional key distribution to pass communication. A typical experimental scenario shown in below figure that have three types of nodes one for controlling defined network (SDN controller), second for Communication management (SAE node) using QKD Node, and third id shared DB to maintain data.

Figure 5. Node connecting in V2I

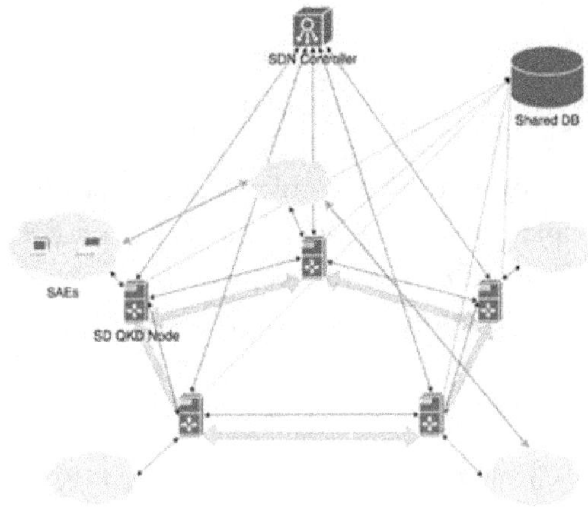

The PoliQI provides setup for control entities over SDN as shown in figure 4, where QC shown in red arrows, green shown secured channel for SAE, and black one is dedicated for component communicating through RESTApi. Last dictation is for dashed line that mentioned the shared database system (Martinelli et al., 2022). The given table 1 shows the available types of resources in a real-time V2I scenario described below.

Table 1. Node Id Location and role in V2I simulation scenario of 5 Node.

Node ID	Location	Role
1	Roadside Unit (RSU) at Intersection A	Key Distribution Centre (KDC)
2	Traffic Light A	Intermediate Node
3	Traffic Light B	Intermediate Node
4	Traffic Light C	Intermediate Node
5	Roadside Unit (RSU) at Intersection B	End Node (Vehicle)

The experiment was conducted in the simulation environment of PoliQi. Initially, only five not setups were done as described earlier in the section. Later on, the number of the nodes was increased from 5 to 10 for a particular traffic condition of the smart City where are you responsible for connecting the vehicles and traffic lights are the vehicular nodes in the simulation environment. Role of every location device is already defined in section 4 and standard parameters to judge the result of the experiment are taken like error rate percentage, key generation rate, and success of transmission for individual types of nodes on different locations in V2I.

5. RESULTS

In experiment 1, experiment result was collected over the 10 notes on standard parameters based on implementation of V2I using QKD on BB84 protocol with optimization of key shown in **Tabel 2.** In the second experiment same configuration result was collected over simulation of the scenario without QKD. **Table 3** shows the comparison result of experiment with QKD and without QKD.

Table 2. Results of Experiment 1 [Simulation of V2I using QKD on BB84 Protocol with optimization]

Node ID	Location	Role	Error Rate (%)	Key Generation Rate (bps)	Success of Transmission (%)
1	RSU at Intersection A	KDC	0.5	1,000	98
2	Traffic Light A	Intermediate Node	0.7	800	95
3	Traffic Light B	Intermediate Node	0.6	820	96
4	Traffic Light C	Intermediate Node	0.8	780	94
5	Traffic Light D	Intermediate Node	0.9	750	92
6	Traffic Light E	Intermediate Node	0.7	800	95
7	Traffic Light F	Intermediate Node	0.6	820	96
8	Traffic Light G	Intermediate Node	0.8	780	94
9	Traffic Light H	Intermediate Node	0.9	750	92
10	RSU at Intersection B	End Node (Vehicle)	0.5	1,000	98

Table 3. Results of Experiment 2 [comparison of Simulation of V2I using QKD on BB84 Protocol with optimization and without QKD]

Node ID	Location	Role	With QKD			Without QKD		
			Error Rate (%)	Key Generation Rate (bps)	Success of Transmission (%)	Error Rate (%)	Key Generation Rate (bps)	Success of Transmission (%)
1	RSU at Intersection A	KDC	0.5	1000	98	0.48	1100	96
2	Traffic Light A	Intermediate Node	0.7	800	95	0.7	900	92
3	Traffic Light B	Intermediate Node	0.6	820	96	0.58	840	96
4	Traffic Light C	Intermediate Node	0.8	780	94	0.7	760	92
5	Traffic Light D	Intermediate Node	0.9	750	92	0.83	742	90
6	Traffic Light E	Intermediate Node	0.7	800	95	0.65	810	93
7	Traffic Light F	Intermediate Node	0.6	820	96	0.6	810	94
8	Traffic Light G	Intermediate Node	0.8	780	94	0.76	762	90
9	Traffic Light H	Intermediate Node	0.9	750	92	0.79	751	90
10	RSU at Intersection B	End Node (Vehicle)	0.5	1000	98	0.48	1100	97

Table 4. Simulation Comparison 1: V2I with QKD vs. Without QKD on quality parameters

Metrics	V2I with QKD and optimization	V2I without QKD
Error Rate (%)	0.5	.48
Key Generation Rate (bps)	1000	800
Success of Transmission (%)	98	90
Security	High	Low
Cost	High	Low
Complexity	High	Low
Resistance to Eavesdropping	Strong	Weak
Reliability	High	Moderate

Table 5. Simulation Comparison 2: Performance Metrics

Metrics	V2I with QKD	V2I without QKD
Latency (ms)	50	30
Throughput (Mbps)	5	8
Computational Overhead	Moderate	Low
Scalability	Moderate	High
Suitable Environments	Highly secure	Less secure

Apart from the numerical drop of packets, key generation, rates in success of transmission, other parameters like securities, cost complexity, Resistance to Eavesdropping and reliability was compared over the existing setup experiments and comparative results are shown in Table 4. The performance of experiments was checked on latency and throughput ad computation overhead, scalability and reliability given in Table 5. Comparative performance of V2I communication implementations using Quantum Key Distribution (QKD) versus traditional methods across various parameters in a real-world environment was furnished over the comparative chart Figure 5, & 6. Note that these results are subject to specific implementation details and environmental conditions.

Figure 6. Comparison of results on performance parameters on respective experiments

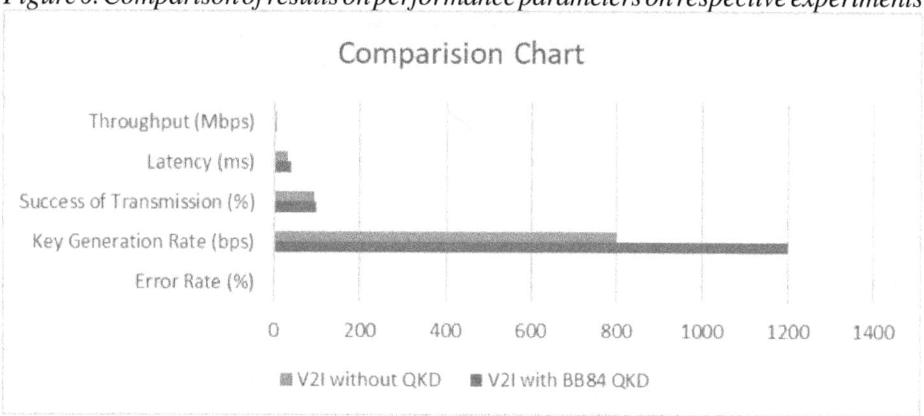

Figure 7. Visualization of result comparison with and without QKD optimization

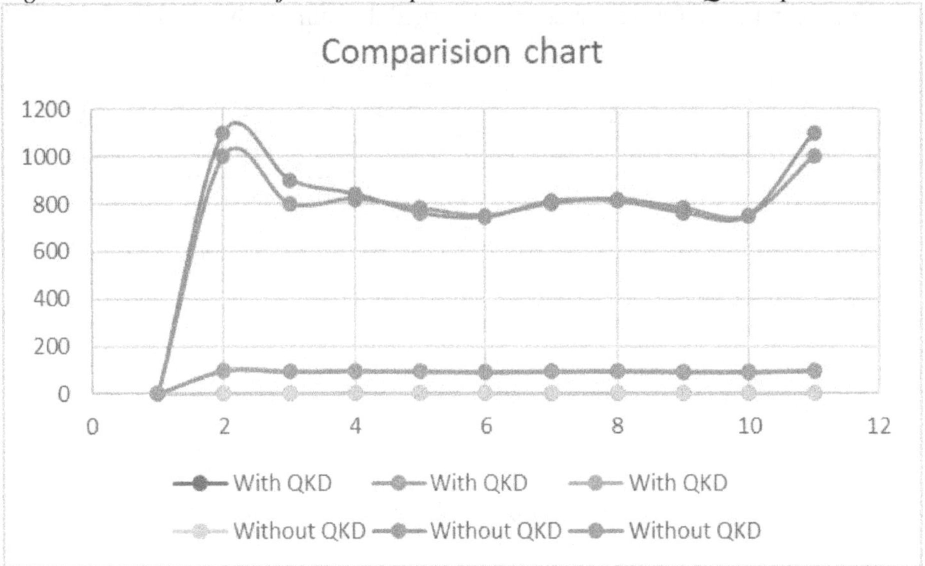

CONCLUSION

As shown in the results collected from experiment 1 and experiment 2 to set up, the present showcases the better performance of the Vehicular to infrastructure system based on the Quantum key distribution method using key optimization technique. Figure 6 in 5 this is visible that with QKD system can be the optimal system for the implementation of V2I in the physical environment. For the application domain of smart projects, it is necessary to deploy secure, channelized, lightweight, software-defined networks that must be easy to manage and up-holding the schemes for better management of traffics and a resource for public interest. This will also help for managing the incidence response mechanism from the government in case of incidents or disaster management.

In this article implementation of a prototype for control architecture of QKD was tested and result was identified for getting the attentions of key management through Quantum key distribution and BB84 protocol proposed optimization on Quantum key. These methods are the latest explorations in the SDN controlled network on V2I infrastructure that may be used for trusted system. Article results demonstrate a new functional environment and shows performance and efficiency of the quantum technology on key distribution methods the compatibility of SDM controller maybe considered as a criterion for reducing error percentage and the resource allocation algorithm can be the future issues of concerned domain for the researchers.

REFERENCES

Aguado, A., Lopez, V., & Lopez, D.. (2019). The engineering of software-defined quantum key distribution. networks. *IEEE Communications Magazine*, *57*(7), 20–26.

Ali, Z., Lagen, S., Giupponi, L., & Rouil, R. (2021). 3GPP NR V2X Mode 2: Overview, Models and System-Level Evaluation. *IEEE Access : Practical Innovations, Open Solutions*, *9*, 89554–89579. DOI: 10.1109/ACCESS.2021.3090855 PMID: 37461405

Alladi, T., Chamola, V., Sahu, N., Venkatesh, V., Goyal, A., & Guizani, M. (2022). A comprehensive survey on the applications of blockchain for securing vehicular networks. *IEEE Communications Surveys and Tutorials*, *24*(2), 1212–1239.

Ansari, K. (2020, December 31). Joint use of DSRC and C-V2X for V2X communications in the 5.9 GHz ITS band. *IET Intelligent Transport Systems*, *15*(2), 213–224. DOI: 10.1049/itr2.12015

Cao, Y., Zhao, Y., Colman-Meixner, C., Yu, X., & Zhang, J. (2017). Key on demand (KoD) for software-defined optical networks secured by quantum key distribution (QKD). *Optics Express*, *25*(22), 26453–26467.

Cao, Y., Zhao, Y., Wang, Q., Zhang, J., Ng, S. X., & Hanzo, L. (2022). The evolution of quantum key distribution networks: On the road to the qinternet. *IEEE Communications Surveys and Tutorials*, *24*(2), 839–894. DOI: 10.1109/COMST.2022.3144219

Chen, T.-Y., Jiang, X., Tang, S.-B., Zhou, L., Yuan, X., Zhou, H., Wang, J., Liu, Y., Chen, L.-K., Liu, W.-Y., Zhang, H.-F., Cui, K., Liang, H., Li, X.-G., Mao, Y., Wang, L.-J., Feng, S.-B., Chen, Q., Zhang, Q., & Pan, J.-W. (2021). Implementation of a 46-node quantum metropolitan area network. *NPJ Quantum Information*, *7*(1), 1–6. DOI: 10.1038/s41534-021-00474-3

Chunduru, A., & Lenka, S. (2024). A Secure Method of Communication Through BB84 Protocol in Quantum Key Distribution. Scalable Computing. *Practice and Experience.*, *25*(1), 21–33. DOI: 10.12694/scpe.v25i1.2152

Costandoiu, A., & Leba, M. (2019, February 18). Convergence of V2X communication systems and next generation networks. *IOP Conference Series. Materials Science and Engineering*, *477*, 012052. DOI: 10.1088/1757-899X/477/1/012052

ETSI, G. (2021, March). 015. Quantum Key Distribution (QKD); Control Interface for Software Defined Networks.

. A. Gatto, M. Brunero, M. Ferrari, et al., "A BB84 QKD Field-Trial in the Turin Metropolitan Area," in PSC, 2021, Tu1A–2.

. Houari, A., & Mazri, T. (2021). Improving V2X-6G network capacity using a new UAV-based approach in a Cloud/ICN architecture, case Study: VANET network. E3S Web of Conferences, 297, 01019. https://doi.org/DOI: 10.1051/e3sconf/202129701019

Ivanov, I., Maple, C., Watson, T., & Lee, S. (2018, March). Cyber security standards and issues in V2X communications for Internet of Vehicles. In *Living in the Internet of Things: Cybersecurity of the IoT-2018* (pp. 1–6). IET.

Joshi, S. K., Aktas, D., Wengerowsky, S., Lončarić, M., Neumann, S. P., Liu, B., Scheidl, T., Lorenzo, G. C., Samec, Ž., Kling, L., Qiu, A., Razavi, M., Stipčević, M., Rarity, J. G., & Ursin, R. (2020). A trusted node–free eight-user metropolitan quantum communication network. *Science Advances*, *6*(36), 73–81. DOI: 10.1126/sciadv.aba0959 PMID: 32917585

Leach, P., Mealling, M., & Salz, R. (2005). A universally unique identifier (uuid) urn namespace (No. rfc4122).

Martinelli, M., Martelli, P., & Gatto, A. (2022). "POLIQI: Milano quantum infrastructure," in ICOP.

Matsuda, K.. (2021). *"Field Demonstration of Real-time 14 Tb/s 220 m FSO Transmission with Class 1 Eye-safe 9-aperture Transmitter," Optical Fiber Communications Conference and Exhibition*. OFC.

Singh, M., & Kim, S. 2017, Introduce reward-based intelligent vehicles communication using blockchain, International SoC Design Conference (ISOCC), Seoul, Korea (South), pp. 15-16

Zhang, Q., Ayoub, O., & Gatto, A.. (2022). Joint routing, channel, and key-rate assignment for resourcie-efficient qkd networking. *IEEE Globecom Workshops*.

Chapter 6
Fuzzy Decision Model for Selecting Core Factors in Vehicular Network Communications

M. Clement Joe Anand
https://orcid.org/0000-0002-1959-7631
Mount Carmel College, Bengaluru, India

M. Santoshi Kumari
https://orcid.org/0000-0002-0391-1171
Chaitanya Bharathi Institute of Technology, India

Devasis Pradhan
https://orcid.org/0000-0002-8201-2210
Acharya Institute of Technology, India

Shaik Khaja Mohiddin
Koneru Lakshmaiah Education Foundation, India

ABSTRACT

The efficacy of the vehicular networks is highly dependent on the choice of communication models. The vehicular network communication models outline the nature of interactions between the entities of vehicles, infrastructure, and the environment. Some of the most common communication models in existence are Vehicle-to-Vehicle, Vehicle-to-Infrastructure, Vehicle-to-Grid, and Vehicle-to-Network. These models are categorized based on the entities involved, direction, and purpose of communication. The choice-making of these communication models are hurdled by the primary challenges of heterogeneity of applications and dynamic network conditions. This

DOI: 10.4018/979-8-3693-6422-2.ch006

chapter proposes solutions to these challenges by constructing a fuzzy-based multi-criteria decision-making method of LBWA (Level Based Weighting Assessment). This fuzzy model considers the factors to be taken into account in selecting the vehicular network communication models and also determines the weightage to each of the factors considered. This model developed in this chapter facilitates identifying the most significant factors in selecting the communication models. The optimal ranking of the factors is validated with other decision methods to determine the consistency of the results. The chapter concludes with future research directions.

1. INTRODUCTION

Vehicular networks are revolutionizing transportation systems and are emerging as a significant component in establishing smooth networking. These wireless networks form the building blocks of the intelligent transportation system and they are crucial in managing traffic congestion and safety measures. The vehicular network communication models (VNC) give directions for the vehicles to interact with other vehicles, infrastructure, and the immediate environment. These VNC models shall be categorized into three groups namely Direct vehicle communication, Vehicle to infrastructure communication, and Vehicle to network communication (Aljeri & Boukerche, 2023). Direct vehicle communication encompasses different communication models that facilitate direct interaction with every vehicle, pedestrians, and electronic devices. This group of models comprises Vehicle-to-Vehicle (V2V), Vehicle-to-Pedestrian (V2P) and Vehicle-to-Device (V2D). Direct communication between the vehicles is governed by V2V kind of communication networks and these networks consider the exchange of data related to the position, speed, acceleration, and other pertinent data (Almeida, 2023). The communication between vehicles and pedestrians is guided by V2P kind of networks (Appaji & Raviraj, 2023). The communication between the entities of the vehicle and external devices is networked by V2D type of communication models. These kinds of communication networks play a crucial role in averting traffic congestion and in maintaining smooth traffic flow.

Vehicle-to-infrastructure (V2I) communication and Vehicle-to-Grid (V2G) communication are network models contributing to the interaction between vehicles and various infrastructure elements. V2I communication deals with direct communication between vehicles and roadside infrastructure, such as traffic lights, road signs, toll booths, and roadside sensors. This type of networking facilitates in receiving real-time data from the elements of infrastructure namely the traffic status, road conditions, and other related information. V2G communication enables bi-directional communication between electric vehicles (EVs) and the power grid infrastructure. Further, this kind of networking facilitates the electric vehicles to emancipate stored energy to the grid on a demand basis. Vehicle-to-network (V2N) communication facilitates the exchange of data between vehicles and centralized servers or cloud-based platforms. This type of network enables access to services such as traffic updates, navigation assistance, and many others. Also, this model is crucial in constructing a well-connected transportation system. This communication model supports over-the-air software updates, remote diagnostics, and vehicle monitoring, which further enhance fleet efficiency and maintenance (Cui & Chen, 2023). V2N communication also reinforces emerging technologies such as autonomous driving and connected mobility services by providing vehicles with access to cloud-based computing resources and AI algorithms. A comparative analysis of these VNC is presented in Table 1 based on certain characteristics.

Table 1. Comparison of Different Vehicular Communications Network

Characteristic	Vehicle-to-Vehicle (V2V)	Vehicle-to-Infrastructure (V2I)	Vehicle-to-Network (V2N)	Vehicle-to-Device (V2D)	Vehicle-to-Grid (V2G)
Communication Entities	Vehicles	Vehicles and Roadside Infrastructure	Vehicles and Central Servers	Vehicles and Electronic Devices	Vehicles and Power Grid
Communication Range	Short-range (few hundred meters)	Medium to long-range (extended coverage)	Long-range (network connectivity)	Short-range (proximity-based)	Medium to long-range (grid connectivity)
Purpose	Cooperative driving, safety	Traffic management, data exchange	Access to services, data exchange	Entertainment, navigation	Energy exchange, grid integration
Typical Applications	Collision avoidance, platooning	Dynamic traffic signal control, ADAS	Traffic updates, infotainment	Media streaming, hands-free calling	Vehicle-to-home, demand response
Key Technologies	DSRC, C-V2X	Dedicated roadside units, cellular networks	Cellular networks, cloud services	Wi-Fi, Bluetooth	Smart charging protocols, energy management systems
Primary Challenges	Latency, reliability	Infrastructure deployment, interoperability	Network coverage, scalability	Device compatibility, data security	Grid stability, interoperability
Safety Considerations	Critical for safety applications	Enhances safety through traffic management	Safety-critical applications	Safety-critical applications	Safety-critical applications

On the whole, these vehicular communication models provide several advantages that contribute to safer, more efficient, and interconnected transportation systems (Doosti-Aref et al., 2023). They facilitate the real-time exchange of information between vehicles, infrastructure, and central servers, offering solutions to the problems caused by traffic collisions, mismanagement, and other challenges of safety. These communication models pave the way for smarter, safer, and more connected transportation systems. The different VNC models are developed based on the requirements, application needs, infrastructure availability, and technological capabilities (Pandey, Gupta, et al., 2023)(Pandey, Perumalla, et al., 2023). There are various factors to be considered in designing and developing vehicular network communication models and the core factors have to be identified in optimal management. This research work presents the factors essential in drawing VNC models and also lays significant focus on choosing the core factors using fuzzy-based decision-making method. The optimal results are sensitively analyzed. The future directions and the concluding remarks together end this chapter.

2. STATE OF ART

This section presents the literature review of the interventions of multi-criteria decision-making methods within the context of vehicle network communication models and the associated framework. In recent years, research in vehicular communication and decision-making has advanced significantly. Purkait and Tripathi (Purkait & Tripathi, 2020) discussed a fuzzy logic-based multi-criteria intelligent routing approach for VANETs, while Nandy et al (Nandy et al., 2020) outlined a multiple-criteria decision analysis for vehicular networks. (Arif et al., 2021) proposed a channel decision model for cooperative CR-enabled Internet on Vehicles using DAHP-TOPSIS, and (Dzemydienė et al., 2021) associated multi-criteria decision support with smart services infrastructure for sustainable freight transportation. Furthermore, Alawi et al (Alawi et al., 2021) designed a multi-criteria prediction approach for vehicular Wi-Fi offloading, while Shah et al (Shah et al., 2021) identified an optimal shortest path selection based on multi-criteria decision-making in Vehicular Ad Hoc Networks. (Evangeline & Kumaravelu, 2020) developed an AHP-FVIKOR-based access network selection method in vehicular communications.

Pradhan et al (Pradhan et al., 2022) discoursed a multi-criteria route selection method for vehicles using genetic algorithms based on driver's preference, and (Bersali et al., 2022) proposed a cooperative clustering approach based on multi-criteria decision-making for the Internet of Vehicles. Singh et al (Singh et al., 2023) sketched out an efficient vehicular-relay selection scheme for vehicular communication, and Taylor et al (Taylor et al., 2023) conducted a comparative analysis of multi-criteria decision methods for secure beacon selection in vehicular platoons. Jahangeer (Jahangeer, 2023) developed a fast multicriteria network selection scheme using a hybrid of cat swarm optimization and TOPSIS algorithm for optimal handover in VANET, while Zhao et al (Zhao et al., 2023) dealt with a multi-criteria contribution evaluation scheme for federated learning in the Internet of Vehicles. Nath et al (Nath et al., 2023) proposed a multi-criteria approach for identifying and ranking key interventions for the seamless adoption of electric vehicle charging infrastructure, and Mekala et al (Mekala et al., 2023) presented an object-aware multi-criteria decision-making approach using heuristic data-driven theory for intelligent transportation systems. Furthermore, (Chai & Ren, 2023) proposed multi-objective evolutionary optimization for spectrum allocation and power control in vehicular communications. Ndashimye (Ndashimye et al., 2021) and Tiwari et al (Tiwari et al., 2022) developed a novel framework for recommending various communication networks using multi-criteria. Various types of fuzzy parameters were discussed in this decision-making in vehicular communications (Varalakshmi et al., 2023)(Bharatraj & Anand, 2019)(Anand & Bharatraj, 2019a)(Anand & Bharatraj, 2018)(Anand & Bharatraj, 2017)(Miriam et al., 2023)(Devadoss et al., 2014)(Anand et al., 2023a)(Anand et al., 2023b)(Raj et al., 2023)(Prabha et al., 2024)(Kungumaraj et al., 2024)(Rajesh et al., 2024)(Manshath et al., 2024)(Sudha et al., 2024).

From the aforementioned literature, it is quite evident that conventional multi-criteria decision methods such as TOPSIS, AHP, and VIKOR are primarily used in ranking the alternatives. However, the method of Level Based Weighting Assessment is not applied. Moreover, the factors that are significant in pertinent to VNC are also not explored to the best of our knowledge. These are a few shortcomings identified from the literature. This has motivated the authors to investigate on the factors related to VNC and to develop fuzzy-based LBWA to determine the core factors.

3. METHODOLOGY OF FUZZY LBWA

This section presents the steps involved in finding the core factors using the method of Fuzzy Level Based Weighting Assessment. This fuzzy-based method is a modification of the method of LBWA developed by Malisa and Pamucar (Žižović & Pamucar, 2019). This method has been applied in several decision-making circumstances such as in evaluating the risks assessments, selecting the alternatives, and performance analysis by integrating with other decision methods. Some of the areas, where the method of LBWA is applied are Logistics, Supplier selection, social enterprises, waste disposal, food industries, and many others (Feng et al., 2023)(Gökhan & Çakir, 2022)(Jahangeer, 2023)(Jokić et al., 2021). The method of LBWA is newly discussed in a fuzzy environment in this section. The steps are presented as a flowchart in Figure 1

Figure 1. Flowchart of Fuzzy LBWA

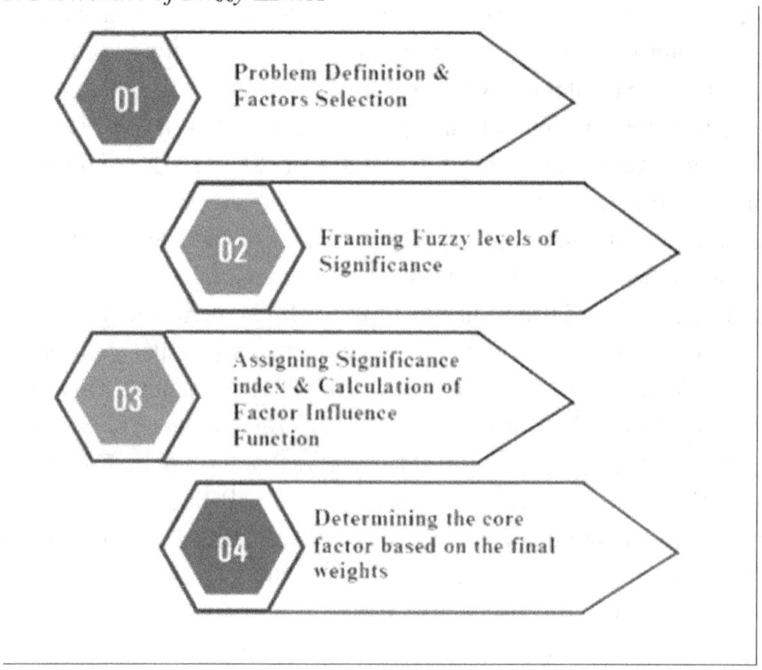

Step 1: Problem Definition and Factor Selection.

This is the first step of this method in which the problem is defined initially and the factors associated with the problem are selected. Let the factors be say CF1,CF2,..CFm.

Step 2: Framing Fuzzy Levels of Significance

In this step, the factor which is considered to be significant is chosen and it is considered to be CF_h. The fuzzy levels of significance say S_L are determined, where each level say S_{Lk} denotes the set of all the factors which is low in significance. In this case L denotes the different linguistic levels of Low.

Step 3: Assigning Significance Index & Calculation of Factor Influence Function

The factors placed in different fuzzy level sets are assigned significance index value say B_h. where h cannot exceed the value of r and r = $\max\{|S_{L1}|, |S_{L2}|, \ldots, |S_{Lp}|\}$ .The factor influence function f: S → R is defined for every criterion as

$$f(CF_j) = \frac{r_0}{i^* r_0 + B_{h_j}}$$

, (1)

here r_0 is the coefficient of elasticity for criterion comparison, j refers to the number of the level set.

Step 4: Determining the Factor Weights

The optimal weight of the significant factor (CF_1) is determined by

$$w_1 = \frac{1}{1 + f(CF_2) + f(CF_3) + f(CF_4) \ldots + f(CF_n)} \quad (2)$$

And the other criterion weights say $w_2, w_3, \ldots w_n$ are calculated as

$$w_j = f(CF_j) * w_1 \quad (3)$$

The process is repeated with different values of r_0 to determine the optimal weights of the factors.

4. FUZZY LBWA IN SELECTING THE CORE FACTORS OF VNC

In this section, the method of Fuzzy LBWA is applied to determine the significant factors to be considered in choice making of vehicular communication networks. The possible factors (Sinha & Saha, 2023) (Suresh, 2023) (Yao et al., 2023) (Yuan et al., 2023) are presented in Table 2.

Step 1: Determination of the Factors associated with VNC

Table 2. Factors and its Description

Factors	Description
Latency (CF1)	Latency refers to the delay between the transmission and reception of data. The latency is expected to be at lower levels to avoid collisions
Reliability (CF2)	Reliability indicates the consistency and dependability of communication. The network is expected to be more reliable to avoid communication errors.
Bandwidth (CF3)	Bandwidth refers to the maximum data transfer rate supported by the communication channel. A large bandwidth is required for maximum coverage.
Coverage (CF4)	Coverage denotes the geographical area within which communication can occur. The extent of coverage is expected to be maximum.
Security (CF5)	Security encompasses measures to protect communication channels, data, and connected devices from unauthorized access, manipulation, and cyber-attacks. The security mechanisms are expected to be strong and shielded.
Scalability (CF6)	Scalability refers to the ability of the communication system to accommodate a growing number of users, devices, and data traffic without compromising performance or efficiency.
Interoperability (CF7)	Interoperability is the ability of different communication systems, devices, and protocols to work together seamlessly.
Energy Efficiency (CF8)	Energy efficiency measures the amount of energy consumed by communication devices and networks.

Step 2: In this case, the factor CF2 i.e Reliability is considered to be the base criterion or the significant

The other criteria are grouped into fuzzy levels of significance say S_L where L refers to the set of linguistic values as stated in section 3.

Let S_{L1} denote the criteria of G that are significantly equal or slightly low in comparison with CF2
Let S_{L2} denote the criteria of G that are significantly somewhat low in comparison with CF2
Let S_{L3} denote the criteria of G that are significantly moderately low in comparison with CFk
Let S_{L4} denote the criteria of G that are significantly low in comparison with CF2
Let S_{L5} denote the criteria of G that are significantly very low in comparison with CF2

These are the five fuzzy levels of significance.

SL1 = {Reliability, Latency}
SL2 = {Bandwidth, Coverage}
SL3 = {Security, Scalability, Interoperability, Energy Efficiency}
SL4 = SL5 = φ

Step 3: The value of r is 4 as $r = \max\{|S_{L1}|, |S_{L2}|, \ldots, |S_{L5}|\}$. The preference to the factors by the decision makers are assumed to be $B_1 = 1$, $B_2 = 0$, $B_3 = 1$, $B_4 = 2$, $B_5 = 1$, $B_6 = 2$, $B_7 = 3$, $B_8 = 4$. By applying factor influence function, the following values are determined. In this case, the coefficient of elasticity is 5.

$$f(C_1) = \frac{5}{1.5+1} = \frac{5}{6} \quad f(C_2) = \frac{5}{1.5+0} = 1 \quad f(C_3) = \frac{5}{2.5+1} = \frac{5}{11}$$

$$f(C_4) = \frac{5}{2.5+2} = \frac{5}{12} \quad f(C_5) = \frac{5}{3.5+1} = \frac{5}{16} \quad f(C_6) = \frac{5}{3.5+2} = \frac{5}{17}$$

$$f(C_7) = \frac{5}{3.5+3} = \frac{5}{18} \quad f(C_8) = \frac{5}{3.5+4} = \frac{5}{19}$$

The above values are tabulated in Table 3

Table 3. Factor Influence values

$f(C_1)$	$f(C_2)$	$f(C_3)$	$f(C_4)$	$f(C_5)$	$f(C_6)$	$f(C_7)$	$f(C_8)$
0.83	1	0.45	0.42	0.31	0.29	0.27	0.26

By using Eq. 2 the value of w_2 is found to be 0.261 and by using the equation 3 the weights of the factors are determined and are presented in Table 4

Table 4. Factor weights

w_1(CF1)	w_2(CF2)	w_3 (CF3)	w_4 (CF4)	w_5(CF5)	w_6(CF6)	w_7 (CF7)	w_8 (CF8)
0.217	0.261	0.117	0.109	0.081	0.075	0.07	0.068

5. DISCUSSIONS

In this table, the weights w_i presented correspond to the weights of the factors CFi. It is very evident that the factor CF2 occupies the highest weight and next to it the factor CF1 occupies the second position followed by other factors. i.e CF2 > CF1> CF3 > CF4 > CF5 > CF6 > CF7 > CF8. This ranking helps the decision makers to make choices of the VNC based on the significance of the factors. Reliability is considered to be the most significant factor followed by latency, bandwidth, and coverage. These are the core factors that are identified to be more important, however, the other factors are also important and they cannot be ignored. The remaining factors are either associated or integrated with these core factors. For instance, though the factor security is considered for ranking, it is an integral component of reliability and it is built with the core factors. Similarly, the factor energy efficiency is an intrinsic aspect of the core factors. The other factors of scalability and interoperability are also the inherent traits of the core factors. Thus the decision makers shall consider or focus on the significant core factors for making choice of vehicular network communications.

5.1 Sensitivity Analysis

The consistency of the rankings is determined by performing a sensitivity analysis of the factor weights concerning different values of the coefficient of elasticity. The factor weights obtained are presented in Table 5.

Table 5. Factor weights with Different values of the coefficient of elasticity

	$r_0 = 5$	$r_0 = 6$	$r_0 = 7$	$r_0 = 8$	$r_0 = 9$	$r_0 = 10$	$r_0 = 11$	$r_0 = 12$	$r_0 = 13$	$r_0 = 14$	$r_0 = 15$
CF1	0.2163	0.2185	0.2202	0.2214	0.2224	0.2232	0.2239	0.2244	0.2249	0.2253	0.2256
CF2	0.2595	0.2550	0.2516	0.2491	0.2471	0.2455	0.2442	0.2431	0.2422	0.24143	0.2407
CF3	0.1179	0.1176	0.1174	0.1172	0.1170	0.1169	0.1168	0.116	0.1166	0.1165	0.1164
CF4	0.1081	0.1092	0.1101	0.1107	0.1112	0.1116	0.1119	0.112	0.1124	0.1126	0.1128
CF5	0.0811	0.0805	0.0800	0.0797	0.0794	0.0792	0.0790	0.0788	0.0787	0.0786	0.0785
CF6	0.0763	0.0765	0.0765	0.0766	0.0767	0.0767	0.0767	0.0767	0.0768	0.0768	0.07683
CF7	0.0721	0.0728	0.0734	0.0738	0.0741	0.0744	0.0746	0.074	0.0749	0.0751	0.0752
CF8	0.0683	0.0695	0.0704	0.071	0.0717	0.0722	0.0726	0.0729	0.0732	0.0734	0.0736

The weights of the factors are influenced by the varying values of the coefficient of elasticity. The weights are found to be converging as the value of r_0 increases. The graphical representations of the same is presented in Figure 2.

Figure 2. Factor Weights versus coefficient of Elasticity

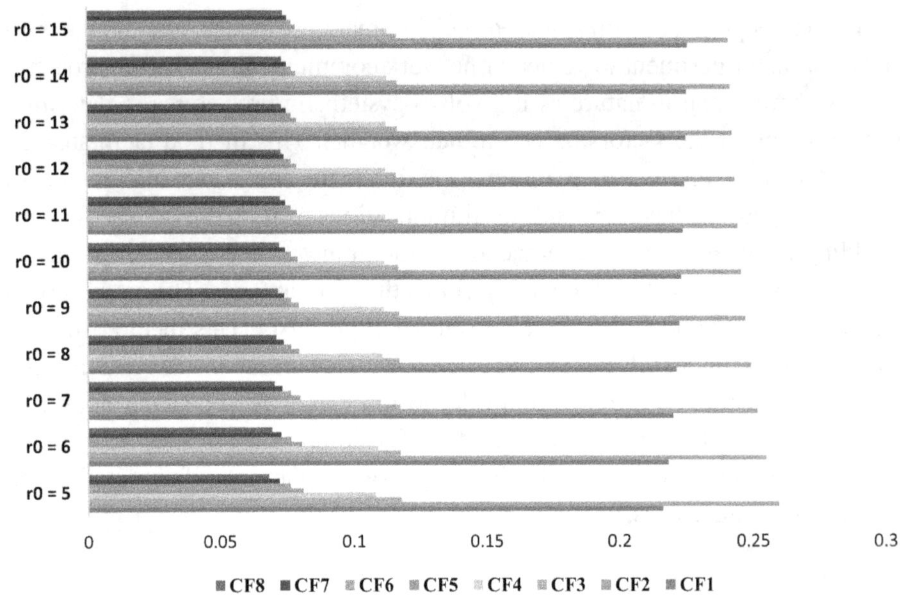

5.2 Industrial Implications

The newly developed Fuzzy based Level Based Weighting Assessment has high potential in selecting vehicular network communication models and this has positive impacts with regard to industrial implications especially in automotive and communication industries. The dynamic networking conditions shall be maintained with the assistance of these communication models. Based on the core factors, the companies shall be able to make feasible decisions on resource allocation, and network performance on par with their competitors. This fuzzy-based selection approach will certainly give directions to the companies in targeting and positioning of their products and services. Moreover, the impreciseness and uncertainty that prevails generally in any decision-making scenario shall be resolved with this fuzzy-based approach. Also, this fuzzy model provides insights for further research and development, facilitating innovation and driving technological advancements in vehicular network communications.

6. CONCLUSION

This chapter proposes a fuzzy-based decision-making model to identify the most significant factor pertinent to vehicular network communication. This approach is a more scientific one in nature as it involves systematic procedures in determining the ranking of the factors based on their weights. This method facilitates the stakeholders in designing a more optimal solution to the choice making problem of VNC. This approach has more industrial implications as it comprises the features of tackling the challenges of resource allocation, improved network performance, and the potential acquisition for gaining competitive advantages. The credibility of the results is also validated by making sensitivity analysis. This method shall be further discussed under different fuzzy extensions. This research will certainly contribute to the creation of different rooms for fostering the theoretical developments of vehicular network communication models. The competency of such fuzzy based decision model will definitely assist the progression in the field of automotive and communication industries.

REFERENCES

Alawi, M., Alsaqour, R., Abdalla, A., Abdelhaq, M., & Uddin, M. (2021). Multi-criteria prediction mechanism for vehicular Wi-fi offloading. *Computers, Materials & Continua, 69*(2), 2313–2337. DOI: 10.32604/cmc.2021.018282

Aljeri, N., & Boukerche, A. (2023, October). Performance Evaluation of Communication Lifetime Prediction Model for Autonomous Vehicular Networks. In *Proceedings of the Int'l ACM Symposium on Performance Evaluation of Wireless Ad Hoc, Sensor, & Ubiquitous Networks* (pp. 17-25).

Almeida, L. G. D. A. (2023). Driver's behavior classification in vehicular communication networks for commercial vehicles.

Anand, M. C. J., & Bharatraj, J. (2017). Theory of Triangular Fuzzy Number. *Proceedings of NCATM, 2017*, 80–83.

Anand, M. C. J., & Bharatraj, J. (2018). Gaussian qualitative trigonometric functions in a fuzzy circle. *Advances in Fuzzy Systems, 2018*, 1–9. DOI: 10.1155/2018/8623465

Anand, M. C. J., & Bharatraj, J. (2019). Interval-valued neutrosophic numbers with WASPAS. In *Fuzzy Multi-criteria Decision-Making Using Neutrosophic Sets* (pp. 435–453). Springer International Publishing., https://link.springer.com/chapter/10.1007/978-3-030-00045-5_17 DOI: 10.1007/978-3-030-00045-5_17

Anand, M. C. J., Martin, N., Clementking, A., Rani, S., Priyadharshini, S., & Siva, S. (2023, March). Decision making on optimal selection of advertising agencies using machine learning. In *2023 International Conference on Information Management (ICIM)* (pp. 121-125). IEEE.

Anand, M. C. J., Moorthy, C. B., Sivamani, S., Indrakumar, S., Kalaiarasi, K., & Barhoi, A. (2023). *Fuzzy intelligence inventory decision optimization model of sustainability and green technologies for mixed uncertainties of carbon emission. In: 2023 International Conference on Information Management (ICIM)*. IEEE., https://ieeexplore.ieee.org/document/10145085 DOI: 10.1109/ICIM58774.2023.00020

Appaji, I., & Raviraj, P. (2023). Framework for simulation of vehicular communication using LSTM-based graph attention networks. *Indian Journal of Science and Technology, 16*(16), 1230–1240. DOI: 10.17485/IJST/v16i16.1777

Arif, M., Kumar, V. D., Jayakumar, L., Ungurean, I., Izdrui, D., & Geman, O. (2021). DAHP–TOPSIS-Based Channel Decision Model for Co-Operative CR-Enabled Internet on Vehicle (CR-IoV). *Sustainability (Basel), 13*(24), 13966. DOI: 10.3390/su132413966

Bersali, M., Rachedi, A., Bouarfa, H., & Badjara, M. E. A. (2022). A novel cooperative clustering approach based on multi-criteria decision-making for IoV. *International Journal of High Performance Systems Architecture*, *11*(1), 36–46. DOI: 10.1504/IJHPSA.2022.121923

Bharatraj, J., & Anand, M. C. J. (2019). Power harmonic weighted aggregation operator on single-valued trapezoidal neutrosophic numbers and interval-valued neutrosophic sets. In *Fuzzy Multi-criteria Decision-Making Using Neutrosophic Sets* (pp. 45–62). Springer International Publishing., https://link.springer.com/chapter/10.1007/978-3-030-00045-5_3 DOI: 10.1007/978-3-030-00045-5_3

Chai, Z., & Ren, J. (2023). Multi-objective evolutionary optimization for spectrum allocation and power control in vehicular communications. *Wireless Personal Communications*, *129*(4), 2767–2789.

Cui, X., & Chen, G. (2023). The application of advanced computational algorithms used for cooperative communication transmission of vehicular networks: A proposed method. *PeerJ. Computer Science*, *9*, e1643. DOI: 10.7717/peerj-cs.1643 PMID: 38077548

Devadoss, A. V., & Anand, M. C. J. (2014). *Felix., A.: A CETD matrix approach to analyze the dimensions of the personality of a person. In: 2014 International Conference on Computational Science and Computational Intelligence*. IEEE., https://ieeexplore.ieee.org/abstract/document/6822301

Doosti-Aref, A., Neelam, S. G., Sahu, P. R., Zhu, X., Basar, E., Coleri, S., & Arslan, H. (2023). Performance Analysis of OTSM under Hardware Impairments in Millimeter-Wave Vehicular Communication Networks. *arXiv preprint arXiv:2309.04161*.

Dzemydienė, D., Burinskienė, A., & Miliauskas, A. (2021). Integration of multi-criteria decision support with infrastructure of smart services for sustainable multi-modal transportation of freights. *Sustainability (Basel)*, *13*(9), 4675. DOI: 10.3390/su13094675

Evangeline, C. S., & Kumaravelu, V. B. (2020). AHP-FVIKOR based access network selection in vehicular communications. In *Inventive Communication and Computational Technologies* [Springer Singapore.]. *Proceedings of ICICCT, 2019*, 829–837.

[] Feng, X., Zibibula, M., & Wei, C. (2023). Evaluation of dynamic technological innovation capability in high-tech enterprises based on pythagorean fuzzy LBWA and MULTIMOORA. Journal of Intelligent & Fuzzy Systems, (Preprint), 1-23.

Gökhan, A. K. E. L. (2022). Prioritization of the Theme Park satisfaction criteria with multi-criteria decision-making method: Level based weight assessment model. *Alphanumeric Journal*, *10*(2), 105–126. DOI: 10.17093/alphanumeric.1028864

Jahangeer, N. (2023). Fast multicriteria network selection scheme using hybrid of cat swarm optimization and TOPSIS algorithm for optimal handover in VANET. *Transactions on Emerging Telecommunications Technologies*, *34*(5), e4743. DOI: 10.1002/ett.4743

Jokić, Ž., Božanić, D., & Pamučar, D. (2021). Selection of fire position of mortar units using LBWA and Fuzzy MABAC model. *Operational Research in Engineering Sciences: Theory and Applications*, *4*(1), 115–135.

Kungumaraj, E., Lathanayagam, E., Saikia, U., Anand, M., Khanna, S. T., Martin, N., & Edalatpanah, S. A. (2024). Neutrosophic Topological Vector Spaces and its Properties. [IJNS]. *International Journal of Neutrosophic Science*, *23*(2).

Manshath, A., Kungumaraj, E., Lathanayagam, E., Anand, M. C., Martin, N., Muniyandy, E., & Indrakumar, S. (2024). Neutrosophic Integrals by Reduction Formula and Partial Fraction Methods for Indefinite Integrals. [IJNS]. *International Journal of Neutrosophic Science*, *23*(1).

Mekala, M. S., Eyad, E., & Srivastava, G. (2023, October). Object-aware Multicriteria Decision-Making Approach using the Heuristic data-driven Theory for Intelligent Transportation Systems. In *2023 IEEE 10th International Conference on Data Science and Advanced Analytics (DSAA)* (pp. 1-8). IEEE. DOI: 10.1109/DSAA60987.2023.10302554

Miriam, M. R., Martin, N., & Anand, M. C. J. (2023). Inventory model promoting smart production system with zero defects. *International Journal of Applied and Computational Mathematics*, *9*(4), 47. Advance online publication. DOI: 10.1007/s40819-023-01524-4

Nandy, T., Idris, M. Y. I. B., Noor, R. M., Ahmedy, I., & Bhattacharyya, S. (2020, December). *A multiple-criteria decision analysis clustering and cluster head selection algorithm in vehicular network. In 2020 IEEE 8th R10 Humanitarian Technology Conference (R10-HTC)*. IEEE.

Nath, A., Rather, Z., Mitra, I., & Sahana, L. (2023). Multi-criteria Approach for Identification and Ranking of Key Interventions for Seamless Adoption of Electric Vehicle Charging Infrastructure. *IEEE Transactions on Vehicular Technology*, *72*(7), 8697–8708. DOI: 10.1109/TVT.2023.3252273

Ndashimye, E., Sarkar, N. I., & Ray, S. K. (2021). A Multi-criteria based handover algorithm for vehicle-to-infrastructure communications. *Computer Networks, 185*, 107652. DOI: 10.1016/j.comnet.2020.107652

Pandey, K., Gupta, A. K., Perumalla, K. R., & Dhillon, H. S. (2023, May). Vehicular Communication Networks with Platooned Vehicles: Modeling and Analysis. In *ICC 2023-IEEE International Conference on Communications* (pp. 3995-4000). IEEE.

Pandey, K., Perumalla, K. R., Gupta, A. K., & Dhillon, H. S. (2023). Fundamentals of vehicular communication networks with vehicle platoons. *IEEE Transactions on Wireless Communications*, *22*(12), 8634–8649. DOI: 10.1109/TWC.2023.3264662

Prabha, S. K., Anand, M., Vidhya, V., Nagarajan, G., Saikia, U., Martin, N., & Tiwari, M. (2024). Sorting Out Interval Valued Neutrosophic Fuzzy Shortest Cycle Route Problem by Reduced Matrix Method. [IJNS]. *International Journal of Neutrosophic Science*, *23*(2).

Pradhan, R., Agarwal, A., & De, T. (2022). A multi-criteria route selection method for vehicles using genetic algorithms based on driver's preference. *Wireless Personal Communications*, *125*(3), 2477–2496. DOI: 10.1007/s11277-022-09670-6

Purkait, R., & Tripathi, S. (2020). Fuzzy logic based multi-criteria intelligent forward routing in VANET. *Wireless Personal Communications*, *111*(3), 1871–1897. DOI: 10.1007/s11277-019-06962-2

Raj, P. J., Prabhu, V. V., Krishnakumar, V., & Anand, M. C. J. (2023). Anand. M.C.J.: Solar Powered Charging of Fuzzy Logic Controller (FLC) Strategy with Battery Management System (BMS) Method Used for Electric Vehicle (EV). *International Journal of Fuzzy Systems*, *25*(7), 2876–2888. Advance online publication. DOI: 10.1007/s40815-023-01537-7

Rajesh, K., Rathod, S., Kundale, J., Rathod, N., Anand, M. C.J., Saikia, U., Tiwari, M,, Martin, N., "A Study on Interval Valued Temporal Neutrosophic Fuzzy Sets", International Journal of Neutrosophic Science, Volume 23, Issue 1, PP: 341-349, (2024). https://doi.org/DOI: 10.54216/IJNS.230129

Shah, A., Hussain, T., & Ali, I. (2021). Optimal Shortest Path Selection Base on Multi-Criteria Decision Making in Vehicular Ad Hoc Network.

Singh, A. K., Pamula, R., Jain, P. K., & Srivastava, G. (2023). An efficient vehicular-relay selection scheme for vehicular communication. *Soft Computing*, *27*(6), 3443–3459. DOI: 10.1007/s00500-021-06106-4

Sinha, S., & Saha, S. (2023). An improved communication strategy in vehicular ad hoc networks: Adaptive game theoretic modelling approach. *International Journal of Mobile Communications*, *21*(2), 273–293. DOI: 10.1504/IJMC.2023.128824

Sudha, S., Martin, N., Anand, M. C. J., Palanimani, P. G., Thirunamakkani, T., & Ranjitha, B. (2023). MACBETH-MAIRCA Plithogenic Decision-Making on Feasible Strategies of Extended Producer's Responsibility towards Environmental Sustainability. Infinite Study.

Suresh, S. (2023). Performance Evaluation of Vehicular Communication in Simulated Environment: A Comprehensive Evaluation of Vehicular Communication in LTE/5G Networks using a Simulated Environment.

Taylor, S. J., Ahmad, F., Nguyen, H. N., Shaikh, S. A., Bryans, J., & Wartnaby, C. E. (2023). A comparative analysis of multi-criteria decision methods for secure beacon selection in vehicular platoons. *Transactions on Emerging Telecommunications Technologies*, *34*(11), e4841. DOI: 10.1002/ett.4841

Tiwari, P. K., Tripathi, A., Srivastava, G., Prakash, S., & Shukla, N. K. (2022). A Novel Framework for Recommendation of Various Communication Networks Using Multi-criteria. In *VLSI, Microwave and Wireless Technologies: Select Proceedings of ICVMWT 2021* (pp. 635-645). Singapore: Springer Nature Singapore.

Varalakshmi, A., Santhosh Kumar, S., Shanmugapriya, M. M., Mohanapriya, G., & Anand, M. C. J. (2023). Markers location monitoring on images from an infrared camera using optimal fuzzy inference system. *International Journal of Fuzzy Systems*, *25*(2), 731–742. DOI: 10.1007/s40815-022-01407-8

Yao, Y., Shu, F., Li, Z., Cheng, X., & Wu, L. (2023). Secure transmission scheme based on joint radar and communication in mobile vehicular networks. *IEEE Transactions on Intelligent Transportation Systems*, *24*(9), 10027–10037. DOI: 10.1109/TITS.2023.3271452

Yuan, W., Du, Z., Meng, X., Liu, F., & Masouros, C. (2023). Integrated Sensing and Communication for Vehicular Networks. In *Integrated Sensing and Communications* (pp. 419–444). Springer Nature Singapore. DOI: 10.1007/978-981-99-2501-8_15

Zhao, Z., Xiang, T., Bi, Y., Zhi, M., & Huang, Z. (2023, July). A Novel Multi-Criteria Contribution Evaluation Scheme for Federated Learning in Internet of Vehicles. In *2023 15th International Conference on Communication Software and Networks (ICCSN)* (pp. 319-325). IEEE. DOI: 10.1109/ICCSN57992.2023.10297351

Žižović, M., & Pamucar, D. (2019). New model for determining criteria weights: Level Based Weight Assessment (LBWA) model. *Decision Making: Applications in Management and Engineering*, *2*(2), 126–137.

Chapter 7
Machine Learning in Estimating Vehicular Efficiency With Special Focus on Customer Acquisition and Retention

R. C. Thivyarathi
RMK College of Engineering and Technology, India

Parul Sharda
https://orcid.org/0000-0003-1188-6903
Medi Caps University, India

B. Ranjitha
https://orcid.org/0000-0003-2138-8430
Mohan Babu University, India

Utpal Saikia
https://orcid.org/0000-0003-0235-6871
Silapathar College, India

M. Clement Joe Anand
https://orcid.org/0000-0002-1959-7631
Mount Carmel College, Bengaluru, India

Charles Robert Kenneth
Loyola College, India

ABSTRACT

The estimation of vehicular efficiency is highly significant in enhancing the sustainability of transportation and optimizing the allocation of resources. Machine learning algorithms are extensively applied in the domain of vehicular networks with a special focus on enriching vehicular potency. In this research work, a machine learning-based estimation model is developed for customer relationship management. The association between vehicular efficiency and customer acquisition cum retention is studied exclusively in this chapter. This work is indeed an intersection of machine learning algorithms, vehicular efficiency calculation, and customer relationship

DOI: 10.4018/979-8-3693-6422-2.ch007

management. The interventions of the customer-oriented strategies with vehicular efficiency are explored in this study by determining the influence of customer-oriented metrics over vehicular efficiency. The results and the insights acquired from this work will certainly facilitate the policymakers and decision-makers in comprehending a comprehensive model integrating customer-centered metrics.

1. INTRODUCTION

The advancing technology is bringing a massive transformation in the automotive industry with an increasing prominence on sustainability. The intervention of machine learning is scaling this industrial sector to elevated levels as it is a significant tool in increasing vehicular efficiency. In the context of an intelligent transportation system vehicular efficiency refers to the effective utilization of the resources with the objective of mitigating emissions, control of traffic congestion, ensuring safety and minimizing the consumption of energy and functional costs. The efficacy of these vehicles is calculated by several factors associated with the characteristics of the vehicles, environmental aspects, operational data, sensor data, and other external factors. As the computation of vehicular efficacy depends on large-scale of data sets, machine learning algorithms are the optimal choice in handling these voluminous data. The competency of the vehicles shall be enhanced by leveraging different algorithms of machine learning which are more competent than the traditional approaches.

Determination of vehicular efficiency is highly a significant phenomenon for the stakeholder's group comprising manufacturers, consumers, and policymakers. The whole framework of the smart transportation system is customer-centered as the choice of the vehicular networks is decided by the satisfaction rate of the consumers. From the perspective of manufacturers, higher efficiency and compliance with environmental standards are key features of vehicular networks. The consumers expect high quality with low-cost benefits and the policy makers are bound to economic and environmental efficiency. The efficiency of the machine learning algorithms has influenced the researchers to employ it in making analysis with respect to several aspects associated with smart intelligent systems. Machine learning is basically a subset of artificial intelligence and it works on the principle of training and testing the data sets to make optimal predictions, estimations and evaluations of vehicular efficiency. These algorithms shall be employed in developing models pertinent to prediction of vehicle failures, estimation of fuel efficiency, carbon emissions and battery usage. This research work intersects not just the machine learning with vehicular efficiency but also explores the possibilities of integrating the aspects of customer acquisition and retention.

Enhancing the vehicular efficiency will significantly contribute to the acquisition and retention of the customers. The present generation customers are highly concerned on environmental sustainability. Hence on applying the machine learning algorithms in enhancing the vehicular efficiency, these automotive industries gain the attention of the customers to a larger extent. In addition, the customers also exercise their loyalty towards the suppliers with superior performances and consistency. The data driven models considers the customers feedback and customer related factors which play a crucial role in comprehending and elevating the vehicular efficiency with special reference to customer acquisition and retention. Personalized recommendations given by the customers, the analysis of the driver's behaviour, service history of the customers, the engagement of the customers and many other aspects related to the customers contribute to the vehicular efficiency in a direct or indirect manner.

This research work attempts in integrating vehicular related characteristics and the customer centric metrics in determining the vehicular efficiency together with the interventions of machine learning algorithms. Customer relationship management and vehicular efficiency are mutually interrelated. The vehicular efficiency influences customer acquisition and retention also, on other hand the customer centered metrics increases vehicular efficiency. This study considers the customer centric features as one of the input components in the development of the ML based prediction model. The remaining contents of the paper is structured as follows. Section 2 sketches the state of art of the related works. Section 3 presents the materials and methods applied in this work. Section 4 applies the ML algorithm of Support vector machine in developing a prediction model. Section 5 discusses the results and the last section concludes the work with future directions.

2. STATE OF ART OF RESEARCH

This section presents a detailed review of the applications of machine learning in the automotive industry; identifies the lacunae concerning the domains of implications; and enumerates the novel contributions made in this chapter.

(Yao & Feng, 2018) describe the extensive role of machine learning in the automotive industry. The advantages of algorithmic framework-based decision-making are well presented. Lopez (Lopez Alduncin, 2018) discussed the ways of employing machine learning to enhance customer satisfaction by predicting customer demands. Peres et al (Peres, Barata, Leitao, & Garcia, 2019) stated the efficacy of machine learning in ensuring quality control in the automotive industry at multiple stages and also presented the key aspects of quality sustenance. Malaca et al (Malaca, Rocha, Gomes, Silva, & Veiga, 2019) illustrated the competency of algorithmic approaches in quality inspection by making optimal detections of faulty items. Bodendorf et al (Bodendorf, Merbele, & Franke, 2019) demonstrated the potency of machine learning in making predictions on the costs associated with the assembling process of the automotive industry. Berges et al (Bergès et al., 2020) employed machine learning strategies in identifying and preventing defects in assembling the components. (Theissler, Pérez-Velázquez, Kettelgerdes, & Elger, 2021) explored the interventions of machine learning techniques in making predictive maintenance of equipment failures. (Bodendorf & Franke, 2021) presented an estimation machine learning model for product designing and development. (Hemmer, Kühl, & Schöffer, 2021) discussed the efficiency of the algorithms in quality checking. (Bergès, Bird, Shroff, Rongen, & Smith, 2021) employed data analytics and machine learning in mitigating the productivity loss and in dealing with the quality checks.

(Ram Seshadri, 2022) elucidated the integration of machine learning tactics with the simulation approaches to optimize the performance. (Pan & Stark, 2022) illustrated the caliber of machine learning in making optimal decisions. (Solke, Shah, Sekhar, & Singh, 2022) explained the applications of learning algorithms in enhancing production efficiency and waste reduction. (Sivabalan & Minu, 2022) forecasted the sales with machine learning tools. (Grube & Berglund, 2022) presented a case study to sketch out the implications of machine learning in test maintenance. (Arakelyan, n.d.) integrated the efficacy of machine learning in making predictions of accident occurrences. (Msakni, Risan, & Schütz, 2023) described the implications of the algorithmic framework in making optimal predictions on quality control. (Gonçalves, Machado, Nara, Dias, & Vaz, 2023) applied in predicting the future prices to compete with the competitors in the automotive market. (Veiga, de Miranda Junior, Silva, Cavalcante, & dos Santos, 2023) explained the efficiency of machine learning in handling customer-quality data. (Kayış, 2023) made demand forecasting with the intervention of both statistical and machine learning methods. (Azouaghe, 2021) discussed the predictions of engine failures. (Bergès, Bird, Shroff, Lumanauw, Raghunathan, & Smith, 2022)(Bergès et al., 2023) tested the potency of machine learning methods in defect identification and hence improving the reliability of the products utilized in the semiconductor industry for automotive applications. (Meinzer, Jensen, Thamm, Hornegger, & Eskofier, 2017) in his research explored the possible ways of enhancing the profits of the automotive industry with machine learning analysis of the customer data and different parameters were discussed. (Varalakshmi, Kumar, Shanmugapriya, Mohanapriya, & Anand, 2022)(Miriam, Martin, & Anand, 2023)(Bharatraj & Anand, 2019)(Anand & Bharatraj, 2019)(Justin Raj, Prabhu, Krishnakumar, & Anand, 2023)(Devadoss, Anand, & Felix, 2014)(Anand & Bharatraj, 2017)(Anand & Bharatraj, 2019)(Sudha, Nivetha, Anand, Palanimani, Thirunamakkani, & Ranjitha, 2023)(Prabha, Anand, Vidhya, Nagarajan, Saikia, Martin, Kumari, Tiwari, 2024)(Kungumaraj, Lathanayagam,

Saikia, Anand, Khanna, Martin, Tiwari, & Edalatpanah, 2024)(Anand et al., 2023)(Rajesh, Rathod, Kundale, Rathod, Anand, Saikia, Tiwari, & Martin, 2024)(Manshath, Kungumaraj, Lathanayagam, Anand, Martin, Muniyandy, & Indrakumar, 2024)(Anand et al., 2023)(Anand, Martin, Clementking, Rani, Priyadharshini, & Siva, 2023)(Anand, Priyadharshini, Martin, Anand, Poorani, Mohiddin, 2023)(Anand, Kalaiarasi, Martin, Ranjitha, Dharshini, & Tiwari, 2023)(Nagaraj, Anand, Priyadharshini, Aparna, 2024)(Prabha, Anand, Vidhya, Nagarajan, Saikia, Martin, Kumari, Tiwari, 2024)(Kungumaraj, Lathanayagam, Saikia, Anand, Khanna, Martin, Tiwari, & Edalatpanah, 2024)(Anand et al., 2023)(Rajesh, Rathod, Kundale, Rathod, Anand, Saikia, Tiwari, & Martin, 2024)(Manshath, Kungumaraj, Lathanayagam, Anand, Martin, Muniyandy, & Indrakumar, 2024)(Anand, Martin, Clementking, Rani, Priyadharshini, & Siva, 2023)(Anand, Priyadharshini, Martin, Anand, Poorani, Mohiddin, 2023)(Anand, Kalaiarasi, Martin, Ranjitha, Dharshini, & Tiwari, 2023)

From the aforementioned literature, it is very evident that machine learning interferences have been in escalation mode over the years. The efficacy of the tools and techniques of machine learning are well demonstrated in the above research works. On intense analysis, these machine learning approaches are leveraged in several areas such as predictions of defects, forecasting of sales and prices, estimation of quality, and related aspects. However, the machine learning interferences with vehicular efficiency are not much explored directly and this is one of the predominant research gaps identified from the literature. In addition to it to the best of the knowledge the machine learning algorithms are not applied in computing the efficiency of the vehicles. This has motivated the authors to make a research contribution by leveraging artificial neural networks to make predictions on vehicular efficiency by considering customer-centric metrics together with vehicle characteristics.

3. MATERIALS AND METHODS

This section presents the procedures of the neural networks in developing a prediction-based decision model. Artificial neural networks are bio-inspired networks that work on the layers. These networks comprise three layers namely input, hidden, and output with several nodes. This input layer comprises neurons signifying input features, hidden layers serve as the intermediate layers acting weight updating with activation function, and the output layer yielding the results. The algorithm of feedforward propagation is applied to determine the predicted outputs and backpropagation is applied to train the neural network and to update the weights. The loss function computes the errors and the recurrent of the process minimizes the errors caused by the differences between the predicted and the actual outputs. The stepwise procedure involved in formulating a machine learning-based model is as follows.

Data Preparation

This is the primary step and it is the most crucial step in preparing the data for modelling. This phase comprises several sub-stages such as a collection of the dataset with input features say X and the target variable say y. The collected data has to be cleansed to remove the missing values, and outliers and to deal with data inconsistency. The features are selected, and extracted and then suitable transformations are made for modeling. Finally, the data sets are split into training, validation, and testing sets.

Selection of the Architecture

The neural network architecture is developed with the optimal number of layers and neurons. The network framework is designed suiting to the nature of the problem, and the characteristics of the data. The choice of the activation function also takes place in this stage.

Hyperparameter Regulation

The tuning of the hyperparameters namely the learning rate, batch size, and many others is an essential step. This is performed by using several optimization techniques to enhance the functioning efficiency of these networks in the validation data sets (X_{val}, y_{val}).

Training the Model

The neural network is subjected to train the data sets (X_{train}, y_{train}) to optimize the loss function (J) with the intervention of backpropagation and gradient descent.

$$\text{Model} = \text{argmin}_{\text{Model}} J\,(X_{train}, y_{train})$$

Evaluating the Model

The competency of the model is evaluated by subjecting the model to the evaluating data sets (X_{val}, y_{val}) and with the suitable metrics of evaluation E.

$$E = \frac{1}{N_{val}} \sum_{i=1}^{N_{val}} L\left(y_{val}, \hat{y}_{val}\right)$$

Selection and Validation of the Model

Based on the performance metrics, the model is selected and validated with the test data (X_{test}, y_{test}) to determine the real efficiency of the model.

Deployment of the Model

The model is deployed to make decisions on new data sets. Based on the results the consistencies of the model are determined. The above-mentioned algorithmic framework is presented in the Fig.1.

Figure 1. ANN Algorithmic Framework

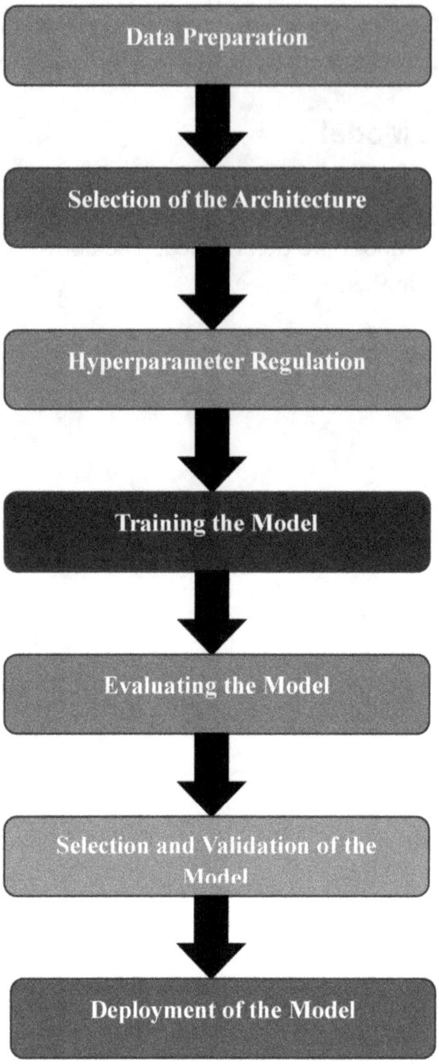

4. APPLICATION OF MACHINE LEARNING IN DEVELOPING PREDICTION MODEL

This section presents a vehicular efficiency prediction model with the intervention of ANN. This model considers different sets of input features which are presented in the Tables 1 and 2. Based on the opinion of the experts, the possible features are presented in the tables. Also, the core categorization of the features is pictorially presented in the Fig.2. The primary sources of the input features are associated with the vehicle characteristics, the external environmental conditions, operational data sets, sensory data and customer related features.

Figure 2. Sources of Input Features

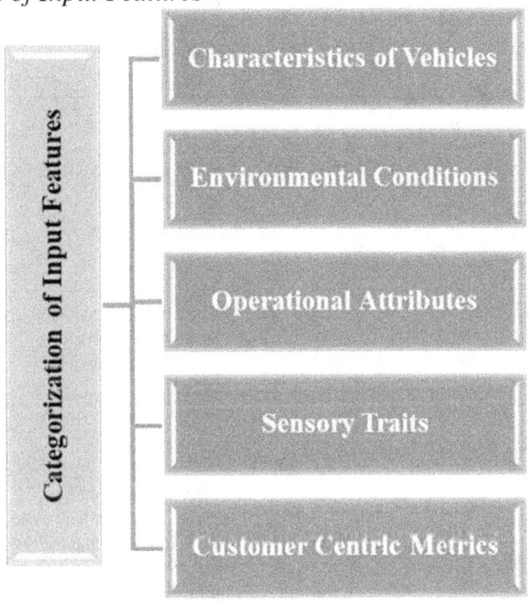

Table 1 presents the input features associated with vehicle nature, environment, operational and sensory.

Table 1. Input Features

Vehicle Characteristics	Environmental Conditions	Operational Attributes	Sensory Traits
• Engine Type • Transmission Type • Vehicle Weight • Aerodynamics • Tire Type and Pressure • Fuel Type • Powertrain Configuration	• Ambient Temperature • Humidity • Altitude • Wind Speed and Direction • Road Type and Condition • Traffic Conditions	• Speed • Acceleration and Deceleration Rates • Engine RPM • Fuel Consumption Rate • Distance Traveled • Idle Time • Gear Shifts • Driving Patterns • Load Weight • Use of Auxiliary Systems • Battery State of Charge • Regenerative Braking Usage	• Oxygen Sensor Readings • Throttle Position • Mass Airflow Rate • Fuel Pressure • Exhaust Gas Recirculation Valve Position • Catalytic Converter Temperature • Tire Pressure Monitoring System Data

Table 2 presents the input features with respect to customer centric metrics

Table 2. Customer Centric Metrics

Customer-centric Metrics	• Customer Demographics (Age, Gender, Income Level) • Purchase History • Service History • Customer Feedback and Reviews • Brand Loyalty Scores • Marketing Campaign Responses • Website Interaction Data (Time on Site, Pages Visited) • Test Drive Data • Sales Interaction Records • Customer Satisfaction Scores (CSAT) • Net Promoter Score (NPS) • Customer Engagement Levels (Service Reminders, Offers) • Subscription to Connected Services • In-car app Usage Data • Social Media Interaction and Sentiment Analysis • Competitive Vehicle Comparisons • Customer Referral Rates • Retention Program Participation • Loyalty Program Membership and Activity

Among the input features presented in Tables 1 and 2, few significant features are determined using the techniques of criterion reduction and they are presented in Table 3 along with their nature.

Table 3. Final Features & Nature

Features	Nature of the Values
Engine Type (F1)	Categorical
Fuel Consumption Rate (F2)	Continuous
Vehicle weight (F3)	Continuous
Aerodynamics (F4)	Continuous
Ambient Temperature (F5)	Continuous
Speed (F6)	Continuous
Distance covered (F7)	Continuous
Customer satisfaction score (F8)	Continuous
Driving patterns (F9)	Categorical
Retention Program Participation (F10)	Categorical
Loyalty Program Membership and Activity (F11)	Continuous

The sample data is presented in Table 4.

Table 4. Sample Data

F1	F2	F3	F4	F5	F6	F7	F8	F9	F10	F11	VE
1	8.3	2501	0.25	18	85	503	4.3	1	1	7	3.654
2	10.6	2203	0.35	20	75	562	4.6	2	0	5	2.158
3	13.9	2845	0.39	23	95	478	3.8	1	1	8	2.984
1	7.6	2763	0.24	25	100	650	4.4	2	0	5	3.356
4	20.1	2004	0.35	10	65	584	3.9	1	1	7	1.056

By subjecting the data to neural network modelling using the neuralnet package with R programming, the following plots and results are obtained. Figure 3 presents ANN plot

Figure 3. ANN Architecture

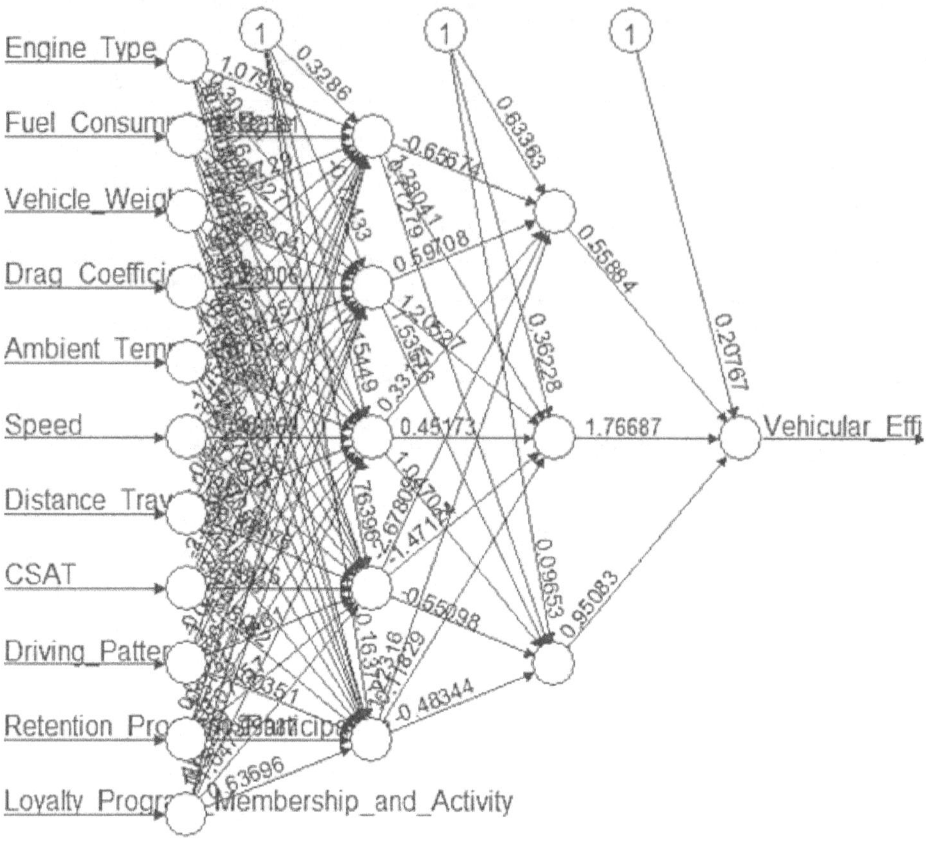

5. SENSITIVE ANALYSIS

This section analyses the efficiency of the prediction model based on the score values of the performance metrics. The measures of the metrics are presented in Table 5.

Table 5. Performance Metrics

MSE	RMSE	MAE	R^2	Adjusted R-squared
0.08	0.2828	0.12	0.95	0.935

From the metric values, the following inferences are obtained

- MSE refers to Mean squared error which measures the average of the squared differences between the actual and the predicted values. The MSE value of 0.08 indicates the predictions made by the model are more accurate.
- RMSE refers to the root mean squared error and it gives a similar expression of the the nature of the model.
- MAE stands for Mean Absolute error which computes the absolute differences between the predicted and the actual values. The value of MAE is the indication of model's high prediction accuracy and consistency.
- R-squared is otherwise referred as the coefficient of determination which computes the proportion of variance between the dependent and independent variables. The high values of R-squared has high explanatory power.
- The adjusted R-squared regulates the R-squared values for the number of predictors in the model. The measure of this metric determines the model performance which indicates the model is well-fitted and the predictors are well justified.

The graphical representation of the performance metrics over the range of epochs is presented in Fig.4.

Figure 4. MSE over Epochs

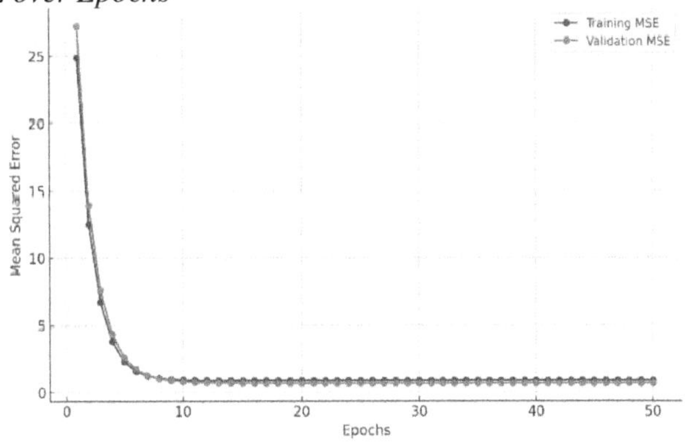

The overall interpretations of the model are as follows. The model has high accuracy, strong predictive power, and suitable model complexity.

6. CONCLUSION

This research work presents the application of neural networks in making optimal predictions of vehicular efficiency. The model proposed in this chapter estimates the efficiency of the vehicles by considering several input features about the domains of vehicle characteristics, environment, operational, sensory traits, and customer-centric metrics. The factors associated with customer relationship management are taken into account in designing this model. The prediction model is more comprehensive as it is an intersection of the domains of the automotive industry, machine learning, and customer-centered factors. The formulation of such an inclusive model will benefit the industries in evolving the strategies of enhancing the efficiency of the vehicles. This model shall be further extended by discussing with other supervised learning algorithms.

REFERENCES

Anand, M. C. J., & Bharatraj, J. (2017). Theory of Triangular Fuzzy Number. *Proceedings of NCATM, 2017*, 80–83.

Anand, M. C. J., & Bharatraj, J. (2018). Gaussian qualitative trigonometric functions in a fuzzy circle. *Advances in Fuzzy Systems, 2018*(1), 8623465.

Anand, M. C. J., & Bharatraj, J. (2019). Interval-valued neutrosophic numbers with WASPAS. In *Fuzzy Multi-criteria Decision-Making Using Neutrosophic Sets* (pp. 435–453). Springer International Publishing.

Anand, M. C. J., Kalaiarasi, K., Martin, N., Ranjitha, B., Priyadharshini, S. S., & Tiwari, M. (2023, September). Fuzzy C-Means Clustering with MAIRCA-MCDM Method in Classifying Feasible Logistic Suppliers of Electrical Products. In *2023 First International Conference on Cyber Physical Systems, Power Electronics and Electric Vehicles (ICPEEV)* (pp. 1-7). IEEE.

Anand, M. C. J., Martin, N., Clementking, A., Rani, S., Priyadharshini, S., & Siva, S. (2023, March). Decision making on optimal selection of advertising agencies using machine learning. In *2023 International Conference on Information Management (ICIM)* (pp. 121-125). IEEE.

Anand, M. C. J., Moorthy, C. B., Sivamani, S., Indrakumar, S., Kalaiarasi, K., & Barhoi, A. (2023). Fuzzy intelligence inventory decision optimization model of sustainability and green technologies for mixed uncertainties of carbon emission. In *In: 2023 International Conference on Information Management (ICIM)*. IEEE. https://ieeexplore.ieee.org/document/10145085

Anand, M. C. J., Priyadharshini, S. S., Martin, N., Anand, M. E., Poorani, B., & Mohiddin, S. K. (2023, September). Decision Trees in the Selection of Electric Vehicles Based on Various Parameters. In *2023 First International Conference on Cyber Physical Systems, Power Electronics and Electric Vehicles (ICPEEV)* (pp. 1-6). IEEE.

Arakelyan, D. Machine Learning-Powered Accident Prediction for the Automotive Insurance Industry.

Azouaghe, Y. (2021). Prediction of engine failures in the automotive industry using machine learning techniques (Doctoral dissertation, Universitat Politècnica de València).

Bergès, C., Bird, J., Shroff, M. D., Lumanauw, E., Raghunathan, S., & Smith, C. (2022, July). Inspection methodologies and machine-learning approaches for defectivity data in semiconductor industry for automotive applications: case study for field-failure prevention. In *2022 IEEE International Symposium on the Physical and Failure Analysis of Integrated Circuits (IPFA)* (pp. 1-7). IEEE. DOI: 10.1109/IPFA55383.2022.9915769

Bergès, C., Bird, J., Shroff, M. D., Rongen, R., & Smith, C. (2021, September). Data analytics and machine learning: root-cause problem-solving approach to prevent yield loss and quality issues in semiconductor industry for automotive applications. In *2021 IEEE International Symposium on the Physical and Failure Analysis of Integrated Circuits (IPFA)* (pp. 1-10). IEEE.

Bergès, C., Céneray, C., Albermann, G., Bird, J., Balakrishnan, B., Mustafa, S. M., & Smith, C. (2020, July). Detection and prevention of assembly defects, by machine learning algorithms, in semiconductor industry for automotive. In *2020 IEEE International Symposium on the Physical and Failure Analysis of Integrated Circuits (IPFA)* (pp. 1-9). IEEE. DOI: 10.1109/IPFA49335.2020.9260582

Bergès, C., Zhang, S., Létévé, A., Anilturk, O., Bird, J., Lumanauw, E., & Smith, C. (2023, July). Test And Reliability Improvement With Defect-Image Classification And Machine-Learning Algorithms In Semiconductor Industry For Automotive Applications. In *2023 IEEE International Symposium on the Physical and Failure Analysis of Integrated Circuits (IPFA)* (pp. 1-9). IEEE.

Bharatraj, J., & Anand, M. C. J. (2019). Power harmonic weighted aggregation operator on single-valued trapezoidal neutrosophic numbers and interval-valued neutrosophic sets. Fuzzy Multi-criteria Decision-Making Using Neutrosophic Sets, 45-62.

Bodendorf, F., & Franke, J. (2021). A machine learning approach to estimate product costs in the early product design phase: A use case from the automotive industry. *Procedia CIRP*, *100*, 643–648. DOI: 10.1016/j.procir.2021.05.137

[] Bodendorf, F., Merbele, S., & Franke, J. (2019). Predictive Cost Analytics of Vehicle Assemblies Based on Machine Learning in the Automotive Industry.

Devadoss, A. V., Anand, M. C. J., & Felix, A. (2014). A CETD matrix approach to analyze the dimensions of the personality of the person. *2014 International Conference on Computational Science and Computational Intelligence*, DOI: 10.1109/CSCI.2014.92

Gonçalves, M. C., Machado, T. R., Nara, E. O. B., Dias, I. C. P., & Vaz, L. V. (2023, November). Integrating Machine Learning for Predicting Future Automobile Prices: A Practical Solution for Enhanced Decision-Making in the Automotive Industry. In Conference on e-Business, e-Services and e-Society (pp. 91-103). Cham: Springer Nature Switzerland.

Grube, T., & Berglund, L. (2022). Test Maintenance for Machine Learning Systems: A Case Study in the Automotive Industry.

Hemmer, P., Kühl, N., & Schöffer, J. (2021). Utilizing Active Machine Learning for Quality Assurance: A case study of virtual car renderings in the automotive industry. arXiv preprint arXiv:2110.09023.

Kayış, E. A. (2023). Demand forecasting using statistical and machine learning methods–a case studyin automotive industry (Master's thesis, İzmir Ekonomi Üniversitesi).

Kungumaraj, E., Lathanayagam, E., Saikia, U., Anand, M., Khanna, S. T., Martin, N., & Edalatpanah, S. A. (2024). Neutrosophic Topological Vector Spaces and its Properties. [IJNS]. *International Journal of Neutrosophic Science*, *23*(2).

Kungumaraj, E., Lathanayagam, E., Saikia, U., Anand, M. C. J., Khanna, S. T., Martin, N., Tiwari, M., & Edalatpanah, S. A. (2024). Khanna, S.T., Martin, N., Tiwari, M., Edalatpanah, S. A.: Neutrosophic Topological Vector Spaces and its Properties.*International Journal of Neutrosophic Science*, *23*(2), 63–76. DOI: 10.54216/IJNS.230206

Lopez Alduncin, M. B. (2018). A customer relationship management, machine learning case study in the automotive service industry.

Malaca, P., Rocha, L. F., Gomes, D., Silva, J., & Veiga, G. (2019). Online inspection system based on machine learning techniques: Real case study of fabric textures classification for the automotive industry. *Journal of Intelligent Manufacturing*, *30*, 351–361.

Manshath, A., Kungumaraj, E., Lathanayagam, E., Anand, M. C., Martin, N., Muniyandy, E., & Indrakumar, S. (2024). Neutrosophic Integrals by Reduction Formula and Partial Fraction Methods for Indefinite Integrals. [IJNS]. *International Journal of Neutrosophic Science*, *23*(1).

Meinzer, S., Jensen, U., Thamm, A., Hornegger, J., & Eskofier, B. M. (2017). Can machine learning techniques predict customer dissatisfaction? A feasibility study for the automotive industry. *Artificial Intelligence Research*, *6*(1), 80–90.

Miriam, M. R., Martin, N., & Anand, M. C. J. (2023). Inventory model promoting smart production system with zero defects. *International Journal of Applied and Computational Mathematics*, *9*(4), 47. DOI: 10.1007/s40819-023-01524-4

Msakni, M. K., Risan, A., & Schütz, P. (2023). Using machine learning prediction models for quality control: A case study from the automotive industry. *Computational Management Science*, *20*(1), 14. PMID: 36942085

Nagaraj, C. T., Anand, M. C. J., Priyadharshini, S. S., & Aparna, P. (2023, December). GCNXG: Detecting Fraudulent Activities in Financial Networks: A Graph Analytics and Machine Learning Fusion. In International Conference on Renewable Energy, Green Computing, and Sustainable Development (pp. 17-32). Cham: Springer Nature Switzerland.

Pan, Y., & Stark, R. (2022). An interpretable machine learning approach for engineering change management decision support in automotive industry. *Computers in Industry*, *138*, 103633.

Peres, R. S., Barata, J., Leitao, P., & Garcia, G. (2019). Multistage quality control using machine learning in the automotive industry. *IEEE Access : Practical Innovations, Open Solutions*, *7*, 79908–79916.

Prabha, S. K., Anand, M., Vidhya, V., Nagarajan, G., Saikia, U., Martin, N., & Tiwari, M. (2024). Sorting Out Interval Valued Neutrosophic Fuzzy Shortest Cycle Route Problem by Reduced Matrix Method. *International Journal of Neutrosophic Science*, *23*(2).

Prabha, S. K., Anand, M. C. J., Vidhya, V., Nagarajan, G., Saikia, U., Martin, N., Kumary, M. S., & Tiwari, M. (2024). Sorting Out Interval Valued Neutrosophic Fuzzy Shortest Cycle Route Problem by Reduced Matrix Method. *International Journal of Neutrosophic Science*, *23*(2), 91–103. DOI: 10.54216/IJNS.230208

Raj, P. J., Prabhu, V. V., Krishnakumar, V., & Anand, M. C. J. (2023). Solar Powered Charging -of Fuzzy Logic Controller (FLC) Strategy with Battery Management System (BMS) Method Used for Electric Vehicle (EV). *International Journal of Fuzzy Systems*, *25*(7), 2876–2888. DOI: 10.1007/s40815-023-01537-7

Rajesh, K., Rathod, S., Kundale, J., Rathod, N., Anand, M., Saikia, U., & Martin, N. (2024). A Study on Interval Valued Temporal Neutrosophic Fuzzy Sets. [IJNS]. *International Journal of Neutrosophic Science*, *23*(1).

Rajesh, K., Rathod, S., Kundale, J., Rathod, N., Anand, M. C. A., Saikia, U., Tiwari, M., & Martin, N. (2024). A Study on Interval Valued Temporal Neutrosophic Fuzzy Sets. *International Journal of Neutrosophic Science*, *23*(1), 341–349. DOI: 10.54216/IJNS.230129

Ram Seshadri, A. (2022). Machine Learning Driven Simulation in the Automotive Industry.

Sivabalan, S., & Minu, R. I. (2022, December). Statistical Sales Forecasting Using Machine Learning Forecasting Methods for Automotive Industry. In International Conference on Communication and Intelligent Systems (pp. 311-319). Singapore: Springer Nature Singapore.

Solke, N. S., Shah, P., Sekhar, R., & Singh, T. P. (2022). Machine learning-based predictive modeling and control of lean manufacturing in automotive parts manufacturing industry. *Global Journal of Flexible Systems Managment*, *23*(1), 89–112.

Sudha, S., Nivetha, M., & Anand, M. C. J. (2023). MACBETH-MAIRCA Plithogenic Decision-Making on Feasible Strategies of Extended Producer's Responsibility towards Environmental Sustainability. *International Journal of Neutrosophic Science*, *22*(2), 114–130. DOI: 10.54216/IJNS.220210

Theissler, A., Pérez-Velázquez, J., Kettelgerdes, M., & Elger, G. (2021). Predictive maintenance enabled by machine learning: Use cases and challenges in the automotive industry. *Reliability Engineering & System Safety*, *215*, 107864.

Varalakshmi, A., Kumar, S. S., Shanmugapriya, M. M., Mohanapriya, G., & Anand, M. C. J. (2022). Markers location monitoring on images from an infrared camera using optimal fuzzy inference system. *International Journal of Fuzzy Systems*.

Veiga, D. T. T. F., de Miranda, A. W., Junior., Silva, L. N., Cavalcante, M. S., & dos Santos, M. D. C. (2023). Applying Machine Learning on automotive customer quality data to improve user experience and increase industry competitiveness (No. 2022-36-0070). SAE Technical Paper.

Yao, B., & Feng, T. (2018). Machine learning in automotive industry. *Advances in Mechanical Engineering*, *10*(10), 1687814018805787.

Chapter 8
Multi-Criteria Approaches in Selecting Optimal Vehicle-to-Vehicle Communication Protocols

N. Anitha
https://orcid.org/0000-0003-0381-2437
Periyar University Centre for Postgraduate and Research Studies, India

Mohit Tiwari
https://orcid.org/0000-0003-1836-3451
Bharat Vidyapeeth's College of Engineering, India

Shaik Moinuddin Imran
Sri Venkateswara Institute of Technology, India

Y. Monikarchana
Mohan Babu University, India

Manish Kumar Thakur
Acharya Institute of Technology, India

M. Clement Joe Anand
https://orcid.org/0000-0002-1959-7631
Mount Carmel College, India

ABSTRACT

Communication protocols are guidelines governing the transmission of data between the entities of the vehicular networks. These protocols play a crucial role in enabling optimal connectivity, ensuring safety and traffic management. The effective functioning of the V2V network depends on the right choice of communication protocols, as the mismatch in selection results in incompatibility, performance degradation, congestion, and other security issues. This chapter focuses primarily

DOI: 10.4018/979-8-3693-6422-2.ch008

on the intervention of a multi-criteria approach in making an optimal selection of the communication protocols used in the V2V network. The decision-making problem comprises the alternatives namely IEEE 802.11p, Cellular-V2X, Dedicated Short-Range Communication (DSRC), LTE-V2X, IEEE 1609.x, and ITS-G5. The criteria considered are Compatibility, Security, Data Rate, Range, Scalability, and Spectral Efficiency. Different MCDM approaches such as AHP, Entropy method, and FUCOM are applied to this linguistic decision matrix to obtain the criterion weights, and methods such as COPRAS, SMART, and MAIRCA are applied in ranking the alternatives. A comparative analysis is made to determine the validity of the criterion weights and ranking results. These combined MCDM approaches shall be applied to other decision-making scenarios of vehicular networking to design optimal solutions.

1. INTRODUCTION

The vehicular communication networks are highly significant in enhancing the smart transportation system. The vehicular network efficiency is purely dependent on the kind of communication protocols. There are several kinds of vehicular communication networks namely V2V, V2I, V2G, V2X which deal with information exchange between the entities namely the vehicles, infrastructure, grid, and other components. The decision-makers have to make the optimal choice of these protocols to ensure the efficient functioning of the vehicular networks. Multi-criteria decision-making methods are the best choice to make an optimal selection of communication protocols. This decision-making framework comprises alternatives criteria and suitable methods of making optimal decisions. This framework comprises two predominant phases one is the computation of criterion weights and the other is the ranking of alternatives.

Multi-criteria decision-making methods comprise several methods to find feasible solutions to the problems. The optimal choice of communication protocols is a basic requisite for enriching and enhancing the network efficiency. The protocols that are generally considered are IEEE 802.11p, Cellular-V2X, Dedicated Short-Range Communication (DSRC), LTE-V2X, IEEE 1609.x and ITS-G5. These protocols possess different attributes and the decision-makers choose either of the protocols based on their requirements. However, the optimal choice of these protocols and rankings of the same is essential at times of drawing decisions on vehicular efficiency. This research work develops a decision-making model using multi-criteria approaches with communication protocols as alternatives and the criteria of Compatibility, Security, Data Rate Range, Scalability, and Spectral Efficiency.

In this research work a combo of decision-making methods are applied. The two sets of methods are labeled as criterion computation and alternative ranking. For the first phase, the methods of the Analytical Hierarchy process, Entropy, and FUCOM are employed. The methods of COPRAS, SMART, and MAIRCA are applied for the second phase. Researchers have used multi-criteria approaches in making decisions but not have explored the decision problem of communication protocols. This has motivated to apply several combinations of methods in computing criterion weights and ranking alternatives of the communication protocols. The remaining of the paper is structured into the following sections. Section 2 sketches the applications of the multi-criteria approach in making decisions with regard to vehicular networks. Section 3 describes the methodology of decision-making methods. Section 4 applies the proposed combination of methods. Section 5 discusses the results and the last section concludes the work.

2. RELATED WORKS

This section presents a detailed description of the state of art of works related to different applications of multi-criteria decision-making methods in obtaining optimal solutions to the problems associated with vehicular networks. Researchers have applied different decision approaches with special focus on multi-criteria and integrated with optimization algorithms and other machine learning frameworks. This literature review sketches out recent trends in research on vehicular networks.

(Chettibi, 2020) formulated a MCDM approach with the conjoin of HF sets in making optimal decisions on VANETs. (Patil & Ragha, 2020) proposed a fuzzy based optimization algorithm in choosing optimal routings. (Ndashimye, Sarkar, & Ray, 2021) framed multi-criteria decision approach in the context of vehicle to-infrastructure communications. (Arif et al., 2021) developed a decision model combining AHP with TOPSIS for choosing competent communication channels. (Malik, 2021) applied AHP-Entropy VIKOR decision method in making optimal ranking of location of road side units. Evangeline and (Evangeline & Kumaravelu, 2022) developed fuzzy based decision approach in selecting communication networks. (Bersali, Rachedi, Bouarfa, & Badjara, 2022) framed clustering based MCDM approach for Internet of Vehicles. (Zeng, Zhang, Zhang, Su, & Carlos, 2022) proposed Pythagorean based decision framework for smart transportation. (Tiwari, Dubey, Prakash, & Tripathi, 2022) developed MAC framework for choosing cognitive vehicular networks. (Chakraborty, Mondal, Alam, Pamucar, & Marinkovic, 2022) framed multi criteria group decision making approach with interval-valued Pythagorean representations in resolving complicated network aspects. (Alawi et al., 2022) employed adaptive QoS decision approach in determining optimal vehicular networks. (Altahrawi, Abdullah, & Nordin, 2022) constructed a MCDM decision model with long short term memory for vehicle-to-infrastructure communication. (Bouakouk, Abdelli, Mokdad, & Othman, 2022) discussed a MCDM model for making routing decisions. (Perla & Ramesh, 2022) framed an integrated decision model with a combination of MCDM and machine learning in assessing the quality of urban road. (Abizar, Farman, Jan, Khan, & Koubaa, 2022) applied the similar approach in assessing IoT-based vehicular ad hoc networks. (Vitianingsih, Baharin, Othman, & Suraji, 2022) applied MCDM approach in classification of traffic accidents. (Magableh & Mumani, 2022) developed simulation based MCDM method to design traffic solutions. (Mamadou & Chalhoub, 2022) proposed a decision model for making selections between LTE-D2D and IEEE 802.11.

(Naderi, Chakareski, & Ghanbari, 2023) developed neutrosophic based AHP for making decisions on vehicular networks. (Malik, Arooj, Jamil, Munir, & Ijaz, 2023) applied MCDM method for making optimal gateway selection. (Yacheur, Ahmed, & Mosbah, 2023) designed deep learning-based decision model for making optimal decisions on hybrid vehicular networks. The adaptability and efficiency are determined through these models. (Naderi & Ghanbari, 2023) applied the decision method for enhancing the routing efficiency. (Okpeh, Okpe, Tekanyi, & Tanko, 2023) developed a supportive multi-criteria handover management scheme for steadfast vehicle-to-infrastructure (V2I) communication. (Tiwari, Prakash, & Tripathi, 2023) formulated metaheuristic-based MAC method in making decisions on cognitive vehicular networks. (Taylor et al., 2023) performed a comparative analysis of the decision methods in determining the vehicular platoons. (Pu, 2023) developed a block chain based MCDM model for VANET. Bhabani and (Bhabani & Mahapatro, 2023) made a performance analysis of hybrid VANET efficiency using AHP-TOPSIS. (Yigit Katanalp, Eren, & Alver, 2023) initiated GIS-based MCDM approach for enhancing traffic safety measures. (Lyu, Yin, Zhou, & Shen, 2023) presented a comprehensive review of the significance of MCDM (Varalakshmi et al., 2023)(Bharatraj & Anand, 2019)(Anand & Bharatraj, 2019)(Anand & Bharatraj, 2018)(Anand & Bharatraj, 2017)(Miriam, Martin, & Anand, 2023)(Devadoss, Anand, & Felix, 2014)(Anand et al., 2023)(Anand et al., 2023)(Raj, Prabhu, Krishnakumar, & Anand, 2023)(Prabha et al., 2024)(Kungumaraj et al., 2024)(Rajesh et al., 2024)(Manshath et al., 2024) (Sudha et al., 2023) in the context of transportation infrastructures. From the afore mentioned literature, some of the shortcomings are identified.

(i) The multi-criteria decision-making models are formulated to make optimal decisions only with regard to limited domains of vehicular networks
(ii) Only very few methods are applied in making optimal decisions
(iii) Decision-making on communication protocols are very limited

.To bridge these gaps, a multi-criteria decision-making model is developed to rank the communication protocols with the integration of methods of AHP, Entropy, FUCOM, COPRAS, SMART, MAIRCA.

3. MCDM METHODOLOGY

This section presents the framework of MCDM methods with special reference to AHP, Entropy, FUCOM, COPRAS, SMART, MAIRCA. The first three methods are used to determine the criterion weights and the later three methods are used in ranking. The algorithmic frameworks are presented for each of the methods (Sriram, Ramachandran, Chinnasamy, & Mathivanan, 2022)(Azhar, Radzi, & Wan Ahmad, 2021).

Analytic Hierarchy Process (AHP)

The method of AHP is initiated by Saaty in the year 1970. This method is a structured technique that uses pairwise comparisons in prioritizing the criteria. In this method the problem is divided into several subproblems for easy comprehension and evaluation. The algorithmic framework of AHP method is presented in Figure 1

Entropy Method

Shannon is the founder of the method of Entropy in the year 1948. This method is based on his information theory. This method calculates the extent of uncertainty existing in the criterion information. The algorithmic framework of the Entropy method is presented in Figure 2

Full Consistency Method (FUCOM)

Dragan Pamucar is the pioneer of FUCOM method and this method is developed in the year 2014. This technique is a systematic approach which calculates the relative significance of criteria. It works on the aspects of ensuring consistency in pairwise comparisons. The algorithmic framework of FUCOM method is presented in Figure 3

Complex Proportional Assessment (COPRAS)

This methos is developed in the year 1996 by Zavadskas. This method works on the principle of considering the relative importance of each alternative by considering both the benefit and non-benefit criteria. The algorithmic framework of COPRAS method is presented in Figure 4.

Simple Multi-Attribute Rating Technique (SMART)

The method of SMART is developed by Bowman in the year 1981. This method is based on utility theory. The additive model is used to associate the ratings and the weights. The algorithmic framework of SMART method is presented in Figure 5.

Multi-Attributive Ideal-Real Comparative Analysis (MAIRCA)

This method is initiated by Neculai Andrei in the year 2013. In this method, the ranking of the alternatives is purely dependent on calculating the deviations from the ideal. The algorithmic framework of MAIRCA method is presented in Figure 6.

Figure 1. AHP Framework

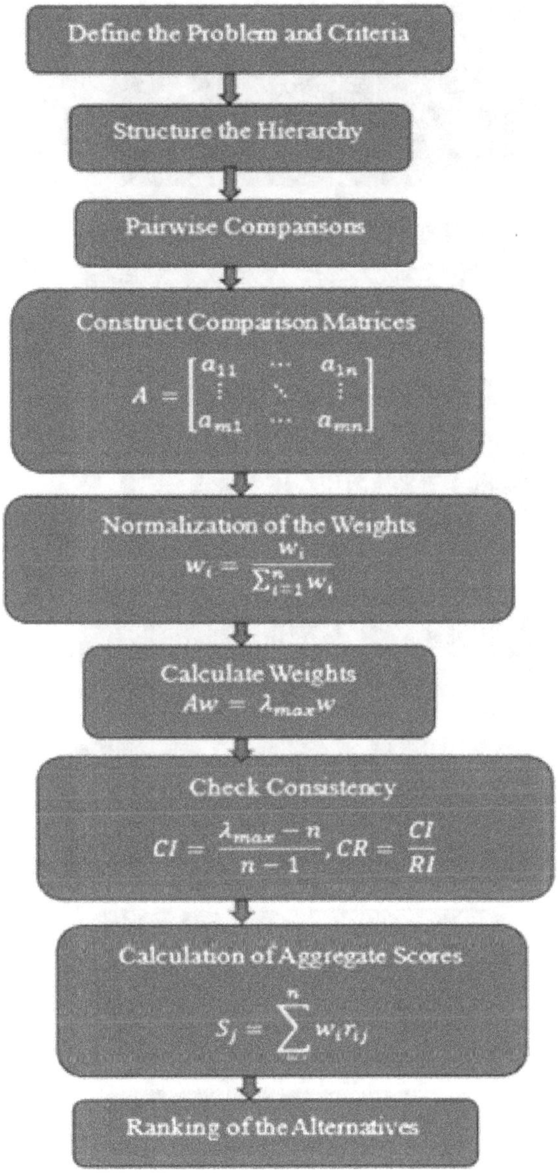

Figure 2. Entropy Framework

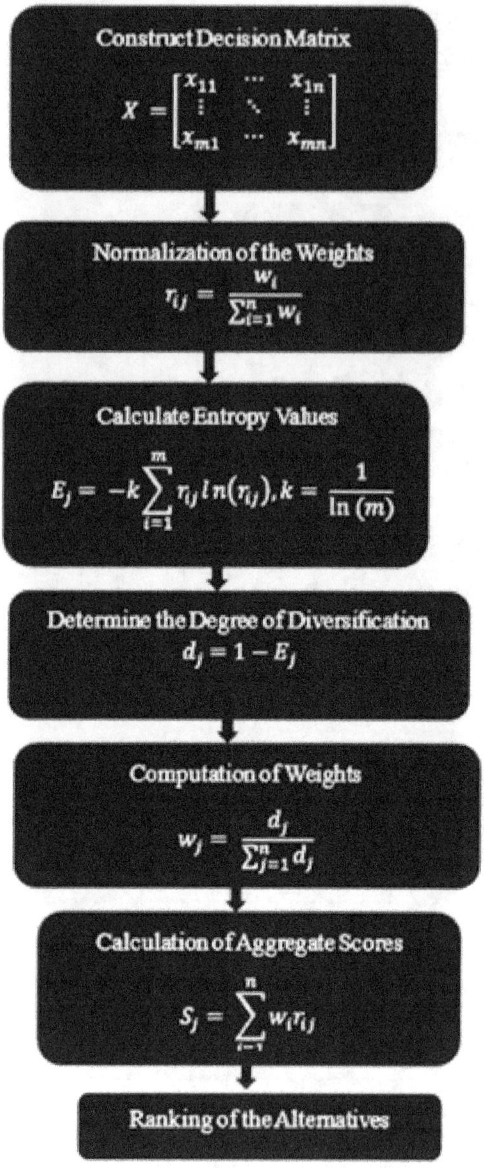

Figure 3. FUCOM Framework

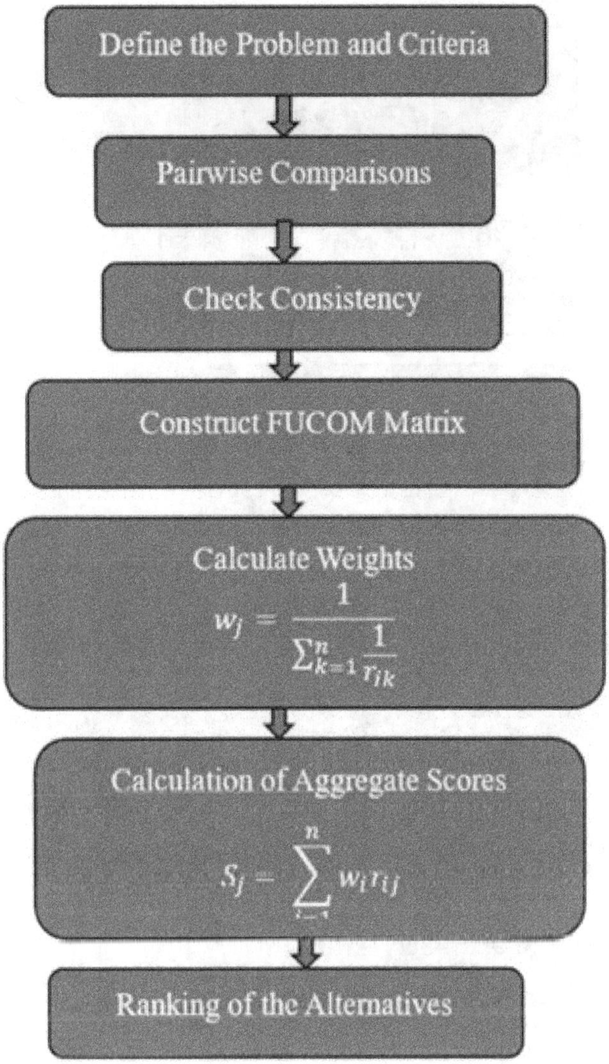

Figure 4 COPRAS Framework

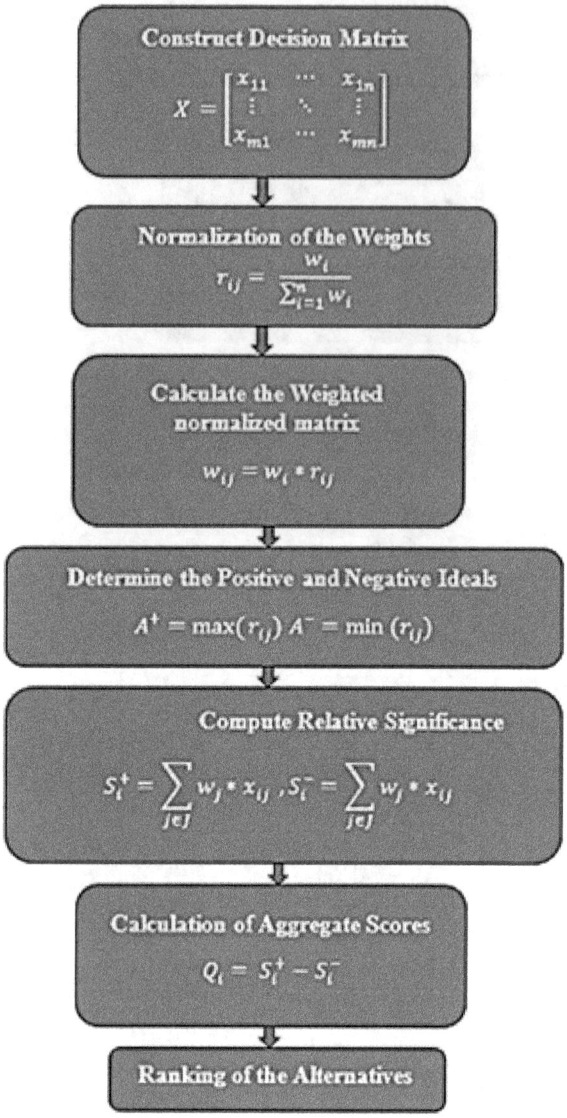

Figure 5. SMART Framework

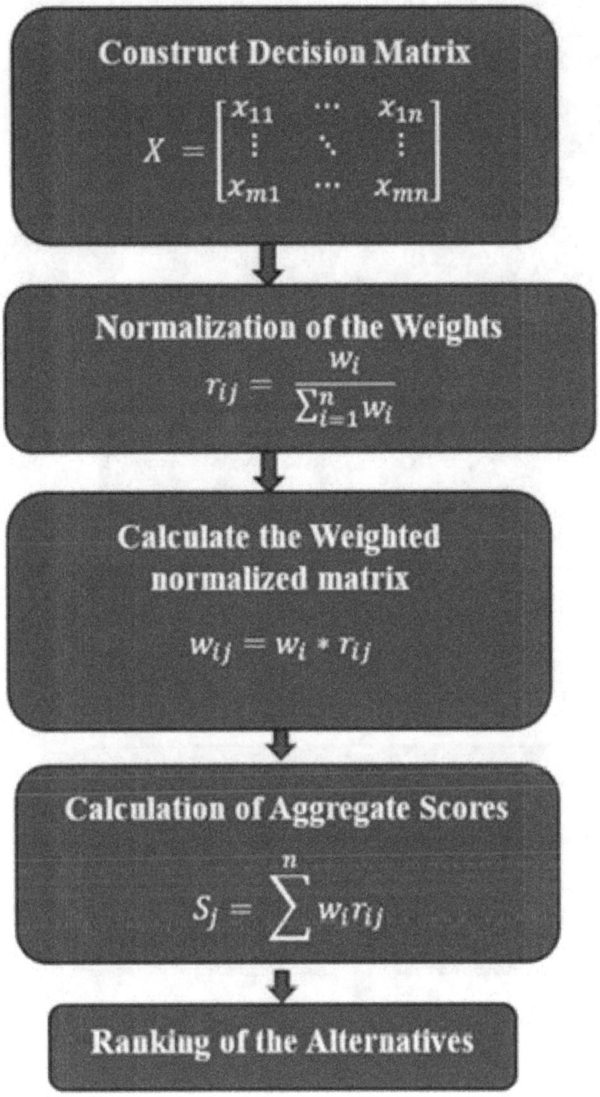

Figure 6. MAIRCA Framework

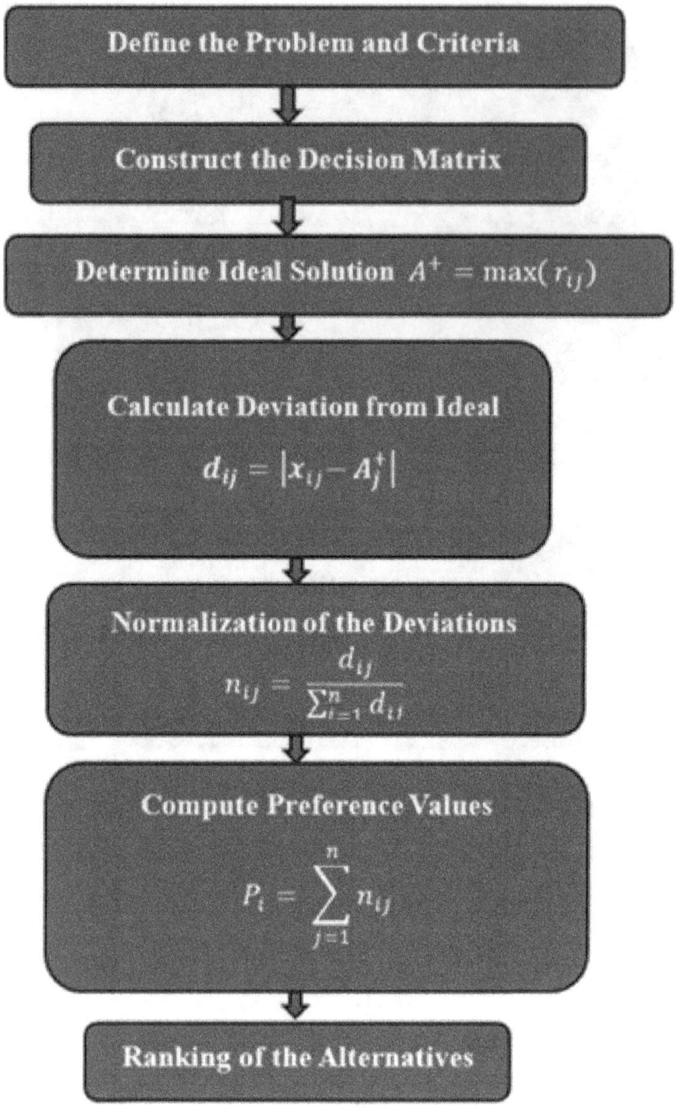

4. MCDM APPLICATION IN RANKING COMMUNICATION PROTOCOLS

In this section the different MCDM methods are applied in computing criterion weights and ranking of the alternatives. The alternatives and the criteria are presented in Table 1.

Table 1. Alternatives & Criteria

Alternatives	Criteria
IEEE 802.11p (WAVE) (A1)	Compatibility (C1)
Cellular-V2X (C-V2X) (A2)	Security (C2)
Dedicated Short-Range Communication DSRC (A3)	Data Rate (C3)
LTE-V2X (A4)	Range (C4)
ITS-G5 (A5)	Scalability (C5)
	Spectral Efficiency (C6)

The decision-making problem considered in this section considers the communication protocols as the alternatives. These protocols are competent in enhancing the vehicular efficiency. However, the ranking of these protocols is significant to make optimal choice of the protocols based on the requirements. The description of the aforementioned communication protocols is presented in Table 2.

Table 2. Description of Communication Protocol

Communication Protocol	Frequency	Latency	Range	Topology	Applications
A1	5.9 GHz	Very low	300 m to 1,000 m	Ad-hoc	Safety-critical applications
A2	5.9 GHz (PC5), Cellular bands (Uu)	Low (PC5), Slightly higher (Uu)	Up to 1,000 m (PC5), Extended via cellular networks	Direct (PC5), Network (Uu)	Safety, autonomous driving, infotainment
A3	5.9 GHz	Low	Up to 1,000 m	Ad-hoc, Infrastructure-based	Safety messages, traffic management, infotainment
A4	5.9 GHz, LTE bands	Low (sidelink)	Up to 1,000 m	Sidelink (PC5), Uplink/Downlink (Uu)	Safety, traffic efficiency, infotainment
A5	5.9 GHz	Low	Up to 1,000 m	Ad-hoc, Infrastructure-based	Safety-critical applications, traffic efficiency, cooperative driving

The initial decision-making matrix is presented as follows. The matrix values are of linguistic variables. The Table 3 presents the construction of the initial matrix.

Table 3. Initial Decision Matrix

Communication Protocol	C1	C2	C3	C4	C5	C6
A1	High	High	Moderate	Short	High	Moderate
A2	High	Very High	High	Short	High	High
A3	Very High	High	Moderate	Medium	Very High	High
A4	High	Moderate	High	Long	Very High	Very High
A5	High	High	Moderate	Medium	High	High

The values presented in the above matrix is converted into numerical values using the rating scale. By using the MCDM methods described in section 3, the criterion weights are computed and the alternatives are ranked. The criterion weights computed using the methods of AHP, Entropy and FUCOM are presented in Table 4.

Table 4. Criterion Weights

Methods	C1	C2	C3	C4	C5	C6
AHP	0.0415	0.0618	0.0983	0.1542	0.2393	0.4049
Entropy	0.166	0.164	0.162	0.168	0.167	0.172
FUCOM	0.1079	0.0973	0.0885	0.1923	0.1282	0.3858

Bu using the criterion weights mentioned in Table 4, the alternatives are ranked based on the score values mentioned in Table 5.

Table 5. Score values of the Alternatives based on AHP Weights

Alternatives	COPRAS	SMART	MAIRCA
A1	0.8370	0.162956	0.5000
A2	0.8300	0.232016	0.5556
A3	0.7739	0.226083	0.4444
A4	0.7743	0.237246	0.4286
A5	0.7490	0.216819	0.4987

The score values obtained using the method of Entropy are presented in Table 6.

Table 6. Score values of the Alternatives based on Entropy Weights

Alternatives	COPRAS	SMART	MAIRCA
A1	0.8290	0.170972	0.5000
A2	0.8037	0.196342	0.5556
A3	0.7946	0.205409	0.4444
A4	0.7816	0.218411	0.4286
A5	0.8107	0.189259	0.4867

The score values obtained using the method of FUCOM are presented in Table 7.

Table 7. Score values of the Alternatives based on FUCOM Weights

Alternatives	COPRAS	SMART	MAIRCA
A1	0.8364	0.163641	0.5000
A2	0.8062	0.193782	0.5556
A3	0.7925	0.207453	0.4444
A4	0.7645	0.235498	0.4286
A5	0.8032	0.196773	0.4753

5. DISCUSSION

The values presented in Table 5,6 and 7 are pictorially represented in Figures 7,8,9 respectively.

Figure 7. AHP Based Scores

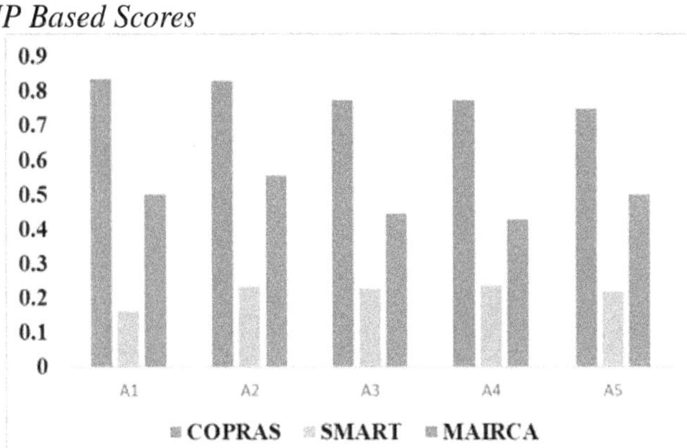

Figure 8. Entropy Based Scores

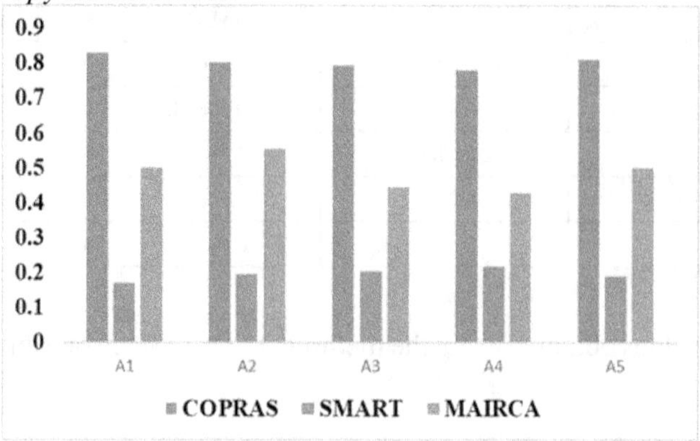

Figure 9. FUCOM Based Scores

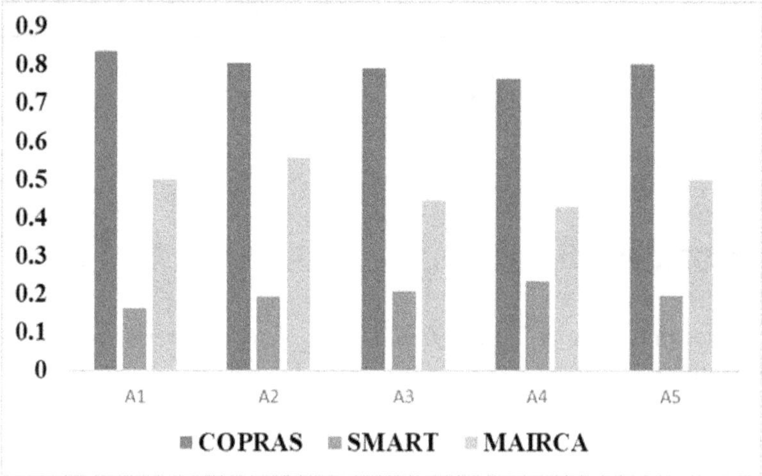

Based on the scores, the alternatives are ranked and the rankings are presented in Table 8.

Table 8. Ranking of the Alternatives

Alternatives	AHP			Entropy			FUCOM		
	COPRAS	SMART	MAIRCA	COPRAS	SMART	MAIRCA	COPRAS	SMART	MAIRCA
A1	1	5	2	1	5	2	1	5	2
A2	2	2	1	2	3	1	2	3	1
A3	4	3	4	4	2	4	4	2	4
A4	5	1	5	5	1	5	5	1	5
A5	3	4	3	3	4	3	3	4	3

The overall rankings stated in Table 8 is graphically presented in Fig.10

Figure 10. Combined Rankings of the Alternatives

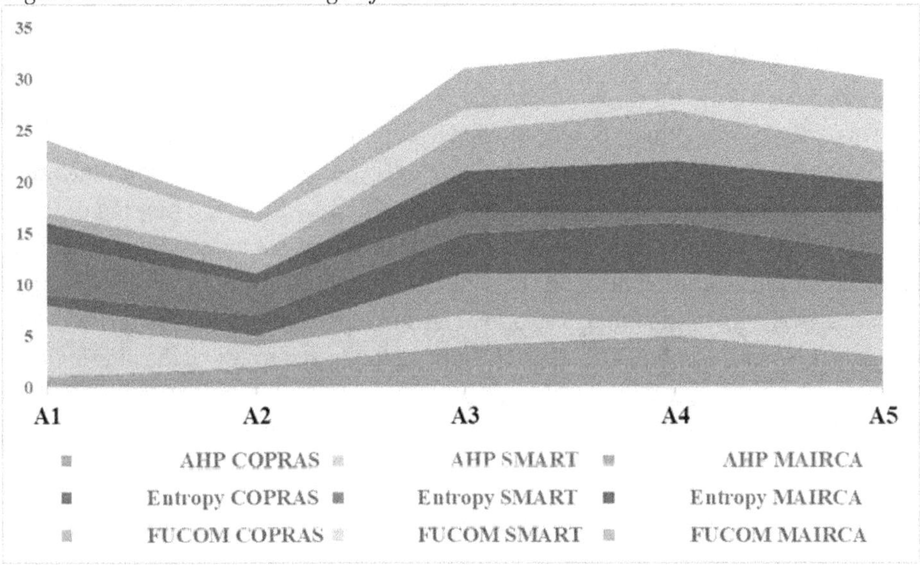

The ranking positions of each of the alternatives is presented in Table 9

Table 9. Ranking Positions of the Alternatives

Alternatives	Ranking Positions				
	I	II	III	IV	V
A1	3	3	0	0	3
A2	3	4	2	0	0
A3	0	2	1	6	0

continued on following page

Table 9. Continued

Alternatives	Ranking Positions				
	I	II	III	IV	V
A4	3	0	0	0	6
A5	0	0	6	3	0

From the ranking positions mentioned in Table 9, it is very evident that the alternative A1 has occupied prime position. Followed by that the alternative A2 has occupied second ranking position. The alternatives A4 and A5 occupies least positions in comparison with all other alternatives and the alternative A3 occupy moderate position.

6. CONCLUSION

This research work presents a comprehensive application of multi-criteria decision-making methods in making optimal decision over the choice of the communication protocols This decision-making problem is more significant as the ranking of these protocols will facilitate the decision makers in the process of enhancing the vehicular efficiency. The combination of these decision-methods shall be applied to other problems pertinent to vehicular networks. The ranking results are more consistent and this research work combines several decision methods in ranking of the alternatives.

REFERENCES

Abizar, F., Farman, H., Jan, B., Khan, Z., & Koubaa, A. (2022). A smart energy-based source location privacy preservation model for Internet of Things-based vehicular ad hoc networks. *Transactions on Emerging Telecommunications Technologies*, *33*(2), e3973. DOI: 10.1002/ett.3973

Alawi, M., Alsaqour, R., Abdelhaq, M., Alkanhel, R., Sharef, B., Sundararajan, E., & Ismail, M. (2022). Adaptive QoS-Aware Multi-Metrics Gateway Selection Scheme for Heterogenous Vehicular Network. *Systems*, *10*(5), 142. DOI: 10.3390/systems10050142

Altahrawi, M., Abdullah, N. F., & Nordin, R. (2022). Service-oriented lstm multi-criteria rat selection scheme for vehicle-to-infrastructure communication. *IEEE Access : Practical Innovations, Open Solutions*, *10*, 110261–110284. DOI: 10.1109/ACCESS.2022.3214852

Anand, M. C. J., & Bharatraj, J. (2017). Theory of Triangular Fuzzy Number. *Proceedings of NCATM*, *2017*, 80–83.

Anand, M. C. J., & Bharatraj, J. (2018). Gaussian qualitative trigonometric functions in a fuzzy circle. *Advances in Fuzzy Systems*, *2018*, 1–9. DOI: 10.1155/2018/8623465

Anand, M. C. J., & Bharatraj, J. (2019). Interval-valued neutrosophic numbers with WASPAS. In *Fuzzy Multi-criteria Decision-Making Using Neutrosophic Sets* (pp. 435–453). Springer International Publishing., https://link.springer.com/chapter/10.1007/978-3-030-00045-5_17 DOI: 10.1007/978-3-030-00045-5_17

Anand, M. C. J., Martin, N., Clementking, A., Rani, S., Priyadharshini, S., & Siva, S. (2023, March). Decision making on optimal selection of advertising agencies using machine learning. In *2023 International Conference on Information Management (ICIM)* (pp. 121-125). IEEE.

Anand, M. C. J., Moorthy, C. B., Sivamani, S., Indrakumar, S., Kalaiarasi, K., & Barhoi, A. (2023). *Fuzzy intelligence inventory decision optimization model of sustainability and green technologies for mixed uncertainties of carbon emission. In: 2023 International Conference on Information Management (ICIM)*. IEEE., https://ieeexplore.ieee.org/document/10145085 DOI: 10.1109/ICIM58774.2023.00020

Arif, M., Kumar, V. D., Jayakumar, L., Ungurean, I., Izdrui, D., & Geman, O. (2021). DAHP–TOPSIS-Based Channel Decision Model for Co-Operative CR-Enabled Internet on Vehicle (CR-IoV). *Sustainability (Basel)*, *13*(24), 13966. DOI: 10.3390/su132413966

[] Azhar, N. A., Radzi, N. A., & Wan Ahmad, W. S. H. M. (2021). Multi-criteria decision making: a systematic review. Recent Advances in Electrical & Electronic Engineering (Formerly Recent Patents on Electrical & Electronic Engineering), 14(8), 779-801.

Bersali, M., Rachedi, A., Bouarfa, H., & Badjara, M. E. A. (2022). A novel cooperative clustering approach based on multi-criteria decision-making for IoV. *International Journal of High Performance Systems Architecture*, *11*(1), 36–46. DOI: 10.1504/IJHPSA.2022.121923

Bhabani, B., & Mahapatro, J. (2023). Performance Evaluation of Priority-based Scheduling in Hybrid VANETs for Different Criteria Weights Using AHP-AHP and AHP-TOPSIS. *Journal of the Institution of Electronics and Telecommunication Engineers*, •••, 1–12.

Bharatraj, J., & Anand, M. C. J. (2019). Power harmonic weighted aggregation operator on single-valued trapezoidal neutrosophic numbers and interval-valued neutrosophic sets. In *Fuzzy Multi-criteria Decision-Making Using Neutrosophic Sets* (pp. 45–62). Springer International Publishing., https://link.springer.com/chapter/10.1007/978-3-030-00045-5_3 DOI: 10.1007/978-3-030-00045-5_3

Bouakouk, M. R., Abdelli, A., Mokdad, L., & Othman, J. B. (2022, May). Dealing with complex routing requirements using an MCDM based approach. In *2022 International Wireless Communications and Mobile Computing (IWCMC)* (pp. 1256-1261). IEEE. DOI: 10.1109/IWCMC55113.2022.9825024

Chakraborty, A., Mondal, S. P., Alam, S., Pamucar, D., & Marinkovic, D. (2022). A new idea to evaluate networking problem and MCGDM problem in parametric interval valued pythagorean arena. *Discrete Dynamics in Nature and Society*, *2022*(1), 2022. DOI: 10.1155/2022/7369045

Chettibi, S. (2020). Combination of HF set and MCDM for stable clustering in VANETs. *IET Intelligent Transport Systems*, *14*(3), 190–195. DOI: 10.1049/iet-its.2019.0283

Devadoss, A. V., & Anand, M. C. J. (2014). *Felix., A.: A CETD matrix approach to analyze the dimensions of the personality of a person. In: 2014 International Conference on Computational Science and Computational Intelligence*. IEEE., https://ieeexplore.ieee.org/abstract/document/6822301

Evangeline, C. S., & Kumaravelu, V. B. (2022). A two-phase fuzzy based access network selection scheme for vehicular ad hoc networks. *Peer-to-Peer Networking and Applications*, *15*(1), 107–133. DOI: 10.1007/s12083-021-01228-w PMID: 36188111

Kungumaraj, E., Lathanayagam, E., Saikia, U., Anand, M., Khanna, S. T., Martin, N., & Edalatpanah, S. A. (2024). Neutrosophic Topological Vector Spaces and its Properties. [IJNS]. *International Journal of Neutrosophic Science*, *23*(2).

Lyu, H. M., Yin, Z. Y., Zhou, A., & Shen, S. L. (2023). MCDM-based flood risk assessment of metro systems in smart city development: A review. *Environmental Impact Assessment Review*, *101*, 107154. DOI: 10.1016/j.eiar.2023.107154

Magableh, G., & Mumani, A. (2022). Simulation Based-MCDM Approach for Evaluating Traffic Solutions. *Promet (Zagreb)*, *34*(1), 117–133. DOI: 10.7307/ptt.v34i1.3842

Malik, K. I., Arooj, H., Jamil, B., Munir, K., & Ijaz, H. (2023). MCDM-Based Optimal Gateway Selection in Mobile Fog Computing. *Technical Journal*, *28*(04), 53–63.

Malik, R. Q. (2021). *A roadside units positioning framework in the context of vehicle-to-infrastructure based on integrated AHP-entropy and group-VIKOR* (Doctoral dissertation, Universiti Tun Hussein Onn Malaysia).

Mamadou, A., & Chalhoub, G. (2022). MCDM-based RAT Selection method between LTE-D2D and IEEE 802.11.

Manshath, A., Kungumaraj, E., Lathanayagam, E., Anand, M. C., Martin, N., Muniyandy, E., & Indrakumar, S. (2024). Neutrosophic Integrals by Reduction Formula and Partial Fraction Methods for Indefinite Integrals. [IJNS]. *International Journal of Neutrosophic Science*, *23*(1).

Miriam, M. R., Martin, N., & Anand, M. C. J. (2023). Inventory model promoting smart production system with zero defects. *International Journal of Applied and Computational Mathematics*, *9*(4), 47. Advance online publication. DOI: 10.1007/s40819-023-01524-4

Naderi, M., Chakareski, J., & Ghanbari, M. (2023). Hierarchical Q-learning-enabled neutrosophic AHP scheme in candidate relay set size adaption in vehicular networks. *Computer Networks*, *235*, 109968. DOI: 10.1016/j.comnet.2023.109968

Naderi, M., & Ghanbari, M. (2023). Adaptively prioritizing candidate forwarding set in opportunistic routing in VANETs. *Ad Hoc Networks*, *140*, 103048. DOI: 10.1016/j.adhoc.2022.103048

Ndashimye, E., Sarkar, N. I., & Ray, S. K. (2021). A Multi-criteria based handover algorithm for vehicle-to-infrastructure communications. *Computer Networks*, *185*, 107652. DOI: 10.1016/j.comnet.2020.107652

Okpeh, A. A., Okpe, J. M., Tekanyi, A., & Tanko, N. (2023). Cooperative Multi-criteria Handover Management For Reliable V2i Communication And Improved Qos In Overlayed Heterogeneous Network Environment. *Annals of the Faculty of Engineering Hunedoara-International Journal of Engineering*, *21*(4).

Patil, S. D., & Ragha, L. (2020). Adaptive fuzzy-based message dissemination and micro-artificial bee colony algorithm optimised routing scheme for vehicular ad hoc network. *IET Communications*, *14*(6), 994–1004. DOI: 10.1049/iet-com.2019.0388

Perla, B., & Ramesh, D. A. (2022). Development of Roughness Index Model For Urban Roads Using Machine Learning Techniques and Prioritizing Using MCDM Techniques. *Available atSSRN* 4160025. DOI: 10.2139/ssrn.4160025

Prabha, S. K., Anand, M., Vidhya, V., Nagarajan, G., Saikia, U., Martin, N., & Tiwari, M. (2024). Sorting Out Interval Valued Neutrosophic Fuzzy Shortest Cycle Route Problem by Reduced Matrix Method. [IJNS]. *International Journal of Neutrosophic Science*, *23*(2).

Pu, C. (2023). Blockchain-Based Trust Management Using Multi-Criteria Decision-Making Model for VANETs. *Authorea Preprints*.

Raj, P. J., Prabhu, V. V., Krishnakumar, V., & Anand, M. C. J. (2023). Anand. M.C.J.: Solar Powered Charging of Fuzzy Logic Controller (FLC) Strategy with Battery Management System (BMS) Method Used for Electric Vehicle (EV). *International Journal of Fuzzy Systems*, *25*(7), 2876–2888. Advance online publication. DOI: 10.1007/s40815-023-01537-7

Rajesh, K., Rathod, S., Kundale, J., Rathod, N., Anand, M., Saikia, U., & Martin, N. (2024). A Study on Interval Valued Temporal Neutrosophic Fuzzy Sets. [IJNS]. *International Journal of Neutrosophic Science*, *23*(1).

Sriram, S., Ramachandran, M., Chinnasamy, S., & Mathivanan, G. (2022). A review on multi-criteria decision-making and its application. REST Journal on Emerging trends in Modelling and Manufacturing, 7(4), 101-107.

Sudha, S., Martin, N., Anand, M. C. J., Palanimani, P. G., Thirunamakkani, T., & Ranjitha, B. (2023). MACBETH-MAIRCA Plithogenic Decision-Making on Feasible Strategies of Extended Producer's Responsibility towards Environmental Sustainability. Infinite Study.

Taylor, S. J., Ahmad, F., Nguyen, H. N., Shaikh, S. A., Bryans, J., & Wartnaby, C. E. (2023). A comparative analysis of multi-criteria decision methods for secure beacon selection in vehicular platoons. *Transactions on Emerging Telecommunications Technologies*, *34*(11), e4841. DOI: 10.1002/ett.4841

Tiwari, J., Dubey, D., Prakash, A., & Tripathi, R. (2022, December). A trustworthy and cooperative MAC protocol for cognitive vehicular networks. In *2022 IEEE 9th Uttar Pradesh Section International Conference on Electrical, Electronics and Computer Engineering (UPCON)* (pp. 1-6). IEEE. DOI: 10.1109/UPCON56432.2022.9986422

Tiwari, J., Prakash, A., & Tripathi, R. (2023). A metaheuristic MAC protocol for safety applications in cognitive vehicular networks. *Vehicular Communications*, *44*, 100686. DOI: 10.1016/j.vehcom.2023.100686

Varalakshmi, A., Santhosh Kumar, S., Shanmugapriya, M. M., Mohanapriya, G., & Anand, M. C. J. (2023). Markers location monitoring on images from an infrared camera using optimal fuzzy inference system. *International Journal of Fuzzy Systems*, *25*(2), 731–742. DOI: 10.1007/s40815-022-01407-8

Vitianingsih, A. V., Baharin, S. S. K., Othman, O., & Suraji, A. (2022). Empirical study of a spatial analysis for prone road traffic accident classification based on MCDM method. *International Journal of Advanced Computer Science and Applications*, *13*(5), 665–679. DOI: 10.14569/IJACSA.2022.0130578

Yacheur, B. Y., Ahmed, T., & Mosbah, M. (2023). Efficient DRL-Based Selection Strategy in Hybrid Vehicular Networks. *IEEE Transactions on Network and Service Management*, *20*(3), 2400–2411. DOI: 10.1109/TNSM.2023.3300653

Yigit Katanalp, B., Eren, E., & Alver, Y. (2023). An integrated solution to identify pedestrian-vehicle accident prone locations: GIS-based multicriteria decision approach. *Journal of Transportation Safety & Security*, *15*(2), 137–176. DOI: 10.1080/19439962.2022.2048760

Zeng, S., Zhang, N., Zhang, C., Su, W., & Carlos, L. A. (2022). Social network multiple-criteria decision-making approach for evaluating unmanned ground delivery vehicles under the Pythagorean fuzzy environment. *Technological Forecasting and Social Change*, *175*, 121414. DOI: 10.1016/j.techfore.2021.121414

Chapter 9

Optimizing Vehicular Network Architecture and Communication Models With Machine Learning Approach

Sakshi Taaresh Khanna
https://orcid.org/0009-0006-2936-8708
Ram Lal Anand College, New Delhi, India

Neeraj Kumar Sharma
Ram Lal Anand College, New Delhi, India

ABSTRACT

Vehicular networks are a digital web that links the entities of vehicles, infrastructure, and information to enable road transportation safer, more efficient, and amicable to the users. It is a more comprehensive system comprising different communication models and architecture, where the latter supplies a supportive framework for the communication models. This chapter proposes a machine learning approach to determine a more suitable combination between the VNA & CM to optimize the efficacy of the VL. The ML algorithm of DL networks with multi-layers is leveraged to analyze factors such as traffic patterns, network topology, communication range, and application needs to identify the most appropriate combination of VNA and CM. The capability of the ML approach is validated using the performance metrics in comparison with other learning algorithms to identify the limitations of the proposed model. This machine learning-based decision model shall also be employed as a predictive model to determine the optimal combination of VNA & CM for optimal

DOI: 10.4018/979-8-3693-6422-2.ch009

planning and implementation of VN.

1. INTRODUCTION

The present modern era of intelligent transportation systems is governed by the emergence of vehicular networks which are the intrinsic components of ensuring smart technology and connectivity. These vehicular networks are termed VANETs i.e. Vehicular Ad hoc NETworks which are embedded with the potency of establishing communications with the vehicles, pedestrians, and infrastructure for sustaining safe transportation. The efficiency of vehicular network communication is solely determined by the vehicular network architecture (VNA) as it is very broader encompassing the suitable infrastructure and framework for communication enabling the communications between the vehicles and other roadside units (Jurczenia & Rak, 2022). VAN assists in building a robust and well-connected network. VAN comprises both onboard units and roadside units, where the former are inherited in the vehicles and the latter are placed to meet the demands of connectivity. VAN considers various factors such as coverage range, scalability, consistency, and latency. These architectural frameworks are vital in traffic management, collision avoidance, automotive driving and other related applications.

Vehicular network communication models (VNC) are of various kinds such as Vehicle-to -vehicle, vehicle-to-pedestrians, vehicle-to-infrastructure, vehicle-to-grid and many other. The choice and the suitability of VNC are based on the available feature of architectural frameworks. The matching of VNC with VAN is very essential to make decisions on the choice of the communication models (Cho, Park, & Paek, 2022). This matching between the entities shall be made optimal with the integration of machine learning algorithms. The intervention of machine learning approaches will alleviate the hurdles in maximizing the competence and effectiveness of communication networks. The functioning of this integrated machine learning approach decreases the manual intervention thereby increasing the efficacy of the automated process of making decisions. The matching model shall be made optimal by considering large data sets to find out the patterns. Also, these algorithms have the potency to accommodate new data sets and provide suitable feedback to make the VNC respond to fluctuating circumstances. In general, these algorithms are applied in evaluating the performances of the VNC, identifying the threatful nodes in networks, and in detecting the hurdles causing the phenomenon. However, machine learning-based algorithms are not applied in building a matching model. Hence, leveraging machine learning algorithms in optimizing the matching of vehicular network architecture and vehicular communication models assists in designing and developing suitable decision models of determining the vehicular network (Jayashree & Kumar, 2023).

The machine learning algorithms are broadly classified into supervised, unsupervised, and reinforcement. Deep learning is characterized as another subset of machine learning and in recent times multilayered neural networks are employed in developing decision models. In addition, other learning approaches are blended with multilayered networks to evolve hybrid models. Deep learning neural networks with multilayers have a single input and output layer with many hidden layers. In general, the input features are many in number which hurdles the decision process, hence algorithms such as principal component analysis, linear discriminant algorithms, non-negative matrix factorization, filter methods, and embedded methods shall be employed in feature reduction. This research work employs multilayered deep neural networks with principal component analysis algorithm in developing a matching model of vehicular architecture framework and vehicular network communication models.

The remaining part of the paper is segmented into the following sections. Section 2 briefs the state of the art of the related works. Section 3 outlines the methods and materials used in this work. Section 4 applies the proposed algorithmic approach in developing the matching model. Section 5 makes sensitivity analysis and the last section concludes the work.

2. RELATED WORKS

This segment of the paper presents the significant applications and interventions of machine learning algorithms with respect to vehicular communication networks. To mention a few, (Eziama, Tepe, Balador, Nwizege, & Jaimes, 2018) studied the utility of deep learning in detecting malicious nodes in VANET. (Sharma & Khanna, 2018) explored the ways of applying the machine learning framework in developing smart and intelligent vehicular networks. (Tang, Kawamoto, Kato, & Liu, 2019)(Fernando, n.d.) discoursed the applicability in the 6G era. (Tayyaba et al., 2020) focused on the management of resources in 5 networks. (Schettler, Buse, Zubow, & Dressler, 2020) explained the machine learning-based training methods for enriching the competency of the intelligent transport system.

Researchers have investigated the advantages of employing machine learning algorithms in the domain of vehicular networks. (Cui, Huang, Wu, & Zheng, 2020) in-network slicing for resource allocations. (Patel & Ajmeri, 2020) in enhancing the transportation services. (Tang, Mao, Kato, & Gui, 2021) in increasing the covering technology. (Hossain et al., 2021) and (Hossain, 2022) in sensing the cooperative spectrum and dynamic segmentation. (do Vale Saraiva, Girardi, & Campos, 2021) in predicting the vehicle context. (Tang, Wen, Zhao, & Kato, 2022) in integrating network assistance in vehicular networks. (Xue, Zhang, Wu, Zhou, & Shen, 2022) in estimating the traffic flow. (Bharti et al., 2022) in controlling traffic congestion. (Kokane et al., 2023) and (Prakash et al., 2024) in developing smart cities with the Internet of vehicles. (Feukeu & Mbuyu, 2023) in link adaptation. (Wijesekara & Gunawardena, 2023) in building a hybrid routing framework for network contention. (Zidi, Alaya, Moulahi, Al-Shargabi, & El Khediri, 2024) in the prediction of faults and in devising recovery techniques. (Ali, Roccotelli, Boggia, & Fanti, 2024) in detecting the network attacks. (Gyawali, Qian, & Hu, 2020)(Gyawali & Qian, 2019) in identifying the instances of misbehavior of networks. (Pandya & Mer, 2020) in avoidance of vehicular collision. (Gautam, Jain, Verma, & Vyas, 2023) in controlling the roadside units. (Nunez et al., 2023) in estimating the path loss in the vehicle-to-infrastructure communication model, various types of parameters were discussed in (Varalakshmi et al., 2022)(Miriam, Martin, & Anand, 2023)(Bharatraj & Anand, 2019)(Anand & Bharatraj, 2018)(Anand & Bharatraj, 2019)(Raj, Prabhu, Krishnakumar, & Anand, 2023)(Devadoss, Anand, & Felix, 2014)(Anand & Bharatraj, 2017)(Anand & Bharatraj, 2019)(Sudha et al., 2023)(Prabha et al., 2024)(Kungumaraj et al., 2024)(Anand et al., 2023)(Rajesh et al., 2024)(Manshath et al., 2024)(Anand et al., 2023)(Anand et al., 2023)(Anand et al., 2023)(Anand et al., 2023)(Anand et al., 2023)(Nagaraj et al., 2023)(Prabha et al., 2024)(Kungumaraj et al., 2024)(Anand et al., 2023)(Rajesh et al., 2024)(Manshath et al., 2024)(Anand et al., 2023).

From the aforementioned literature citing the interventions of machine learning in vehicular network communication, it is very evident that these algorithms are employed in detecting faulty modes, predicting traffic congestion, estimation network flow and evaluation of the network efficiency in the context of vehicular networks. However, these algorithms are not employed in any of the matching models to the best of our knowledge. This has motivated the authors to develop a decision model to optimize the choice of vehicular network communication as the output with several input features of vehicular network architecture. This research work contributes to the development of a matching model with the intervention of multilayered deep neural networks with principal component analysis.

3. MATERIALS AND METHODS

This section presents the sequence of steps involved in developing a decision-making model with machine learning interferences. The procedure shall be characterized as a phased approach. The first phase is defining the input parameters. The second phase is feature reduction using Principal component analysis and the third phase is the applicability of neural networks and the last phase is model training, evaluation and validation and enhancement. The overall algorithmic procedure is presented graphically in Figure 1

Figure 1. Algorithmic Framework (Phase I)

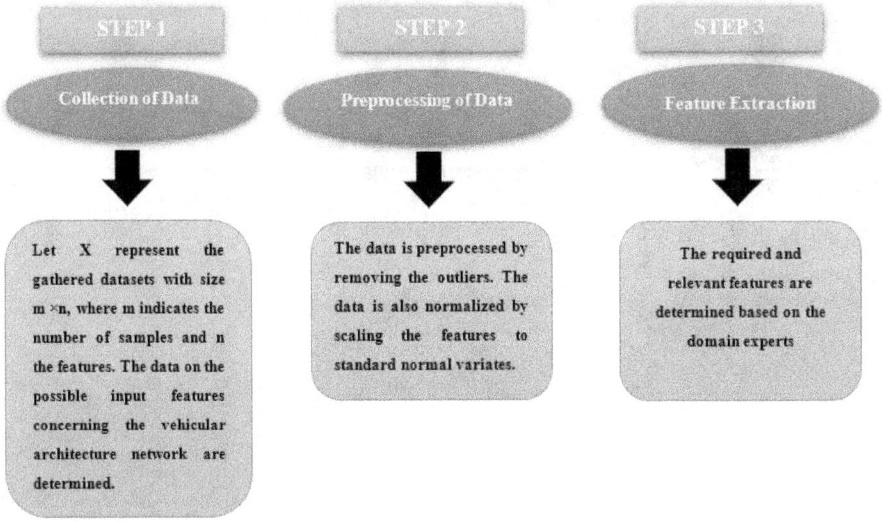

Figure 2. Algorithmic Framework (Phase II)

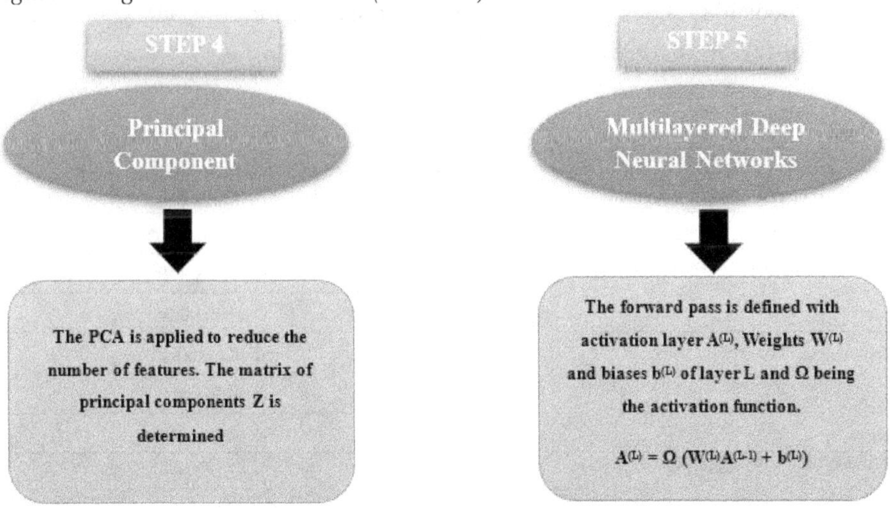

Figure 3. Algorithmic Framework (Phase III)

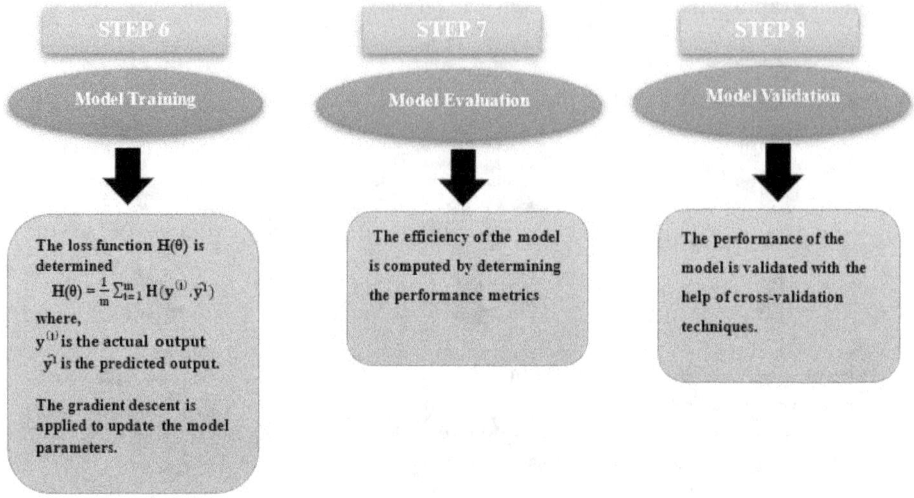

4. OPTIMAL MATCHING WITH MACHINE LEARNING INTERFERENCES

This section employs the methodology discussed in section 3 in devising the matching decision model. The input features considered for this study are categorized as follows as in Table 1

Table 1. Categorization of the Input Features

Category	Features
Vehicular Network Architecture	Density of Roadside units
	Network coverage
	Scalability
Vehicle and Mobility	Vehicular mobility patterns
	Traffic density
	Road conditions

continued on following page

Table 1. Continued

Category	Features
Communication Model	Range of communication
	Latency
	Reliability
	Adaptability
Environmental Factors	Weather conditions

By using Principal component Analysis, the following features are determined to be significant F1Density of roadside units, F2 Traffic density, F3 Range of communication, F4 Reliability, and F5 Latency. The validity is tested by determining the performance metrics and the accuracy of this model is found to be 0.9 in comparison with other feature reduction algorithms.

Figure 4. ROC Curve of PCA

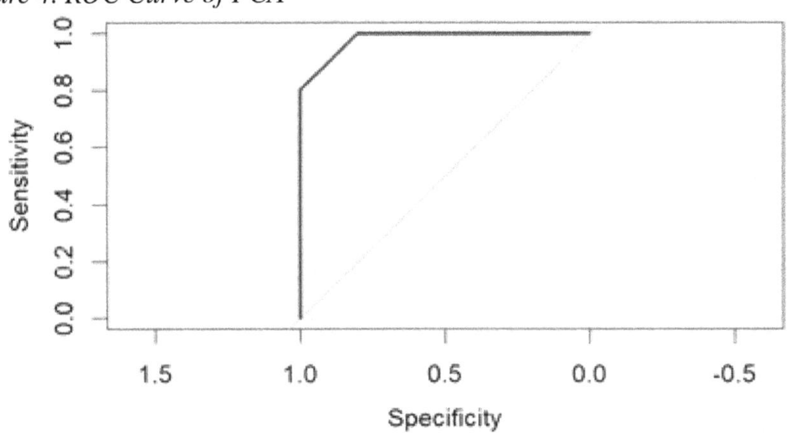

The ROC curve maps the sensitivity and specificity and the area under this curve yields a scalar measure closer to 1 indicating the efficiency of the model in feature reduction. The multilayered deep neural networks are applied to determine the optimal matching of the communicational models with the above input features and the graphical representation of the neural networks is depicted in Figure 3.

Figure 5. Multilayered Neural Networks

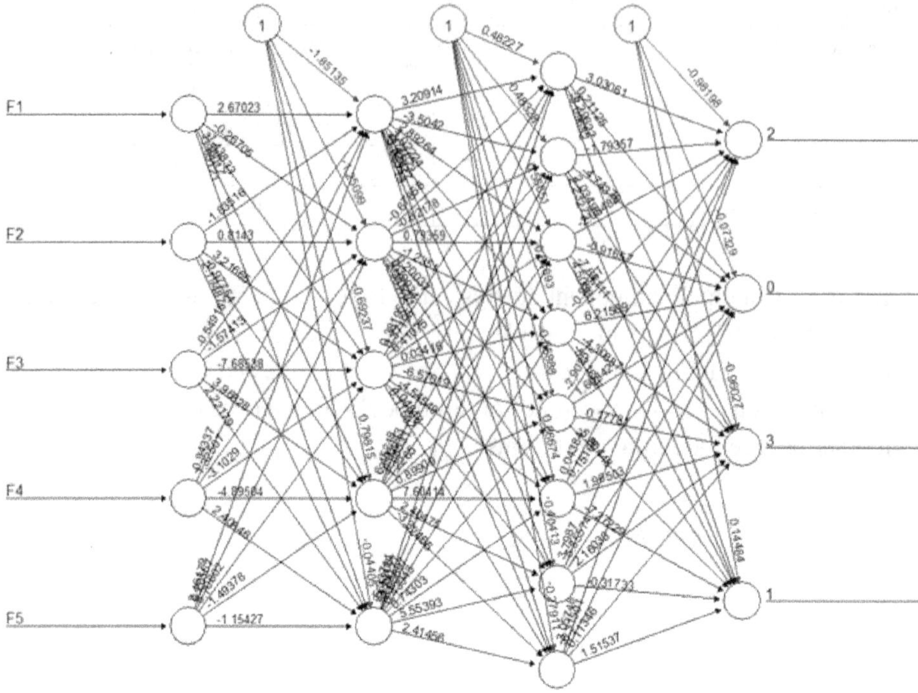

The performance metrics are determined and presented in Table 2.

Table 2. Performance Metrics of the Model

Accuracy	Precision	Recall	F1 Score	Specificity
0.9	0.875	0.942	0.908	0.854

5. SENSITIVITY ANALYSIS

The efficiency of this blended model of PCA and multilayered neural networks (M_0) is compared with other combinations of machine learning algorithms. The performance metric values are presented in Table 3.

Table 3. Aggregate Performance Metric Values

Methods	Accuracy	Precision	Recall	F1 Score	Specificity
Principal Component Analysis & Support Vector Machines (M_1)	0.7700	0.7647	0.7800	0.7720	0.7600
Linear Discriminant Analysis & Multilayered Neural networks (M_2)	0.7150	0.7000	0.7143	0.7078	0.7059
Linear Discriminant Analysis & Support Vector Machines (M_3)	0.6200	0.6000	0.5227	0.5581	0.7037
Principal Component Analysis & KNN (M_4)	0.5250	0.5882	0.5556	0.5714	0.5435
Linear Discriminant Analysis & KNN (M_5)	0.7600	0.7255	0.7879	0.7558	0.7358

The efficiency of the blended method M_0 shall be compared with other methods in terms of the performance metric values of accuracy, precision, recall, F1 score and specificity. The accuracy comparison is presented in Figure 4

Figure 6. Accuracy Comparison

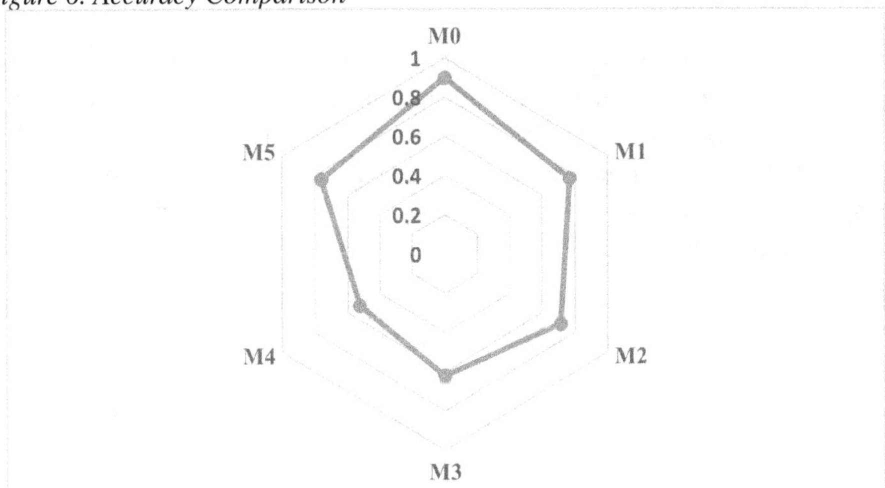

The method M_0 has high accuracy and the other methods possess low accuracy indicating the less effectiveness in classification. On comparing the precision values, the method M_0 has a low rate of false positives in comparison with other methods and it is presented in Figure 5.

Figure 7. Precision Comparison

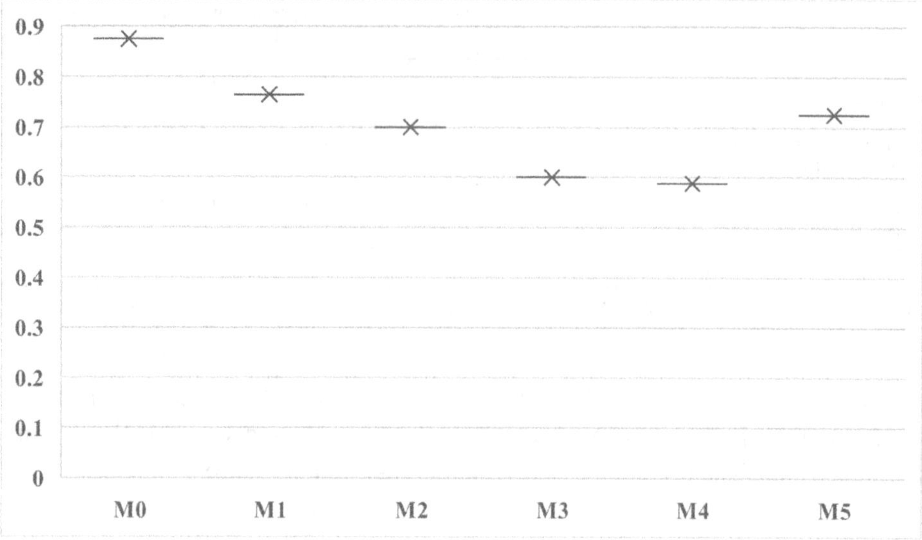

Similarly, on comparing other performance metric values, it is very evident that the method M_0 ranks high in comparison with other methods. Also next to the method of M_0, the other methods shall be ranked as follows. M5 > M1 > M2 > M3 > M4. These inferences are obtained from this data sets and it is subjected to change based on the nature of the input data. The overall comparison is depicted graphically in Figure 6.

Figure 8. Comparison of the Model Efficiencies

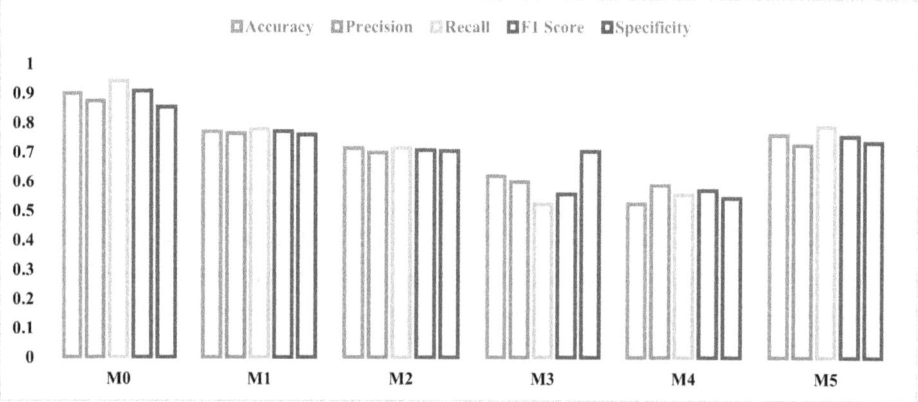

6. CONCLUSION

This chapter presents a combination of the machine learning approach and its utility in making a matching between the vehicular communication model with that of the input features about vehicular architecture. The machine learning-based decision models are more efficient in matching the vehicular networks. The optimal choice-making with different machine learning approaches facilitates the decision-makers in deriving feasible decisions. The results are quite promising for the combination of Principal component analysis and multilayered neural networks. As a part of the extension, the same approach shall be applied to other decision-making scenarios and also other machine learning algorithms shall be employed to determine the reliability and consistency of these model combinations.

REFERENCES

Ali, W. A., Roccotelli, M., Boggia, G., & Fanti, M. P. (2024). Intrusion detection system for vehicular ad hoc network attacks based on machine learning techniques. *Information Security Journal: A Global Perspective*, 1-19.

Anand, M. C. J., & Bharatraj, J. (2017). Theory of Triangular Fuzzy Number. *Proceedings of NCATM, 2017*, 80–83.

Anand, M. C. J., & Bharatraj, J. (2018). Gaussian qualitative trigonometric functions in a fuzzy circle. *Advances in Fuzzy Systems, 2018*, 1–9. DOI: 10.1155/2018/8623465

Anand, M. C. J., & Bharatraj, J. (2019). Interval-valued neutrosophic numbers with WASPAS. In *Fuzzy Multi-criteria Decision-Making Using Neutrosophic Sets* (pp. 435–453). Springer International Publishing. DOI: 10.1007/978-3-030-00045-5_17

Anand, M. C. J., Kalaiarasi, K., Martin, N., Ranjitha, B., Priyadharshini, S. S., & Tiwari, M. (2023, September). Fuzzy C-Means Clustering with MAIRCA-MCDM Method in Classifying Feasible Logistic Suppliers of Electrical Products. In *2023 First International Conference on Cyber Physical Systems, Power Electronics and Electric Vehicles (ICPEEV)* (pp. 1-7). IEEE.

Anand, M. C. J., Martin, N., Clementking, A., Rani, S., Priyadharshini, S., & Siva, S. (2023, March). Decision making on optimal selection of advertising agencies using machine learning. In *2023 International Conference on Information Management (ICIM)* (pp. 121-125). IEEE.

Anand, M. C. J., Moorthy, C. B., Sivamani, S., Indrakumar, S., Kalaiarasi, K., & Barhoi, A. (2023). Fuzzy intelligence inventory decision optimization model of sustainability and green technologies for mixed uncertainties of carbon emission. In *In: 2023 International Conference on Information Management (ICIM)*. IEEE, https://ieeexplore.ieee.org/document/10145085

Anand, M. C. J., Moorthy, C. B., Sivamani, S., Indrakumar, S., Kalaiarasi, K., & Barhoi, A. (2023). Fuzzy intelligence inventory decision optimization model of sustainability and green technologies for mixed uncertainties of carbon emission. In *In: 2023 International Conference on Information Management (ICIM)*. IEEE., https://ieeexplore.ieee.org/document/10145085

Anand, M. C. J., Moorthy, C. B., Sivamani, S., Indrakumar, S., Kalaiarasi, K., & Barhoi, A. (2023, March). Fuzzy intelligence inventory decision optimization model of sustainability and green technologies for mixed uncertainties of carbon emission. In *2023 International Conference on Information Management (ICIM)* (pp. 78-82). IEEE.

Anand, M. C. J., Priyadharshini, S. S., Martin, N., Anand, M. E., Poorani, B., & Mohiddin, S. K. (2023, September). Decision Trees in the Selection of Electric Vehicles Based on Various Parameters. In *2023 First International Conference on Cyber Physical Systems, Power Electronics and Electric Vehicles (ICPEEV)* (pp. 1-6). IEEE.

Bharatraj, J., & Anand, M. C. J. (2019). Power harmonic weighted aggregation operator on single-valued trapezoidal neutrosophic numbers and interval-valued neutrosophic sets. In *Fuzzy Multi-criteria Decision-Making Using Neutrosophic Sets* (pp. 45–62). Springer International Publishing. DOI: 10.1007/978-3-030-00045-5_3

Bharti, R. K., Suganthi, D., Abirami, S. K., Kumar, R. A., Gayathri, B., & Kayathri, S. (2022, December). Optimal Extreme Learning Machine-based Traffic Congestion Control System in Vehicular Network. In *2022 6th International Conference on Electronics, Communication and Aerospace Technology* (pp. 597-603). IEEE.

Cho, Y., Park, M., & Paek, J. (2022, October). Broadcasting in the vehicular ad-hoc network: Survey on recent advances. In 2022 13th International Conference on Information and Communication Technology Convergence (ICTC) (pp. 813-816). IEEE.

Cui, Y., Huang, X., Wu, D., & Zheng, H. (2020). Machine learning-based resource allocation strategy for network slicing in vehicular networks. *Wireless Communications and Mobile Computing*, *2020*, 1–10.

Devadoss, A. V., Anand, M. C. J., & Felix, A. (2014, March). A CETD matrix approach to analyze the dimensions of personality of person. In *2014 International Conference on Computational Science and Computational Intelligence* (Vol. 2, pp. 40-45). IEEE.

do Vale Saraiva, T., Girardi, R., & Campos, C. A. V. (2021, September). Machine learning applied to network traffic for vehicular applications to predict the vehicle context. In *2021 IEEE 94th Vehicular Technology Conference (VTC2021-Fall)* (pp. 1-5). IEEE. DOI: 10.1109/VTC2021-Fall52928.2021.9625538

Eziama, E., Tepe, K., Balador, A., Nwizege, K. S., & Jaimes, L. M. (2018, December). Malicious node detection in vehicular ad-hoc network using machine learning and deep learning. In *2018 IEEE Globecom Workshops (GC Wkshps)* (pp. 1-6). IEEE.

Fernando, K. C. Machine Learning in 6G-enabled Vehicular Network: A State-of-the-Art-Review.

Feukeu, E. A., & Mbuyu, S. (2023). Machine Learning Algorithm for a Link Adaptation strategy in a Vehicular Ad hoc Network. *Inteligencia Artificial: Revista Iberoamericana de Inteligencia Artificial, 26*(72).

Gautam, K., Jain, V. K., Verma, S. S., & Vyas, S. (2023, February). Vehicular Communication Strategy Using Machine Learning and Image Processing to Enhance Observations and Control on the Road Side Area. In *Proceedings of 3rd International Conference on Recent Trends in Machine Learning, IoT, Smart Cities and Applications: ICMISC 2022* (pp. 397-404). Singapore: Springer Nature Singapore. DOI: 10.1007/978-981-19-6088-8_34

Gyawali, S., & Qian, Y. (2019, May). Misbehavior detection using machine learning in vehicular communication networks. In *ICC 2019-2019 IEEE International Conference on Communications (ICC)* (pp. 1-6). IEEE. DOI: 10.1109/ICC.2019.8761300

Gyawali, S., Qian, Y., & Hu, R. Q. (2020). Machine learning and reputation-based misbehavior detection in vehicular communication networks. *IEEE Transactions on Vehicular Technology*, *69*(8), 8871–8885. DOI: 10.1109/TVT.2020.2996620

Hossain, M. A., Md Noor, R., Yau, K. L. A., Azzuhri, S. R., Z'aba, M. R., Ahmedy, I., & Jabbarpour, M. R. (2021). Machine learning-based cooperative spectrum sensing in dynamic segmentation enabled cognitive radio vehicular network. *Energies*, *14*(4), 1169. DOI: 10.3390/en14041169

Jayashree, S., & Kumar, S. S. (2023). A comprehensive survey on effective key management schemes for secured authentication in vehicular ad hoc network. *International Journal of Information and Computer Security*, *22*(3-4), 354–376.

Jurczenia, K., & Rak, J. (2022). A survey of vehicular network systems for road traffic management. *IEEE Access : Practical Innovations, Open Solutions*, *10*, 42365–42385.

Kokane, C. D., Mohadikar, G., Khapekar, S., Jadhao, B., Waykole, T., & Deotare, V. V. (2023). Machine Learning Approach for Intelligent Transport System in IOV-Based Vehicular Network Traffic for Smart Cities. *International Journal of Intelligent Systems and Applications in Engineering, 11*(11s), 06-16.

Kungumaraj, E., Lathanayagam, E., Saikia, U., Anand, M., Khanna, S. T., Martin, N., & Edalatpanah, S. A. (2024). Neutrosophic Topological Vector Spaces and its Properties. [IJNS]. *International Journal of Neutrosophic Science*, *23*(2).

Kungumaraj, E., Lathanayagam, E., Saikia, U., Anand, M. C. J., Khanna, S. T., Martin, N., Tiwari, M., & Edalatpanah, S. A. (2024). Khanna, S.T., Martin, N., Tiwari, M., Edalatpanah, S. A.: Neutrosophic Topological Vector Spaces and its Properties. *International Journal of Neutrosophic Science*, *23*(2), 63–76. DOI: 10.54216/IJNS.230206

Manshath, A., Kungumaraj, E., Lathanayagam, E., Anand, M. C., Martin, N., Muniyandy, E., & Indrakumar, S. (2024). Neutrosophic Integrals by Reduction Formula and Partial Fraction Methods for Indefinite Integrals. [IJNS]. *International Journal of Neutrosophic Science*, *23*(1).

Miriam, M. R., Martin, N., & Anand, M. C. J. (2023). Inventory model promoting smart production system with zero defects. *International Journal of Applied and Computational Mathematics*, *9*(4), 47. DOI: 10.1007/s40819-023-01524-4

Mohammad Asif, H. (2022). *Cooperative spectrum sensing based on machine learning in cognitive radio vehicular network/Mohammad Asif Hossain* (Doctoral dissertation, Universiti Malaya).

Nagaraj, C. T., Anand, M. C. J., Priyadharshini, S. S., & Aparna, P. (2023, December). GCNXG: Detecting Fraudulent Activities in Financial Networks: A Graph Analytics and Machine Learning Fusion. In International Conference on Renewable Energy, Green Computing, and Sustainable Development (pp. 17-32). Cham: Springer Nature Switzerland.

Nuñez, Y., Lovisolo, L., da Silva Mello, L., Ramos, G., Leonor, N. R., Faria, S., & Caldeirinha, R. F. (2023, November). Path Loss Prediction for Vehicular-to-Infrastructure Communication Using Machine Learning Techniques. In *2023 IEEE Virtual Conference on Communications (VCC)* (pp. 270-275). IEEE. DOI: 10.1109/VCC60689.2023.10474798

Pandya, D., & Mer, H. (2020, December). Inter-Vehicular Communication for Intelligent Collision Avoidance Using Machine Learning: An Overview. In *International Conference on Advanced Informatics for Computing Research* (pp. 151-159). Singapore: Springer Singapore.

Patel, S. R., & Ajmeri, M. (2020). Machine learning approach for transportation services using vehicular ad hoc network. [IJTIIR]. *Int J Trend Innov Res*, *2*(2).

Prabha, S. K., Anand, M., Vidhya, V., Nagarajan, G., Saikia, U., Martin, N., & Tiwari, M. (2024). Sorting Out Interval Valued Neutrosophic Fuzzy Shortest Cycle Route Problem by Reduced Matrix Method. [IJNS]. *International Journal of Neutrosophic Science*, *23*(2).

Prabha, S. K., Anand, M. C. J., Vidhya, V., Nagarajan, G., Saikia, U., Martin, N., Kumary, M. S., & Tiwari, M. (2024). Sorting Out Interval Valued Neutrosophic Fuzzy Shortest Cycle Route Problem by Reduced Matrix Method. *International Journal of Neutrosophic Science*, *23*(2), 91–103. DOI: 10.54216/IJNS.230208

Prakash, J., Murali, L., Manikandan, N., Nagaprasad, N., & Ramaswamy, K. (2024). A vehicular network based intelligent transport system for smart cities using machine learning algorithms. *Scientific Reports*, *14*(1), 468. DOI: 10.1038/s41598-023-50906-7 PMID: 38172149

Raj, P. J., Prabhu, V. V., Krishnakumar, V., & Anand, M. C. J.Justin Raj. (2023). P., Prabhu. V. V., Krishnakumar. V., Anand. M. C. J.: Solar Powered Charging -of Fuzzy Logic Controller (FLC) Strategy with Battery Management System (BMS) Method Used for Electric Vehicle (EV). *International Journal of Fuzzy Systems*, *25*(7), 2876–2888. DOI: 10.1007/s40815-023-01537-7

Rajesh, K., Rathod, S., Kundale, J., Rathod, N., Anand, M., Saikia, U., & Martin, N. (2024). A Study on Interval Valued Temporal Neutrosophic Fuzzy Sets. [IJNS]. *International Journal of Neutrosophic Science*, *23*(1).

Rajesh, K., Rathod, S., Kundale, J., Rathod, N., Anand, M. C. A., Saikia, U., Tiwari, M., & Martin, N. (2024). A Study on Interval Valued Temporal Neutrosophic Fuzzy Sets. *International Journal of Neutrosophic Science*, *23*(1), 341–349. DOI: 10.54216/IJNS.230129

Schettler, M., Buse, D. S., Zubow, A., & Dressler, F. (2020, December). How to train your ITS? integrating machine learning with vehicular network simulation. In *2020 IEEE Vehicular Networking Conference (VNC)* (pp. 1-4). IEEE.

Sharma, M., & Khanna, H. (2018). Intelligent and Secure Vehicular Network using Machine Learning. *JETIR-International Journal of Emerging Technologies and Innovative Research (*www. jetir. org*), ISSN*, 2349-5162.

Sudha, S., Nivetha, M., & Anand, M. C. J. (2023). Palanimani, P.G., Thirunamakkani, T., Ranjitha, B.: MACBETH-MAIRCA Plithogenic Decision-Making on Feasible Strategies of Extended Producer's Responsibility towards Environmental Sustainability. *International Journal of Neutrosophic Science*, *22*(2), 114–130. DOI: 10.54216/IJNS.220210

Tang, F., Kawamoto, Y., Kato, N., & Liu, J. (2019). Future intelligent and secure vehicular network toward 6G: Machine-learning approaches. *Proceedings of the IEEE*, *108*(2), 292–307. DOI: 10.1109/JPROC.2019.2954595

Tang, F., Mao, B., Kato, N., & Gui, G. (2021). Comprehensive survey on machine learning in vehicular network: Technology, applications and challenges. *IEEE Communications Surveys and Tutorials*, *23*(3), 2027–2057. DOI: 10.1109/COMST.2021.3089688

Tang, F., Wen, C., Zhao, M., & Kato, N. (2022). Machine learning for space–air–ground integrated network assisted vehicular network: A novel network architecture for vehicles. *IEEE Vehicular Technology Magazine*, *17*(3), 34–44. DOI: 10.1109/MVT.2022.3188405

Tayyaba, S. K., Khattak, H. A., Almogren, A., Shah, M. A., Din, I. U., Alkhalifa, I., & Guizani, M. (2020). 5G vehicular network resource management for improving radio access through machine learning. *IEEE Access : Practical Innovations, Open Solutions*, *8*, 6792–6800.

Varalakshmi, A., Kumar, S. S., Shanmugapriya, M. M., Mohanapriya, G., & Anand, M. C. J. (2022). Markers location monitoring on images from an infrared camera using optimal fuzzy inference system. *International Journal of Fuzzy Systems*.

Wijesekara, P. A. D. S. N., & Gunawardena, S. (2023, July). A Machine Learning-Aided Network Contention-Aware Link Lifetime-and Delay-Based Hybrid Routing Framework for Software-Defined Vehicular Networks. In *Telecom* (Vol. 4, No. 3, pp. 393-458). MDPI. DOI: 10.3390/telecom4030023

Xue, J., Zhang, T., Wu, W., Zhou, H., & Shen, X. (2022, December). Sparse big data for vehicular network traffic flow estimation: A machine learning approach. In *GLOBECOM 2022-2022 IEEE Global Communications Conference* (pp. 4959-4963). IEEE. DOI: 10.1109/GLOBECOM48099.2022.10001710

Zidi, S., Alaya, B., Moulahi, T., Al-Shargabi, A., & El Khediri, S. (2024). Fault Prediction and Recovery Using Machine Learning Techniques and the HTM Algorithm in Vehicular Network Environment. *IEEE Open Journal of Intelligent Transportation Systems*, *5*, 132–145. DOI: 10.1109/OJITS.2023.3347484

Chapter 10
Strategic Integration of WASPAS Method for Effective Traffic Management and Congestion Control

M. Clement Joe Anand
https://orcid.org/0000-0002-1959-7631
Mount Carmel College, Bengaluru, India

C. Manoj Kumar
https://orcid.org/0000-0002-2150-3331
Cambridge Institute of Technology, K.R. Puram, Bengaluru, India

P. S. Devana
https://orcid.org/0009-0008-5438-3541
Mount Carmel College, Bengaluru, India

R. Myna
Erode Sengunthar Engineering College, India

A. P. Pushpalatha
https://orcid.org/0000-0002-0799-7961
Velammal College of Engineering and Technology, India

Chitra Ramaprakash
https://orcid.org/0000-0002-0421-2479
Dayananda Sagar College of Engineering, India

ABSTRACT

Effective traffic management and congestion control are pivotal for sustainable urban development. This study proposes a Multi-Criteria Decision-Making (MCDM) approach, employing the Weighted Aggregated Sum Product Assessment (WASPAS) method, to address these obstacles. Integrating 30 diverse sub-criteria, including infrastructure capacity, environmental impact, and socio-economic elements, the model offers a comprehensive framework for decision-makers. By categorizing

DOI: 10.4018/979-8-3693-6422-2.ch010

criteria through WASPAS, informed decisions can be made to optimize traffic flow and mitigate congestion. Here we highlight the innovative application of MCDM in transportation planning, emphasizing its potential to enhance urban mobility and promote sustainable development in increasingly congested urban environments.

1. INTRODUCTION

Traffic management and congestion control are critical aspects of urbanization and transportation engineering, aimed at ensuring the seamless flow of vehicles and minimizing delays on road highway systems. With the rapid increase in urban sprawl and vehicle ownership, cities worldwide face notable challenges in managing traffic congestion. Effective traffic management involves an inclusive approach that considers multiple factors contributing to congestion, including high vehicle density, inadequate public transportation systems, and behavioural factors of drivers. Fuzzy logic is increasingly being used in traffic management and congestion control to handle the uncertainty and fluctuation of traffic conditions. It helps in dynamically adjusting traffic signals based on real-time data, improving flow efficiency. For instance, fuzzy logic controllers can evaluate traffic density at intersections and adjust signal timings to optimize traffic flow, reducing wait times and improving overall efficiency. This method of using fuzzy logic for traffic control has been successfully implemented in various cities, demonstrating significant reductions in congestion and improvements in traffic flow. One of the primary advantages of using fuzzy logic in traffic management is its ability to handle imprecise and changing data, which is a common characteristic of real-world traffic scenarios. Traditional traffic control systems often rely on fixed signal timings and established guidelines, which may not be effective in dealing with fluctuating traffic volumes and unpredictable conditions. In contrast, fuzzy logic systems can continuously adapt to fluctuating circumstances, providing a more responsive and efficient solution to traffic management. To effectively address the elements of traffic congestion, multi-criteria decision-making (MCDM) techniques are applied to analyse and prioritize the various elements that impact traffic management. MCDM methods are used to make well-informed and organized resolutions when multiple conflicting alternatives are involved. This multi-criterion decision-making approach helps traffic planners recognize the most impactful factors and implement targeted interventions to enhance traffic flow and reduce congestion. They help to analyse and prioritize criteria by methodically considering various factors, enhancing transparency and objectivity. MCDM approach facilitates stakeholder involvement, resolves conflicts, and assists a strategic approach across diverse fields like project management, environmental management, and healthcare. By incorporating both quantitative and qualitative data,

MCDM methods improve decision quality and ensure systematic analysis, leading to enhanced outcomes in complex decision-making scenarios. In traffic management, these methods can be used to evaluate a wide range of criteria, including traffic volume, road density, public transportation availability, and environmental impact.

One common MCDM method is the Analytical Hierarchy Process (AHP), which involves dissecting a complex decision into a hierarchy of simpler decisions, each of which can be analysed independently. By assigning weights to different criteria based on their relative significance, AHP helps decision-makers prioritize their actions and allocate resources more effectively. Another widely used MCDM method is the Technique for Order Preference by Similarity to Ideal Solution (TOPSIS), which determines the best option by comparing the distance of each alternative to an ideal solution. One such method is the Weighted Aggregated Sum Product Assessment (WASPAS), which unifies the strengths of the weighted sum model (WSM) and the weighted product model (WPM). WASPAS allows for an in-depth analysis of multiple criteria, helping decision-makers identify and select the most critical factors affecting traffic congestion. This method is especially useful in scenarios where multiple criteria need to be considered simultaneously, and where the relative importance of these criteria may vary. By using the WASPAS method, urban designers and traffic management authorities can systematically evaluate the relative significance of different criteria such as vehicle density, public transportation quality—which comprises how well the transport system meets the needs and expectations of its users—and driver behaviour. This structured approach enables the development of targeted strategies to alleviate traffic congestion and improve urban mobility. For instance, by analyzing the relative importance of public transportation quality, designers can identify key areas for improvement, such as increasing the frequency of buses or enhancing the connectivity of transit networks. On practical terms, implementing fuzzy logic and MCDM methods like WASPAS can lead to significant advancements in traffic management systems. For example, cities that have enacted these technologies have seen reductions in average travel times and improvements in air quality due to decreased vehicle idling. Furthermore, these strategies can be integrated with smart city initiatives, using certain devices to gather real-time traffic data and further enhance decision-making processes. Such as smart traffic lights equipped with sensors can collect data on traffic flow and adjust signal timings in real-time, improving traffic movement and reducing congestion. The integration of fuzzy logic and MCDM methods into traffic management also supports the development of additional sustainable urban environments. By optimizing traffic flow and reducing congestion, these approaches can contribute to diminish emissions and reduced fuel consumption, supporting broader environmental goals. Additionally, by improving the efficiency of public transportation systems, cities can promote greater use of these systems, further reducing the reliance on private vehicles and alleviating traffic congestion.

Traffic management and congestion control are vital components of urban development and transportation engineering. By leveraging advanced technologies such as fuzzy logic and multi-criteria decision-making methods like WASPAS, cities can develop additional effective and responsive traffic management systems that address the complex and dynamic nature of urban traffic. These approaches support a systematic and comprehensive evaluation of the various factors that contribute to traffic congestion, supporting the development of targeted strategies to enhance traffic movement and improve urban mobility. As cities continue to grow and develop, the integration of these advanced methodologies with emerging technologies such as artificial intelligence and machine learning, will be critical to ensuring sustainable and efficient urban transportation systems.

2. LITERATURE REVIEW

This section delves into the current domain pertaining to our research topic, offering a wide-ranging review of the existing literature. (Bellman & Zadeh, 1970) introduce the idea of fuzzy sets and their application to decision-making processes. It provides a foundation for handling uncertainty and imprecision in various domains, including traffic management, where flexible systems can benefit from fuzzy logic. (Cavallaro, 2010) applies the fuzzy TOPSIS method to evaluate energy storage systems, displaying its flexibility in handling multi-criteria decision-making in various fields, including traffic management for evaluating and optimizing various criteria. Applications of the WASPAS Method in Manufacturing Decision Making (Chakraborty & Zavadskas, 2014) illustrate the effectiveness of the WASPAS method in manufacturing, accentuating its ability to combine the strengths of WSM and WPM. It proposes how WASPAS can be adapted for traffic management to evaluate multiple factors affecting congestion. (Chen & Hwang, 1992) offer a comprehensive overview of fuzzy MCDM methods, detailing their applications and benefits. (Cheng & Li, 2001) examine the use of AHP for resource allocation, highlighting its effectiveness in prioritizing insights and tools. The methodology can be applied to traffic management to prioritize interventions and improve resource use. Chiu and colleagues (Chiu & Chand, 1992) present an adaptive traffic signal control system using fuzzy logic, exhibiting its potential to improve traffic flow by adjusting signal timings based on real-time conditions. Dion, Rakha, and Kang (Dion, Rakha, & Kang, 2004) contrasts delay estimates at signalized intersections, providing valuable data on the effectiveness of different traffic control strategies. It showcases the need for adaptive systems that can handle varying traffic conditions. Figueira and colleagues

(Figueira, Greco, & Ehrgott, 2005) offers a extensive study of MCDM methods, including their theoretical foundations and practical applications. It serves as a vital reference for understanding the application of MCDM in traffic management. (Hobbs & Meier, 2000) explore the application of MCDM methods in environmental and energy decisions, explaining how these methods can balance environmental and mobility goals in traffic management. Hwang and Yoon's book (Hwang & Yoon, 1981) is a foundational text on MCDM methods, specifying various approaches and their applications. It provides essential insights into methodologies applicable to traffic management for assessing and prioritizing different criteria. (Kamarianakis & Prastacos, 2005) study focuses on the space-time modeling of traffic flow, emphasizing the importance of contemplating temporal and spatial factors in traffic management. It supports the need for dynamic and adaptive management systems. (Köksalan, Wallenius, & Zionts, 2011) book traces the development of MCDM methods, providing a historical framework for their use in traffic management. It highlights the evolution and increasing complexity of decision-making processes. (Lee & Lee, 2001) present a case study on the development and evaluation of a fuzzy logic-based traffic signal control system. Their findings illustrate the practical benefits of using fuzzy logic to improve traffic flow and reduce congestion.

(Mardle, Pascoe, & Herrero, 2004) explores the use of AHP in evaluating management objectives in fisheries, showcasing its adaptability. The methodology and insights can be applied to traffic management to prioritize objectives and optimize strategies. (Mishra, Rani, & Gunasekaran, 2017) combine fuzzy AHP and TOPSIS to evaluate cloud-based service providers, demonstrating a hybrid approach. This study illustrates the benefits of integrating multiple MCDM methods for robust decision-making in traffic management. (Niittymäki & Pursula, 2000) discusses the use of fuzzy logic for traffic signal control, demonstrating its ability to deal with uncertainties in traffic flow and improve signal timings dynamically. This approach is crucial for managing congestion and enhancing traffic efficiency. (Papageorgiou, Diakaki, Dinopoulou, Kotsialos, & Wang, 2003) comprehensive review covers various road traffic control strategies, highlighting their effectiveness and challenges. The paper underscores the need for sophisticated control systems like those using fuzzy logic and MCDM for improved traffic management. (Saaty, 2008) elaborates on the operational implementations of AHP in decision-making processes across various domains. It provides insights into how AHP can be used in traffic management to evaluate and prioritize different factors impacting traffic flow. Uddin and colleagues (Uddin, Imran, & Qazi, 2018) apply fuzzy logic to multi-criteria decision-making in urban transportation, demonstrating its effectiveness in addressing complex and uncertain traffic conditions. The approach helps in developing more adaptive and efficient traffic management strategies. (Vahidi & Eskandarian, 2003) highlights advancements in intelligent collision exclusion and adaptive cruise control systems, emphasizing the role of real-time data and adaptive control strategies. These technologies are integral to modern traffic management systems. (Yannis, Golias, & Antoniou, 2006) investigate the impact of real-time traffic information on driver behaviour and traffic flow, highlighting the importance of providing precise and timely information to drivers to manage congestion and improve traffic efficiency

3. BASIC DEFINITIONS

Definition 3.1: Membership Function

A membership function defines the degree to which a given element is part of a fuzzy set. The value ranges from 0 (not a member) to 1 (full member), with values in between representing partial membership.

Definition 3.2: Alternative

These are the different factors or attributes that decision-makers consider when evaluating criteria. Alternatives can be quantitative or qualitative and may have different levels of importance.

Definition 3.3: Criteria

These are the options or choices available for selection. Each crietria is evaluated based on its performance or characteristics with respect to the defined alternative.

Definition 3.4: Fuzzy Weight

Weights are assigned to alternatives to reflect their relative importance in the decision-making process. These weights represent the significance of each alternative in the overall evaluation. Instead of fixed numerical weights, fuzzy weights are often used to capture uncertainty and imprecision in the decision-maker's judgment.

Definition 3.5: Fuzzy Arithmetic

Calculations with fuzzy numbers allow the incorporation of uncertainty on parameters, properties, geometry, initial conditions etc. The arithmetic calculations on fuzzy numbers are implemented using fuzzy arithmetic operators.

Definition 3.6: Fuzzy Decision Matrix

A fuzzy decision matrix extends the classical decision matrix to incorporate fuzzy numbers or linguistic variables. Each element x_{ij} in the matrix represents the performance of alternative j with respect to criterion i, expressed as a fuzzy number.

Definition 3.7: Aggregation of Fuzzy Score

Aggregation methods combine fuzzy information from different alternatives to form a composite performance score for each criterion. Common aggregation techniques include fuzzy weighted sum, fuzzy weighted product, and fuzzy max-min operations.

Definition 3.8: Defuzzification

Defuzzification is the process of converting fuzzy numbers into crisp scores. This step is essential for making actionable decisions based on fuzzy MCDM results. Common defuzzification methods contain the centroid method, mean of maxima, and the alpha-cut method.

Definition 3.9: Alpha-Cut

The alpha-cut or alpha-level cut or cut worthy set of a fuzzy set A of the set X is the following crisp (i.e., conventional) set given by:

$A = \{x \in X: \mu_A(x) \geq \ \}$ where $\ \in [0,1]$

Definition 3.10: Fuzzy Analytic Hierarchy Process (Fuzzy AHP)

Fuzzy AHP is an extension of the Analytic Hierarchy Process that incorporates fuzzy logic to handle the uncertainty in one-on-one comparisons of criteria and alternatives. It allows for a more flexible and realistic depiction of decision-maker preferences.

4. PROPOSED METHOD

Effective solutions require an in-depth approach that considers multiple factors, including vehicle density, public transportation efficiency, and driver behavior. The Weighted Aggregated Sum Product Assessment (WASPAS) method, resilient multi-criteria decision-making (MCDM) technique, offers a structured and systematic approach to tackle these challenges.

Step 1: Construction of Decision Matrix and Weight Matrix

In a multiple criteria decision-making (MCDM) problem, the decision matrix D is a vital tool for organizing and comparing various criteria through multiple alternatives. This matrix is structured with m rows and n columns, where each element x_{ij} represents the value of the ith criterion for the jth alternative. The rows are according to different criteria, such as cost, quality, and time, meanwhile the columns represent the different alternatives available for decision-making. By using this matrix, decision-makers can systematically assess and compare the relative importance and productivity of each criterion for all alternatives. The decision matrix D serves as the foundation for applying various MCDM methods, such as AHP, TOPSIS, and WASPAS, to derive optimal decisions based on multiple conflicting criteria.

$$\mathbf{D} = \begin{vmatrix} x_{11} & x_{12}\cdots & x_{1n} \\ x_{21} & x_{22}\cdots & x_{2n} \\ x_{m1} & x_{m2}\cdots & x_{mn} \end{vmatrix}$$

The weight vector can be expressed as, $w_j = [w_1, w_2, \ldots w_n]$, where $\sum_{j=1}^{n} \left(w_1, w_2, \ldots w_n\right) = 1$

Step 2: Normalized Decision Matrix

$$n_{ij} = \begin{cases} \dfrac{x_{ij}}{\max x_{ij}}, j \in B \\ \dfrac{\min x_{ij}}{x_{ij}}, j \in C \end{cases}$$

The variable n_{ij} represents the normalized value of the ith criterion concerning the jth alternative. The terms max x_{ij} and min x_{ij} signify the maximum and minimum values of x_{ij} within the jth column for the benefit (B) and cost (C) criteria, respectively.

Step 3: Weighted Normalized Decision Matrix

Each alternative in a multiple criteria decision-making (MCDM) problem is allocated a weight that represents its relative importance compared to other alternatives. These weights reflect the decision-maker's preferences or priorities. To assimilate these weights into the decision matrix, we first normalize the decision matrix to ensure comparability across different scales. Then, we evaluate the product of the assigned weights with the normalized decision matrix. This process adjusts each element in the decision matrix by its corresponding weight, effectively integrating the importance of each alternative into the overall evaluation. The outcome is a weighted decision matrix, which better represents the decision-makers priorities and facilitates a more accurate analysis of the criterion

$$W_{ij} = w_j n_{ij}$$

Step 4: Weighted Sum Model

Ordering or arranging alternatives based on their relative merits or performance

$$S_i^{WSM} = \sum_{j=1}^{n} w_j n_{ij}$$

In the weighted sum model (WSM), S_i^{WSM} is the ranking score of the ith criteria and w_j is the weight of the jth criterion and then summing up these weighted scores. The criteria are then arranged in descending order based on their total scores, with the criteria having the highest total score being ranked the highest.

Step 5: Weighted Product Model

By using WPM, the ranking score is calculated as,

$$S_i^{WPM} = \prod_{j=1}^{n} n_{ij}^{w_j}$$

The criteria are organized in a sequence based on the decreasing order of their ranking scores calculated using the WPM (Weighted Performance Model). This entails arranging the criteria according to the diminishing values of their WPM ranking scores, with the highest-ranking alternative having the highest score.

Step 6: Weighted Aggregated Sum Product Assessment

The Weighted Sum Performance Assessment (WASPAS) method combines two multicriteria decision-making (MCDM) approaches, namely the Weighted Sum Model (WSM) and the Weighted Performance Measure (WPM). Finally, the WASPAS score is computed by integrating the results from both WSM and WPM. This comprehensive approach allows decision-makers to consider both the weighted importance of alternatives and the direct performance of criteria.

$$S_i^{WASPAS} = \lambda\, S_i^{WSM} + (1 - \lambda)\, S_i^{WPM}$$

$$S_i^{WASPAS} = \lambda \sum_{j=1}^{n} w_j n_{ij} + \left(1 - \lambda\right) \prod_{j=1}^{n} n_{ij}^{\ w_j}$$

Step 7: Weighted Aggregated Sum Product Assessment for Group-Wise Ranking

By looking at the WASPAS preference score, the group-wise ranking is allocated for all the sub-criteria that come under the major seven criteria, looking forward to which sub-criteria is ranked the highest. The sub-criteria are then ranked from highest to lowest based on these scores within the particular group. The sub-criteria with the highest WASPAS score is considered the most important, reflecting its overall preference within the decision-making framework.

Step 8: Weighted Aggregated Sum Product Assessment for Global Ranking.

To evaluate the highest-ranked sub-criteria among the 30 sub-criteria based on WASPAS preference scores, we first calculate the WASPAS scores for each sub-criteria, and thus repeating the steps of computing the weighted sum model (WSM), and the weighted product model (WPM). We then combine these to get the WASPAS score for each sub-criteria. By comparing these WASPAS scores, we establish a global ranking. The sub-criteria with the highest WASPAS score is considered the best overall, reflecting their superior performance and preference across all evaluated aspects in the deliberation.

Analysis of Major Criteria and Sub-criteria:

$$\text{C1=High Vehicle density=}\begin{cases} C11 = \textit{Population growth} \\ C12 = \textit{Economic prosperity} \\ C13 = \textit{Inadequate parking facilities} \\ C14 = \textit{Lack of public transport} \\ C15 = \textit{Vehicle independent infrastructure} \\ C16 = \textit{Urbanization} \end{cases}$$

$$\text{C2=Poor public transportation system=}\begin{cases} C21 = \textit{Limited coverage} \\ C22 = \textit{Infrequent service} \\ C23 = \textit{Unreliable Scheludes} \\ C24 = \textit{Comfort and safety issues} \end{cases}$$

$$C3=\text{Inadequate Road infrastructure}=\begin{cases} C31 = Narrow\ roads \\ C32 = Inadequate\ traffic\ signals \\ C33 = Poor\ road\ maintenance \end{cases}$$

$$C4=\text{Traffic management issues}=\begin{cases} C41 = Inefficient\ traffic\ signals \\ C42 = Lack\ of\ traffic\ enforcement \\ C43 = Inadequate\ planning \end{cases}$$

$$C5=\text{Inefficient Road capacity}=\begin{cases} C51 = Overloaded\ road\ infrastructure \\ C52 = Limited\ expansion\ options \end{cases}$$

$$C6=\text{Urban design flaws}=\begin{cases} C61 = Sprawling\ development \\ C62 = Inadequate\ mass\ transit\ integration \end{cases}$$

$$C7=\text{Behavioural factors }=\begin{cases} C71 = Aggressive\ driving \\ C72 = Distracted\ driving \\ C73 = Poor\ driving\ skills \\ C74 = Non-adherence\ to\ traffic\ rules \\ C75 = Lack\ of\ car\ pooling \\ C76 = Inconsistent\ driving\ speed \\ C77 = Over\ reliance\ on\ GPS\ navigation \\ C78 = Blocking\ intersections \\ C79 = Impatient\ driving \\ C80 = Poor\ merging\ behaviour \end{cases}$$

5. NUMERICAL CALCULATION

In this section, we discuss the numerical example to test the proposed method.

Table 1. Construction of Decision Matrix

Criteria	E1	E2	E3	E4
C1				
C11	9.7	7	8.5	8.1
C12	7.3	8	7.2	6.6
C13	5.5	6.2	5.4	6.2
C14	8.9	7.7	7.4	8.4
C15	9.6	7.6	8	8.1
C16	9.8	8.1	7.3	7.8
C2				
C21	3	3.5	3.1	2.7
C22	2.4	3.3	3.5	2.6
C23	2.1	2.7	3.2	3.1
C24	3.3	2.9	3.6	3.8
C3				
C31	1.3	2.3	2.5	1.9
C32	2.1	1.6	1.42	2.8
C33	1.8	2.7	1.1	2.2
C4				
C41	2.6	3.5	3.1	2.7
C42	3.6	2.4	2.7	2.9
C43	2.8	3.9	3.2	3.3
C5				
C51	8.6	7.9	8.1	9
C52	7.7	8	8.3	8.9
C6				
C61	4.7	5.4	3.9	4.8
C62	5.6	4.9	4.2	5
C7				
C71	7.5	9.1	8.1	7.2
C72	8.4	8.6	7.4	7.7
C73	9.9	8	8.5	9.7

continued on following page

Table 1. Continued

Criteria	E1	E2	E3	E4
C74	9.2	7.6	9.3	9.4
C75	8.7	9.6	9.8	9.9
C76	8	7.8	9.2	8.9
C77	7.1	7.6	8.9	9.7
C78	8.6	8.3	7.8	9.2
C79	9.3	7.7	7.9	9
C80	8.8	7.4	9.5	7.5

Table 2. Normalized value of Decision Matrix

Criteria	E1	E2	E3	E4
C1				
C11	0.98979	0.86419	0.63529	0.76543
C12	0.74489	0.98765	0.75	0.93939
C13	0.56122	0.76543	1	1
C14	0.90816	0.95061	0.72973	0.73809
C15	0.97959	0.93827	0.675	0.76543
C16	1	1	0.73972	0.79487
C2				
C21	0.90909	1	1	0.96296
C22	0.72727	0.94285	0.88571	1
C23	0.63634	0.77142	0.96875	0.83871
C24	1	0.82857	0.86111	0.68421
C3				
C31	0.61904	0.85185	0.44	1
C32	1	0.59259	0.78571	0.67857
C33	0.85714	1	1	0.86363
C4				
C41	0.72222	0.89743	0.87096	1
C42	1	0.61538	1	0.93103
C43	0.77777	1	0.84375	0.81818
C5				
C51	1	0.9875	1	0.98888

continued on following page

Table 2. Continued

Criteria	E1	E2	E3	E4
C52	0.89534	1	0.97590	1
C6				
C61	0.83928	1	1	1
C62	1	0.90740	0.92857	0.96
C7				
C71	0.75757	0.94791	0.91358	1
C72	0.84848	0.89583	1	0.93506
C73	1	0.83333	0.87058	0.74226
C74	0.92929	0.79166	0.79569	0.76595
C75	0.87878	1	0.75510	0.72727
C76	0.80808	0.8125	0.80434	0.80898
C77	0.71717	0.79166	0.83146	0.74226
C78	0.86868	0.86458	0.94871	0.78260
C79	0.93939	0.80208	0.93670	0.8
C80	0.88888	0.77083	0.77894	0.96

Table 3. Weighted Sum Model

Criteria	E1	E2	E3	E4	Preference score	Rank
C1						
C11	0.35632	0.23333	0.17152	0.07654	0.83773	4
C12	0.26816	0.26666	0.2025	0.09393	0.83126	5
C13	0.20204	0.20666	0.27	0.1	0.77870	6
C14	0.32693	0.25666	0.19702	0.07381	0.85444	3
C15	0.35653	0.25333	0.18225	0.07654	0.86477	2
C16	0.36	0.27	0.19972	0.07948	0.90921	1
C2						
C21	0.32727	0.27	0.27	0.09629	0.96356	1
C22	0.26181	0.25471	0.23914	0.1	0.85553	3
C23	0.22909	0.20828	0.26156	0.08387	0.78281	4
C24	0.36	0.22371	0.2325	0.06842	0.88463	2
C3						
C31	0.22285	0.23	0.1188	0.1	0.67165	3

continued on following page

Table 3. Continued

Criteria	E1	E2	E3	E4	Preference score	Rank
C32	0.36	0.16	0.21214	0.06785	0.8	2
C33	0.30857	0.27	0.27	0.08636	0.93493	1
C4						
C41	0.26	0.24230	0.23516	0.1	0.83746	3
C42	0.36	0.16615	0.27	0.09310	0.88925	1
C43	0.28	0.27	0.26349	0.08181	0.85963	2
C5						
C51	0.36	0.26662	0.27	0.09888	0.99551	1
C52	0.32232	0.27	0.26349	0.1	0.95581	2
C6						
C61	0.30214	0.27	0.27	0.1	0.94214	2
C62	0.36	0.245	0.25071	0.096	0.95171	1
C7						
C71	0.27272	0.25593	0.25666	0.1	0.87533	5
C72	0.30545	0.24187	0.27	0.09350	0.91083	1
C73	0.36	0.225	0.23505	0.07422	0.89428	2
C74	0.33454	0.21375	0.21483	0.076595	0.83972	7
C75	0.31636	0.21375	0.21483	0.07659	0.86296	6
C76	0.29090	0.21937	0.21717	0.08089	0.80835	9
C77	0.25818	0.21375	0.22449	0.07422	0.77065	10
C78	0.31272	0.23343	0.25615	0.07826	0.88057	4
C79	0.33818	0.21656	0.25291	0.08	0.88765	3
C80	0.32	0.20812	0.21031	0.096	0.83444	8

Table 4. Weighted Product Model

Criteria	E1	E2	E3	E4	Preference score	Rank
C1						
C11	0.99631	0.96135	0.88471	0.97362	0.82504	4
C12	0.89940	0.99665	0.92526	0.99376	0.82423	5
C13	0.81224	0.93036	1	1	0.75569	6
C14	0.96591	0.98641	0.91844	0.97008	0.84891	3
C15	0.99260	0.98294	0.89931	0.97362	0.85429	2

continued on following page

Table 4. Continued

Criteria	E1	E2	E3	E4	Preference score	Rank
C16	1	1	0.92182	0.97730	0.90090	1
C2						
C21	0.96627	1	1	0.99623	0.96263	1
C22	0.89168	0.98423	0.96776	1	0.84933	3
C23	0.84983	0.93233	0.99146	0.98256	0.77186	4
C24	1	0.95049	0.96043	0.96276	0.87888	2
C3						
C31	0.84143	0.95763	0.80118	1	0.645582	3
C32	1	0.86824	0.93696	0.99288	0.78257	2
C33	0.94601	1	1	0.98013	0.93225	1
C4						
C41	0.88945	0.97120	0.96338	1	0.83221	3
C42	1	0.87714	1	0.99288	0.87089	1
C43	0.94601	1	0.955164	0.98013	0.85520	2
C5						
C51	1	0.99660	1	0.99888	0.99549	1
C52	0.96098	1	0.99343	1	0.95467	2
C6						
C61	0.93887	1	1	1	0.93887	2
C62	1	0.97410	0.98019	0.99592	0.95092	1
C7						
C71	0.94088	0.98566	0.97589	1	0.87040	5
C72	0.94256	0.97073	1	0.99330	0.90886	1
C73	1	0.95196	0.96327	0.97063	0.89007	2
C74	0.97394	0.93887	0.94016	0.97063	0.83443	7
C75	0.95454	1	0.92696	0.96865	0.85709	6
C76	0.92615	0.94548	0.94290	0.97902	0.80835	9
C77	0.88720	0.93887	0.95138	0.97063	0.76920	10
C78	0.95058	0.96147	0.982588	0.97793	0.87821	4
C79	0.97774	0.94219	0.98250	0.97793	0.88512	3
C80	0.95848	0.92313	0.93477	0.93492	0.83176	8

Table 5. Weighted Aggregate Sum Product Model for Group-Wise Ranking

Criteria	WSM	WPM	WASPAS	Groupwise ranking
C1				
C11	0.83773	0.82504	0.83138	4
C12	0.83126	0.82423	0.82775	5
C13	0.77870	0.75569	0.76719	6
C14	0.85444	0.84891	0.85168	3
C15	0.86477	0.85429	0.85953	2
C16	0.90921	0.90090	0.90505	1
C2				
C21	0.96356	0.96263	0.9631	1
C22	0.85553	0.84933	0.85243	3
C23	0.78281	0.77186	0.77733	4
C24	0.88463	0.87888	0.88176	2
C3				
C31	0.67165	0.645582	0.65861	3
C32	0.8	0.78257	0.79128	2
C33	0.93493	0.93225	0.93359	1
C4				
C41	0.83746	0.83221	0.83484	3
C42	0.88925	0.87089	0.88007	1
C43	0.85963	0.85520	0.85741	2
C5				
C51	0.99551	0.99549	0.99550	1
C52	0.95581	0.95467	0.95524	2
C6				
C61	0.94214	0.93887	0.94050	2
C62	0.95171	0.95092	0.95131	1
C7				
C71	0.87533	0.87040	0.87287	5
C72	0.91083	0.90886	0.90984	1
C73	0.89428	0.89007	0.89218	2
C74	0.83972	0.83443	0.83840	7
C75	0.86296	0.85709	0.86003	6
C76	0.80835	0.80835	0.80835	9
C77	0.77065	0.76920	0.76993	10

continued on following page

Table 5. Continued

Criteria	WSM	WPM	WASPAS	Groupwise ranking
C78	0.88057	0.87821	0.87991	4
C79	0.88765	0.88512	0.88639	3
C80	0.83444	0.83176	0.83310	8

Table 6. Weighted Aggregate Sum Product Model for all the 30 criteria

Criteria	WSM	WPM	WASPAS	Global ranking
C1				
C11	0.60800	0.45397	0.53098	9
C12	0.56049	0.45353	0.50701	14
C13	0.46002	0.41581	0.43791	19
C14	0.60295	0.46711	0.53503	7
C15	0.62342	0.47007	0.54674	3
C16	0.64922	0.49571	0.57246	1
C2				
C21	0.37370	0.36163	0.36766	21
C22	0.33801	0.31907	0.32854	28
C23	0.30640	0.28996	0.29818	29
C24	0.33406	0.33017	0.33211	27
C3				
C31	0.33076	0.26228	0.29652	30
C32	0.40136	0.31794	0.35965	24
C33	0.49775	0.37875	0.43825	18
C4				
C41	0.35915	0.34347	0.35131	26
C42	0.37392	0.35944	0.36668	23
C43	0.36189	0.35296	0.35743	25
C5				
C51	0.59269	0.45027	0.52148	11
C52	0.56213	0.43180	0.49697	15
C6				
C61	0.43852	0.42403	0.43127	20
C62	0.41416	0.31993	0.36704	22

continued on following page

Table 6. Continued

Criteria	WSM	WPM	WASPAS	Global ranking
C7				
C71	0.59172	0.45535	0.52353	10
C72	0.61214	0.47547	0.54380	4
C73	0.63952	0.46564	0.55258	2
C74	0.60044	0.43791	0.51917	12
C75	0.63586	0.44838	0.54212	5
C76	0.56391	0.42288	0.49340	16
C77	0.52489	0.40241	0.46365	17
C78	0.60489	0.45997	0.53243	8
C79	0.61345	0.46305	0.53825	6
C80	0.58472	0.43513	0.50992	13

6. RESULT AND DISCUSSION

Figure 1. Ranking of sub-criteria under high vehicle density.

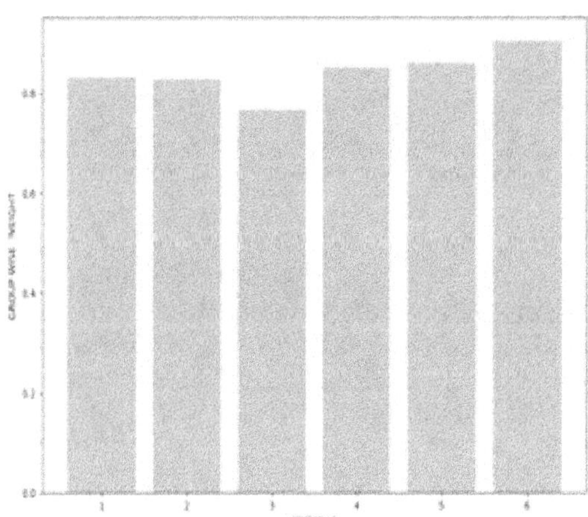

Figure 2. Ranking of sub-criteria under poor public transport system.

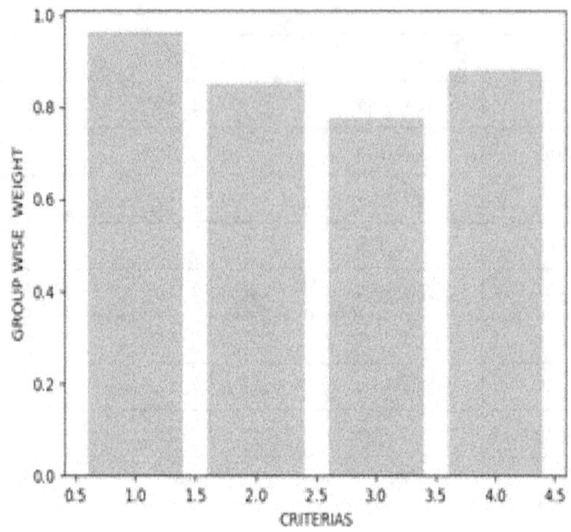

Figure 3. Ranking of sub-criteria under inadequate road infrastructure.

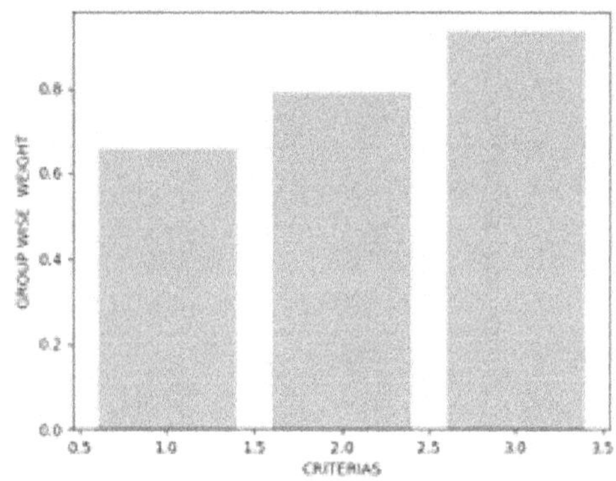

Figure 4. Ranking of sub-criteria under traffic management issues.

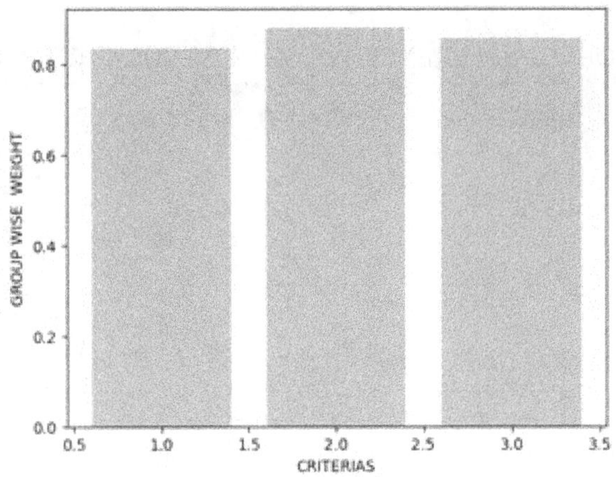

Figure 5. Ranking of sub-criteria under behavioral factors.

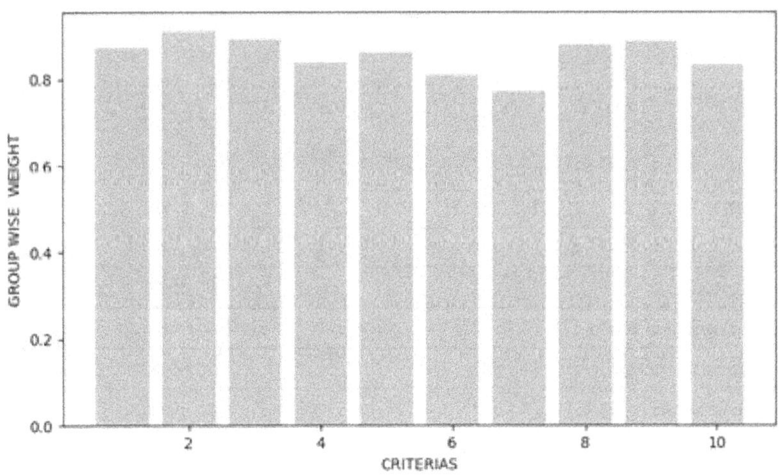

Figure 6. Global preference weight of sub-criteria of traffic management and congestion control.

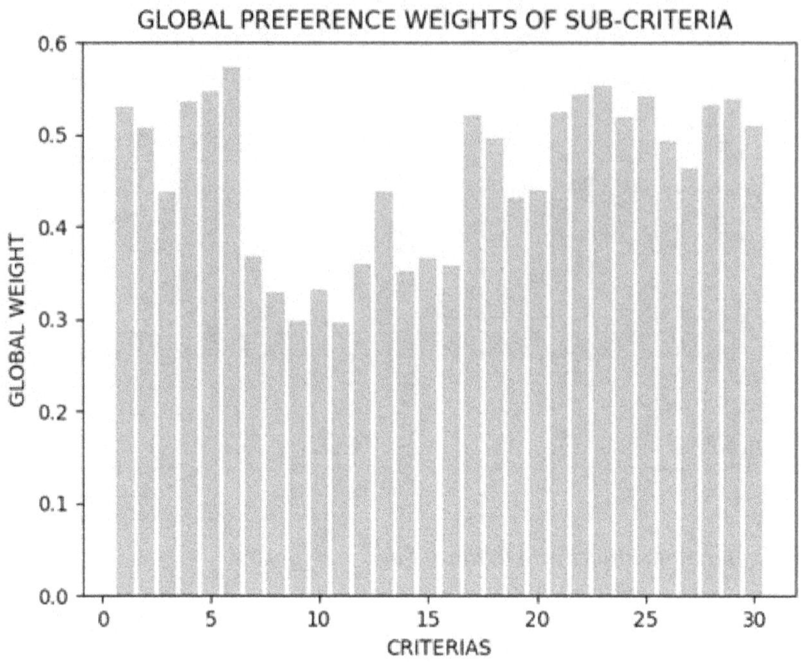

Table 7. Major criteria ranking based on the WASPAS preference score

Criteria	Scores	Rank
High Vehicle density	5.04258	2
Poor public transportation system	3.47462	3
Inadequate road infrastructure	2.38348	5
Traffic management issues	2.55971	4
Inefficient road capacity	1.95074	6
Urban design flaws	1.89181	7
Behavioural factors	8.551	1

7. CONCLUSION

The implementation of the Weighted Aggregated Sum Product Assessment (WASPAS) method has provided a robust framework for evaluating and addressing traffic congestion. Our study has clearly identified behavioural factors and high vehicle density as the primary contributors to urban traffic congestion. Behavioural factors, such as aggressive driving, distracted driving, and poor merging practices, especially disrupt traffic flow and increase accident rates, leading to frequent stop-and-go conditions that increase congestion. Targeted interventions, including driver education programs, stricter enforcement of traffic laws, and stimulants for safe driving, are essential to mitigate these issues and improve overall traffic dynamics. High vehicle density exacerbates congestion due to the sheer volume of vehicles exceeding road capacity. This is further composited by inadequate public transportation systems and limited road infrastructure. Strategic infrastructure improvements are necessary to accommodate the growing number of vehicles. The WASPAS model's combining approach has been instrumental in prioritizing these factors, enabling urban planners to devise targeted and effective strategies for congestion control.

REFERENCES

Anand, M. C. J., Kalaiarasi, K., Martin, N., Ranjitha, B., Priyadharshini, S. S., & Tiwari, M. (2023, September). Fuzzy C-Means Clustering with MAIRCA-MCDM Method in Classifying Feasible Logistic Suppliers of Electrical Products. In *2023 First International Conference on Cyber Physical Systems, Power Electronics and Electric Vehicles (ICPEEV)* (pp. 1-7). IEEE.

Anand, M. C. J., Martin, N., Clementking, A., Rani, S., Priyadharshini, S., & Siva, S. (2023, March). Decision making on optimal selection of advertising agencies using machine learning. In *2023 International Conference on Information Management (ICIM)* (pp. 121-125). IEEE.

Anand, M. C. J., Priyadharshini, S. S., Martin, N., Anand, M. E., Poorani, B., & Mohiddin, S. K. (2023, September). Decision Trees in the Selection of Electric Vehicles Based on Various Parameters. In *2023 First International Conference on Cyber Physical Systems, Power Electronics and Electric Vehicles (ICPEEV)* (pp. 1-6). IEEE.

Bellman, R. E., & Zadeh, L. A. (1970). Decision Making in a Fuzzy Environment. *Management Science, 17*(4), B141–B164. DOI: 10.1287/mnsc.17.4.B141

Bharatraj, J., & Clement Joe Anand, M. (2019). Power harmonic weighted aggregation operator on single-valued trapezoidal neutrosophic numbers and interval-valued neutrosophic sets. In *Fuzzy Multi-criteria Decision-Making Using Neutrosophic Sets* (pp. 45–62). Springer International Publishing. DOI: 10.1007/978-3-030-00045-5_3

Cavallaro, F. (2010). Fuzzy TOPSIS Approach for Assessing Thermal-Energy Storage in Concentrating Solar Power (CSP) Systems. *Applied Energy, 87*(2), 496–503. DOI: 10.1016/j.apenergy.2009.07.009

Chakraborty, S., & Zavadskas, E. K. (2014). Applications of WASPAS Method in Manufacturing Decision Making. *Informatica (Vilnius), 25*(4), 524–540. DOI: 10.15388/Informatica.2014.01

Chen, S. J., & Hwang, C. L. (1992). *Fuzzy Multiple Attribute Decision Making: Methods and Applications. Lecture Notes in Economics and Mathematical Systems, 375*. Springer., DOI: 10.1007/978-3-642-46768-4

Cheng, E. W. L., & Li, H. (2001). Information Priority-Setting for Better Resource Allocation Using Analytic Hierarchy Process (AHP). *Information Management & Computer Security, 9*(2), 61–70. DOI: 10.1108/09685220110388827

Chiu, S., & Chand, S. (1992). Adaptive Traffic Signal Control Using Fuzzy Logic. *Proceedings of the Second IEEE International Conference on Fuzzy Systems, 3,* 1371-1376. https://doi.org/DOI: 10.1109/FUZZY.1992.258699

Clement Joe Anand, M., & Bharatraj, J. (2017). Theory of Triangular Fuzzy Number. *Proceedings of NCATM, 2017*, 80–83.

Clement Joe Anand, M., & Bharatraj, J. (2018). Gaussian qualitative trigonometric functions in a fuzzy circle. *Advances in Fuzzy Systems, 2018*, 1–9. DOI: 10.1155/2018/8623465

Clement Joe Anand, M., & Bharatraj, J. (2019). Interval-valued neutrosophic numbers with WASPAS. In *Fuzzy Multi-criteria Decision-Making Using Neutrosophic Sets* (pp. 435–453). Springer International Publishing. DOI: 10.1007/978-3-030-00045-5_17

Clement Joe Anand, M., & Bharatraj, J. (2019). Interval-valued neutrosophic numbers with WASPAS. In *Fuzzy Multi-criteria Decision-Making Using Neutrosophic Sets* (pp. 435–453). Springer International Publishing. DOI: 10.1007/978-3-030-00045-5_17

Clement Joe Anand, M., Moorthy, C. B., Sivamani, S., Indrakumar, S., Kalaiarasi, K., & Barhoi, A. (2023). Fuzzy intelligence inventory decision optimization model of sustainability and green technologies for mixed uncertainties of carbon emission. In *In: 2023 International Conference on Information Management (ICIM).* IEEE. [REMOVED HYPERLINK FIELD], https://ieeexplore.ieee.org/document/10145085

Clement Joe Anand, M., Moorthy, C. B., Sivamani, S., Indrakumar, S., Kalaiarasi, K., & Barhoi, A. (2023). Fuzzy intelligence inventory decision optimization model of sustainability and green technologies for mixed uncertainties of carbon emission. In *In: 2023 International Conference on Information Management (ICIM).* IEEE., https://ieeexplore.ieee.org/document/10145085

Devadoss, A. V., Anand, M. C. J., & Felix, A. (2014, March). A CETD matrix approach to analyze the dimensions of personality of person. In *2014 International Conference on Computational Science and Computational Intelligence* (Vol. 2, pp. 40-45). IEEE.

Dion, F., Rakha, H., & Kang, Y. S. (2004). Comparison of Delay Estimates at Under-Saturated and Over-Saturated Pre-Timed Signalized Intersections. *Transportation Research Part B: Methodological, 38*(2), 99–122. DOI: 10.1016/S0191-2615(03)00003-1

Figueira J, Greco S and Ehrgott M (2005). Multiple Criteria Decision Analysis: State of the Art Surveys. *Springer, 78.* https://doi.org/DOI: 10.1007/b100605

Hobbs, B. F., & Meier, P. (2000). *Energy Decisions and the Environment: A Guide to the Use of Multicriteria Methods. International Series in Operations Research & Management Science*. Springer., DOI: 10.1007/978-1-4615-4477-7

Hwang, C. L., & Yoon, K. (1981). *Multiple Attribute Decision Making: Methods and Applications. Lecture Notes in Economics and Mathematical Systems, 186*. Springer., DOI: 10.1007/978-3-642-48318-9

Kamarianakis, Y., & Prastacos, P. (2005). Space–time Modeling of Traffic Flow. *Computers & Geosciences*, *31*(2), 119–133. DOI: 10.1016/j.cageo.2004.05.012

Köksalan, M., Wallenius, J., & Zionts, S. (2011). *Multiple Criteria Decision Making: From Early History to the 21st Century*. World Scientific., DOI: 10.1142/8042

Kungumaraj, E., Lathanayagam, E., Saikia, U., Anand, M., Khanna, S. T., Martin, N., & Edalatpanah, S. A. (2024). Neutrosophic Topological Vector Spaces and its Properties. [IJNS]. *International Journal of Neutrosophic Science*, *23*(2).

Kungumaraj, E., Lathanayagam, E., Saikia, U., Clement Joe Anand, M., Khanna, S. T., Martin, N., Tiwari, M., & Edalatpanah, S. A. (2024). Neutrosophic Topological Vector Spaces and its Properties. *International Journal of Neutrosophic Science*, *23*(2), 63–76. DOI: 10.54216/IJNS.230206

Lee, C., & Lee, H. (2001). Development and Evaluation of Fuzzy Logic-Based Traffic Signal Control. *Transportation Research Record: Journal of the Transportation Research Board*, *1748*(1), 82–88. DOI: 10.3141/2663-11

Manshath, A., Kungumaraj, E., Lathanayagam, E., Anand, M. C., Martin, N., Muniyandy, E., & Indrakumar, S. (2024). Neutrosophic Integrals by Reduction Formula and Partial Fraction Methods for Indefinite Integrals. [IJNS]. *International Journal of Neutrosophic Science*, *23*(1).

Mardle, S., Pascoe, S., & Herrero, I. (2004). Management Objective Importance in Fisheries: An Evaluation Using the Analytic Hierarchy Process (AHP). *Environmental Management*, *33*(1), 1–11. DOI: 10.1007/s00267-003-3070-y PMID: 14994158

Miriam, M. R., Martin, N., & Clement Joe Anand, M. (2023). Inventory model promoting smart production system with zero defects. *International Journal of Applied and Computational Mathematics*, *9*(4), 47. DOI: 10.1007/s40819-023-01524-4

Mishra, A. R., Rani, P., & Gunasekaran, M. (2017). A Multi-Criteria Decision-Making Approach for Cloud Service Provider Selection Based on Fuzzy AHP and TOPSIS. *Journal of Enterprise Information Management*, *30*(4), 586–610. DOI: 10.1108/JEIM-12-2016-0201

Nagaraj, C. T., Anand, M. C. J., Priyadharshini, S. S., & Aparna, P. (2023, December). GCNXG: Detecting Fraudulent Activities in Financial Networks: A Graph Analytics and Machine Learning Fusion. In International Conference on Renewable Energy, Green Computing, and Sustainable Development (pp. 17-32). Cham: Springer Nature Switzerland.

Niittymäki, J., & Pursula, M. (2000). Signal Control Using Fuzzy Logic. *Fuzzy Sets and Systems*, *116*(1), 11–22. DOI: 10.1016/S0165-0114(99)00034-2

Papageorgiou, M., Diakaki, C., Dinopoulou, V., Kotsialos, A., & Wang, Y. (2003). Review of Road Traffic Control Strategies. *Proceedings of the IEEE*, *91*(12), 2043–2067. DOI: 10.1109/JPROC.2003.819610

Prabha, S. K., Anand, M., Vidhya, V., Nagarajan, G., Saikia, U., Martin, N., & Tiwari, M. (2024). Sorting Out Interval Valued Neutrosophic Fuzzy Shortest Cycle Route Problem by Reduced Matrix Method. [IJNS]. *International Journal of Neutrosophic Science*, *23*(2).

Prabha, S. K., Clement Joe Anand, M., Vidhya, V., Nagarajan, G., Saikia, U., Martin, N., Kumari, M. S., & Tiwari, M. (2024). Sorting Out Interval Valued Neutrosophic Fuzzy Shortest Cycle Route Problem by Reduced Matrix Method. *International Journal of Neutrosophic Science*, *23*(2), 91–103. DOI: 10.54216/IJNS.230208

Raj, P. J., Prabhu, V. V., Krishnakumar, V., & Anand, M. C. J. (2023). P., Prabhu. V. V., Krishnakumar. V., M. Clement Joe Anand: Solar Powered Charging -of Fuzzy Logic Controller (FLC) Strategy with Battery Management System (BMS) Method Used for Electric Vehicle (EV). *International Journal of Fuzzy Systems*, *25*(7), 2876–2888. DOI: 10.1007/s40815-023-01537-7

Rajesh, K., Rathod, S., Kundale, J., Rathod, N., Anand, M., Saikia, U., & Martin, N. (2024). A Study on Interval Valued Temporal Neutrosophic Fuzzy Sets. [IJNS]. *International Journal of Neutrosophic Science*, *23*(1).

Rajesh, K., Rathod, S., Kundale, J., Rathod, N., Clement Joe Anand, M., Saikia, U., Tiwari, M., & Martin, N. (2024). A Study on Interval Valued Temporal Neutrosophic Fuzzy Sets. *International Journal of Neutrosophic Science*, *23*(1), 341–349. DOI: 10.54216/IJNS.230129

Saaty, T. L. (2008). Decision Making with the Analytic Hierarchy Process. *International Journal of Services Sciences*, *1*(1), 83–98. DOI: 10.1504/IJSSCI.2008.017590

Sudha, S., Nivetha, M., Clement Joe Anand, M., Palanimani, P. G., Thirunamakkani, T., & Ranjitha, B. (2023). MACBETH-MAIRCA Plithogenic Decision-Making on Feasible Strategies of Extended Producer's Responsibility towards Environmental Sustainability. *International Journal of Neutrosophic Science*, *22*(2), 114–130. DOI: 10.54216/IJNS.220210

Uddin, M., Imran, M., & Qazi, A. (2018). Multi-Criteria Decision Making Based on Fuzzy Logic for Urban Transportation System. *International Journal of Transportation Engineering*, *6*(3), 257–270. DOI: 10.22119/IJTE.2018.88293

Vahidi, A., & Eskandarian, A. (2003). Research Advances in Intelligent Collision Avoidance and Adaptive Cruise Control. *IEEE Transactions on Intelligent Transportation Systems*, *4*(3), 143–153. DOI: 10.1109/TITS.2003.821292

Varalakshmi, A., Kumar, S. S., Shanmugapriya, M. M., Mohanapriya, G., & Clement Joe Anand, M. (2022). Markers location monitoring on images from an infrared camera using optimal fuzzy inference system. *International Journal of Fuzzy Systems*.

Yannis, G., Golias, J., & Antoniou, C. (2006). Effects of Real-Time Traffic Information on Driver Behaviour and Traffic Flow. *Transportation Research Part C, Emerging Technologies*, *14*(5), 402–420. DOI: 10.1016/j.trc.2006.09.001

Chapter 11
Traffic Management and Congestion Control in Vehicular Adhoc Networks

Rajesh Kanna Rajendran
https://orcid.org/0000-0001-7228-5031
Christ University, Bangalore, India

P. Nivetha
Dr. N.G.P. Institute of Technology, India

Chippy Mohan
https://orcid.org/0000-0001-8589-4944
Christ University, Bangalore, India

Athira Aneesh
https://orcid.org/0000-0002-2856-7186
Christ University, Bangalore, India

V. Resmi
Mar Ivanios College, India

ABSTRACT

Urban traffic congestion is a growing concern worldwide. Vehicular Adhoc Networks (VANETs) offer a glimpse into a future with smoother traffic flow and reduced congestion. These networks enable real-time communication between vehicles and infrastructure, creating a dynamic traffic management system. Imagine traffic signals that adjust based on real-time data, congestion being predicted and alleviated

DOI: 10.4018/979-8-3693-6422-2.ch011

before it builds, and emergency services receiving faster response times. This is the potential of VANETs. Ensuring reliable communication and data integrity among constantly moving vehicles is crucial. Researchers are developing protocols and algorithms to address this, focusing on efficient routing, data dissemination, and network stability. The integration of emerging technologies like 5G, edge computing, and artificial intelligence holds promise for further enhancing network performance and robustness. While significant progress has been made, widespread adoption of VANETs faces hurdles. Scalability, security, privacy, and infrastructure development costs are significant concerns. T

INTRODUCTION

Urbanization and fast pace of vehicular traffic have exacerbated the problem of congestion in cities all around the world, causing great time loss, more fuel intake, and increased emissions of greenhouse gases. Conventional systems for traffic management fail to deal with the dynamic nature and complexity of modern urban traffic. The challenge has brought advanced technologies, including Vehicular Adhoc Networks (VANETs), to come forward as a viable answer. VANETs enable real-time communication between vehicles and infrastructures, thus providing an interconnected ecosystem for dynamic traffic management, predictive analytics, and efficient resource utilization.

Whereas the VANETs concept brings some excitement into its practice, it poses important problems like efficient data dissemination, dynamic routing, and robust stability in the network in highly mobile and volatile environments. Among such complex, dynamic optimization problems, reinforcement learning has really proven to be a suitable fit in dealing with them. With continuous interaction of systems with their environment through which they learn optimal actions, this enables a framework for adaptive, intelligent decision-making.

The main area in which reinforcement learning plays a revolutionary role for VANETs includes: Traffic Signal Optimization. Dynamic and real-time traffic signal scheduling may be carried out with RL-based agents, avoiding congested traffic conditions and smoothing out the traffic flow. Route Optimization. Vehicles carrying an onboard computer with an RL algorithm could optimize the route for travel on a self-driving basis, which saves more fuel consumption and time while trying to avoid bottlenecks and congested traffic areas.

Congestion Prediction and Management: RL models can predict traffic congestion and propose proactive actions that would be able to counter it.

Resource Allocation in Edge Computing: The allocation of the edge computing resources among devices in VANETs can be optimized by using RL to ensure effective real-time data processing and communication.

Emergency Vehicle Prioritization: RL frameworks can ensure faster and safer routing for emergency vehicles, reducing response times and improving urban resilience.

Advances in 5G, edge computing, and AI further enhance the integration of reinforcement learning with VANETs. The efficiency of RL models gets improved with high-speed low-latency communication offered by 5G. Edge computing allows for real-time data processing closer to the vehicles, not depending on the centralized infrastructure. Combining RL with deep learning techniques helps in modeling complex traffic scenarios, making VANETs more robust and adaptable.

This chapter explores the intersection of reinforcement learning and VANETs, taking into account the critical challenges associated with scalability, security, and infrastructure development. It suggests novel RL-based solutions for such a dynamic nature of vehicular networks, providing insight into potential advancements and future research directions. By using RL, VANETs can evolve into a backbone of smart urban transportation systems to provide safer, more efficient, and environmentally sustainable mobility solutions for cities of the future.

Related Work

Vehicular Adhoc Networks (VANETs) has been an area of much research in the efforts to reduce traffic congestion and improve vehicular communication systems. Real time communication and data sharing between vehicles and infrastructure have been at the center of improving traffic flow and safety. The earlier approaches to traffic management and congestion control were static and based on heuristics, while recent advancements in machine learning, particularly Reinforcement Learning (RL), have presented significant promise in developing adaptive and intelligent solutions for VANETs.

The initial efforts towards congestion control were based on conventional techniques, such as adjusting transmission power and message frequencies, based on predefined thresholds. For example, Kihl et al. (2012) have reported studies focusing on adaptive message dissemination schemes to overcome congestion in denser traffic scenarios. Those methods failed to dynamically respond to network conditions changing with a faster pace. That is why researchers started to investigate more flexible solutions. As RL appears as a mechanism of dynamic learning, this allows the vehicle to decide data-driven in real time, greatly improving the mechanism of congestion control.

In awareness control, it has been a prime concern to prioritize safety-critical messages. Liu et al. (2016) proposed clustering-based approaches that reduce redundant transmissions by having groups of vehicles, ensuring critical information is disseminated in an efficient manner. RL-based systems, as have been proposed by Wang et al. (2019), have introduced multi-agent reinforcement learning frameworks that enable cooperative decision-making among vehicles toward effective message prioritization. Such techniques surpassed traditional methods as they considered the dynamic nature of VANETs and took into account changing traffic and communication loads.

The hybridization of RL with newly developing technologies has further enhanced VANET-based congestion and awareness control mechanisms. Edge computing has been used in studies for reducing latency in RL-based frameworks by processing data near vehicles. Similarly, the adoption of 5G networks, as discussed by Zhang et al. (2021), has improved the responsiveness of RL models, thus enabling faster adaptation to traffic dynamics. Blockchain technology has also received attention, providing secure communication channels and ensuring data integrity in RL-driven VANET systems. These developments point to the potential of hybrid approaches that combine RL with modern computational and communication technologies.

Despite all these advancements, the scalability, security, and high computational demands still pose serious challenges. Recent works suggest hybrid frameworks combining RL with deep learning and predictive analytics for overcoming these challenges. However, further research and large-scale real-world validations are essential to realize the full potential of these RL-based frameworks in VANETs. This growing body of related work lays a strong foundation for novel solutions to optimize congestion control and awareness mechanisms, thus paving the way for smarter and more reliable vehicular networks. The application of reinforcement learning (RL) in VANETs has gained much research interest due to its ability to adapt to complex and dynamic vehicular environments. Researchers have studied a wide range of RL algorithms that may optimize traffic flow and prevent congestion. For example, Ye et al. (2020) applied deep reinforcement learning (DRL) in optimizing traffic signal control in cities and significantly reduced average travel times and congestion levels relative to traditional rule-based systems.

Another significant area is routing optimization in VANETs. Wang et al. (2018) proposed an RL-based protocol for routing, where the automobiles learn and select optimal routes dynamically based on real-time network conditions. This approach showed enhanced rates of packet delivery and decrease in delay, thereby nullifying the challenges in this highly mobile vehicular environment. According to Chen et al. (2021), by allowing vehicles to learn optimally together in shared routing protocols, multi-agent reinforcement learning has furthered the improvement of routing protocols. RL-based methods have also been beneficial for the dissemination of critical safety messages. For instance, Xie et al. (2019) proposed an RL framework that focuses on prioritizing emergency notifications in congested networks to ensure timely and reliable message delivery. These methods outperform static prioritization techniques by adapting to varying network loads and traffic densities.

Hybrid frameworks combining RL with other techniques are emerging as a promising direction. For example, Zhang et al. (2020) integrated RL with predictive analytics to predict congestion hotspots, proactively mitigating such congestion hotspots before occurrence. The approach used historical data and real-time inputs with better performance in preventing bottlenecks. Edge computing has greatly contributed to the success of RL-based VANET systems. According to Li et al. (2022), edge devices can host RL algorithms to reduce latency and allow real-time decision-making for congestion control. Edge-based RL frameworks have been proven to outperform centralized systems, especially in high-density traffic scenarios.

Further focus was put on the integration of 5G technology with RL frameworks. For instance, Singh et al. (2021) demonstrate how 5G capabilities increase the efficiency of RL-based VANET systems through high speeds and low latency. Faster feedback loops allow vehicles to respond to rapidly changing network conditions through adjustments in behavior. Blockchain has been studied as a supporting technology for RL in VANETs to achieve safe and reliable communication. In Kumar et al. (2022), a blockchain-based RL framework is presented with the objective of preserving integrity and privacy during congestion and awareness control. This framework is of particular interest as it solves several critical security issues in VANETs, which is an essential aspect of real-world performance.

Scalability is one of the challenges in RL-based VANET systems, and researchers have been working to address this challenge by proposing hierarchical RL. For instance, Gao et al. (2023) proposed a multi-level RL approach in which vehicles operate under localized policies but with higher-level coordination in order to manage broader network-wide objectives. This framework showed improved scalability without losing any efficiency. The role of MARL in VANETs has been further expanded to include dynamic task allocation for emergency services. Huang et al. (2021) implemented MARL to prioritize routing ambulances through congested areas while minimizing disturbance to regular traffic. This approach highlighted the potential of MARL for managing complex, multi-objective scenarios in VANETs.

DRL has been used for optimizing resource allocation in VANETs, such as spectrum management and bandwidth usage. Zhao et al. (2023) proposed a DRL-based system that efficiently allocated communication resources based on traffic density and application priorities. This method ensured reliable communication for critical applications while minimizing resource wastage, addressing one of the key limitations of VANET deployments. This stream of work collectively points out the power of RL, along with its integration into emerging technologies, to counter the problems of congestion and awareness control in VANETs. Still, a lot of research is still required to refine these techniques further for increased scalability and effectiveness in practical deployments.

The application of RL in VANETs has been under focus because of its adaptability in dynamic vehicular environments. Rahman et al. (2018) proposed a Q-learning-based mechanism for congestion control that dynamically adjusts transmission parameters based on traffic density, which resulted in reduced delays and higher packet delivery ratios compared to static approaches. Wang et al. (2019) presented a DRL model for dynamic spectrum management in VANETs. The framework was able to efficiently allocate bandwidth by preventing network bottlenecks, even in dense traffic scenarios, through the use of neural networks. The results obtained showed improved communication reliability.

Multi-agent RL frameworks, as proposed by Zhou et al. (2020), exhibited better performance in collaborative dissemination of awareness messages. In this system, vehicles learn together to discover optimal transmission strategies that help in the timely delivery of safety-critical information while minimizing network congestion. Chen et al. (2020) proposed the integration of RL with game theory for congestion control in VANETs. In this work, the authors modeled interactions between vehicles as a game, where RL agents optimized their strategies to balance individual and network-wide objectives, leading to improved fairness and efficiency. Edge-assisted RL frameworks also have received attention. Xu et al. proposed in 2021 a system wherein the edge servers processed the RL algorithms locally to enable real-time congestion mitigation and reduce the computational load on individual vehicles. This architecture has been found to effectively manage high-density urban traffic.

Li et al. study 6G's role in promoting RL-based VANETs (2022). Ultra-low latency and massive connectivity from the 6G networks may allow the improvement of the RL performance for the congestion control in scenarios having a high mobility of vehicular units. Pandey et al. 2022 explored blockchain-augmented RL systems. In the study, security concerns related to VANETs were addressed. Such a framework ensured integrity, privacy in communication, and robustness against malicious attacks of congestion control strategies based on reinforcement learning. Sun et al. (2023) developed a hybrid RL model combining supervised learning with RL to predict traffic congestion and adjust routing strategies dynamically. The hybrid approach leveraged historical data for prediction while using RL for real-time decision-making, achieving higher adaptability and accuracy.

Wang et al. (2023) proposed a MARL-based approach for optimizing traffic signals in smart cities. This system enhanced the efficiency of traffic flow by allowing the smart traffic lights to learn adaptive timing strategies based on the real-time vehicular density and patterns.Gupta et al. (2023) proposed an RL-driven priority management system for mixed vehicular environments, including autonomous and human-driven vehicles. The system prioritized safety-critical messages and adjusted communication parameters dynamically, improving both efficiency and safety.

Traffic and Congestion in India

India is one of the most densely populated countries in the world, and cities here often face acute traffic congestion. The rapid process of urbanization combined with inefficient public transport mechanisms in many cities has led to a very sharp increase in the number of vehicles on roads. This congestion not only causes lengthy travel times and fuel consumption but also leads to significant air pollution and road accidents. Most of the major cities in India, such as New Delhi, Mumbai, and Bengaluru, are usually marred by traffic jams that last for hours during peak hours. Limited road infrastructure, insufficient parking facilities, and poor traffic management create further problems, leading to great losses in terms of economy and a fall in overall quality of life among its residents. With the numbers rising day by day, a need for more efficient solutions has been felt for effective management of traffic.

Fixed time, as well as manual, modes of traffic management are failing miserably when there are demands for more complex and dynamic models to run a city's traffic. Now in India, advanced technologies of AI and machine learning in their various modules find new grounds within the traffic management system. The adoption of AI in traffic systems has shown significant potential in alleviating congestion and improving the flow of traffic in Indian cities. Adaptive AI-based traffic management systems use real-time data from sensors, cameras, and GPS devices to monitor traffic conditions continuously.

These systems employ machine learning algorithms that learn from traffic patterns and dynamically adjust traffic signal timings, lane usage, and route suggestions based on current traffic conditions. Unlike traditional systems, which are based on static timers, AI-powered systems can predict congestion patterns and optimize traffic flow, thus reducing bottlenecks and overall efficiency. For instance, AI systems can analyze historical traffic data and use it to predict congestion hotspots during different times of the day or special events. Based on these predictions, the system can adjust signal timings in advance or suggest alternative routes to drivers, making it harder for congestion to occur. AI may also assimilate data from other sources such as weather conditions, accidents, and road closures to enhance decision-making. In India, AI-powered traffic systems have been introduced in cities such as Delhi and Bengaluru, where in real time, the traffic signals are adjusted to ease congestion, minimize delays, and reduce pollution. The use of AI is part of the broader move toward smart cities in India, creating a more sustainable, efficient, and livable urban environment.

Traffic Management of VANETs with Advancements

The main application of VANETs is to relieve congestion, enhance safety, and optimize urban mobility by ensuring traffic management. In the field of VANETs, V2V and V2I are two types of communications: Vehicle-to-Vehicle, which allows vehicles to communicate with each other, and Vehicle-to-Infrastructure, in which a vehicle communicates with the surrounding infrastructure. This means vehicles are highly connected and very dynamic. The VANET-based traffic management system makes use of real-time data provided by vehicles, sensors, and traffic infrastructure.

Core Elements of Traffic Management in VANETs

1. Dynamic Traffic Signal Control

The ordinary traffic lights work on a static schedule, which usually produces inefficient traffic flow. VANET-based traffic light control adjusts in real time to the changing traffic patterns and conditions by analyzing data transmitted from vehicles and roadside units (RSUs). This approach dynamically optimizes idle times at intersections and reduces general congestion.

2. Smart Traffic Routing

VANETs allow real-time optimization of routes based on traffic density, accidents, and road conditions. Vehicles receive new route recommendations that avoid congested areas and ensure smoother traffic flow across the network.

3. Incident Detection and Management

VANETs are crucial for rapid incident detection, such as accidents or road blockages. Vehicles and RSUs communicate the occurrence of incidents to nearby vehicles and traffic authorities, thus allowing timely interventions and rerouting of traffic.

4. Emergency Vehicle Prioritization

VANETs help prioritize emergency vehicles such as ambulances and fire trucks by changing traffic light dynamically in the environment and communicating through adjacent vehicles to leave paths free.

Advances Improving VANET-Based Traffic Management

1. 5G Integration

Enhanced by high speed low-latency communication, VANET's capabilities are greatly expanded in 5G. Thus data-exchange and processing get streamlined efficiently to ensure quicker decisions and management of traffic applications are attained.

2. Edge Computing and Fog Networking

It brings the distribution of computational resources closer to vehicles and RSUs, causing faster analysis times of data and decreased time before giving a response. This decentralization reduces dependence on centralized cloud infrastructure, allowing uninterrupted service even in highly dynamic traffic scenarios.

3. Artificial Intelligence and Machine Learning

AI and machine learning algorithms, in particular, reinforcement learning, are being integrated into VANET-based traffic management systems. These learn from historical and real-time data for predicting traffic trends, optimizing signal timings, and making recommendations for personalized routes.

4. Predictive Analytics

Using historical traffic pattern analysis along with current conditions, predictive models predict bottlenecks in advance and recommend how to avoid them. Besides, it also helps plan cities better through long-term traffic trends.

5. Blockchain for Security and Privacy

Blockchain technology makes the VANETs' data more reliable, secure, and private. The blockchain guarantees tamper-proofing of the traffic data shared among the vehicles and parts of the infrastructure. Misuse of data and unauthorised access will not take place with blockchain.

6. Internet of Things (IoT) Integration

IoT devices, such as smart sensors and connected cameras, enrich VANET-based traffic management systems with detailed and diverse data. IoT integration ensures a comprehensive understanding of traffic dynamics and supports more accurate decision-making.

The advancements taking place in the VANET-based traffic management systems are transformative for urban transport. With emerging technologies of 5G, AI, edge computing, and blockchain, these systems are transforming into more efficient, secure, and scalable entities. So, as the VANETs continue their journey through evolution, this promises that the future would be really smarter, safer, and more sustainable, thereby ameliorating the quality of lives of city inhabitants. Traffic Management in VANET with Advancements

Congestion Control Mechanisms in VANET

VANETs provide dynamic communication infrastructure for vehicles and infrastructure but are prone to network congestion as vehicles move fast and cause dense traffic conditions. The demand for communication exceeds the network capacity, resulting in packet collisions, increased delay, and decreased reliability, which constitute congestion in VANET. One of the effective mechanisms for ensuring that communications be conducted in an efficient manner for VANET applications, like traffic management, accident prevention, and navigation, is to develop congestion control mechanisms. A type of congestion control for VANETs is through adaptive transmission control wherein the vehicle adjusts the parameters used for transmission, like transmission power, rate, or frequency, dynamically according to the state of the network.

For instance, in high-traffic density scenarios, vehicles can reduce their transmission power to minimize interference or switch to alternate communication channels to balance the load. These methods, often supported by algorithms like Reinforcement Learning (RL), enable vehicles to make intelligent decisions, enhancing overall network performance. Another important mechanism is priority-based data dissemination, where critical messages, such as accident alerts or emergency vehicle notifications, are transmitted with a higher priority over non-critical information. This method utilizes Quality of Service (QoS) protocols to classify and prioritize messages, which reduces congestion for high-priority data. Clustering techniques can also be applied in which vehicles are grouped together, and a cluster head manages the communication within the group, reducing redundant transmissions and network congestion. Emerging technologies such as edge computing and artificial intelligence are also playing a pivotal role in congestion control.

Edge devices near the vehicles process data locally, reducing the volume of information transmitted over the network and alleviating congestion. AI-driven predictive analytics further enhance congestion control by forecasting traffic density and communication load, enabling preemptive measures. Combined with 5G networks, these technologies ensure low-latency, high-throughput communication, thus guaranteeing robust congestion management even in the most densely populated urban environments. With these mechanisms in place, VANETs are becoming more efficient, setting the ground for reliable and scalable vehicular communication networks.

Traffic Density-Based Congestion Control Method for VANETs

Traffic density-based congestion control methods are designed to manage communication and vehicular traffic in Vehicular Ad Hoc Networks (VANETs) by dynamically adapting to the density of vehicles on the road. Such methods generally depend on real-time data from vehicle sensors and roadside units (RSUs) to estimate traffic density in a given area. Based on the density, adaptive mechanisms are in place to regulate the messages transmitted to ensure that high-priority safety information prevails over network overhead. Techniques such as dynamic message scheduling and selective broadcasting prevent message collisions and reduce congestion in high-density scenarios, thereby keeping the communication network reliable. More developed traffic density-based methods often employ machine learning algorithms, like reinforcement learning (RL), in order to enhance adaptability.

RL models enable vehicles and RSUs to learn optimal transmission strategies in various traffic conditions, which dynamically adapt parameters like transmission power, frequency, and message prioritization. Such approaches also promote network efficiency while enhancing road safety through the timely dissemination of important information. Additional methods have also been introduced to process traffic density data in real-time using emerging technologies like 5G and edge computing, which reduces latency and enhances the scalability of congestion control systems in VANETs.

RL-Based Framework for Congestion and Awareness Control

RL offers an effective framework to address the issues involved in congestion and awareness control within VANETs. A vehicle can learn and adjust to dynamic network conditions in RL-based approaches through experience and optimization of its own behavior through feedback with its environment. By continuously monitoring the state of the network in terms of traffic density, channel availability, and communication quality, RL agents can make wise decisions to combat congestion without compromising the timely dissemination of critical information. In this context, in congestion control, RL algorithms enable the vehicles to make dynamic adaptations in their transmission parameters.

vehicles can balance the load of the network by varying their transmission power, message frequency, or data packet size. The ability to do so would be very handy in highly dense traffic, where extensive message broadcasting creates packet collisions and delays. Through reinforcement learning, RL-based systems learn patterns about congestion from past interactions and adjust the communication strategy ahead of time to ensure that data will flow smoothly and efficiently. For awareness control, RL-based frameworks are primarily focused on the importance of transmitting safety-critical messages with minimal non-essential communication. Vehicles employ RL algorithms in determining optimal timing and frequency for the transmission of awareness messages so that timely and vital information such as accidents, road hazards, or emergency vehicle notifications are effectively communicated. Moreover, the multi-agent RL system ensures collaborative decision-making among the vehicles to enhance the efficiency of awareness message dissemination within the network. The integration of emerging technologies such as edge computing and 5G further enhances RL-based frameworks for congestion and awareness control.

Edge devices provide real-time processing capabilities, reducing communication delays and improving the responsiveness of RL algorithms. The high-speed, low-latency features of 5G networks enable faster feedback loops for RL agents, allowing them to adapt quickly to changing network conditions. With these advances, these frameworks based on RL not only work against congestion but also ensure reliable communications for critical awareness applications, enhancing safety and efficiency in the vehicular network.

Design and Analysis of Intelligent Vehicular Networks and Applications

The design of intelligent vehicular networks involves creating systems that allow for communication between vehicles (V2V), between vehicles and infrastructure (V2I), and between vehicles and pedestrians (V2P). These rely on advanced communication technologies including 5G, DSRC, and V2X protocols to ensure efficient and real-time data exchange. A critical feature of IVNs is incorporating intelligent traffic management systems sensitive to real-time road conditions; these systems utilize data coming through various sensors, cameras, and GPS devices installed in vehicles and in infrastructure to make quick decisions relating to changing traffic signal timetables, route suggestions for drivers, and hazard indication. The design of these networks also entails secure and scalable communication protocols to handle high mobility and the large number of vehicles on the road. They have to guarantee reliable communication in densely populated urban areas. Low latency and high throughput are one of the key issues to be addressed while designing IVNs, because vehicles will always be on the move in dynamic traffic environments.

To address this, intelligent vehicular networks integrate edge computing. Edge computing allows processing to be closer to the data source. Thus, it reduces the amount of long-distance communication required with centralized servers. This improves faster decision-making and reduces the probability of network congestion. Another benefit of using machine learning algorithms, especially reinforcement learning (RL), in IVNs is the ability to adapt to changes in traffic conditions. For example, RL can optimize traffic signal control and route planning by learning from past traffic data, improving the efficiency of vehicle routing and reducing congestion. The analysis of these networks requires modeling the communication protocols, traffic patterns, and network dynamics to ensure that the system performs optimally under varying conditions. The applications of intelligent vehicular networks extend beyond just congestion control. IVNs play a critical role in improving road safety, energy efficiency, and environmental sustainability.

Real-time vehicle-to-vehicle communication enables features such as collision avoidance, lane-change assistance, and emergency vehicle prioritization, thereby reducing accidents and fatalities. In addition, IVNs can contribute to greener transportation by optimizing fuel consumption and reducing emissions. For instance, AI-powered adaptive traffic signal systems can direct the movement of vehicles through intersections in an efficient way, thus eliminating the time lost in idle situations and the fuel consumed during the process. The EVs incorporated into the IVN system will enjoy smart charging stations and route planning for optimized travel distance and efficiency. It also looks at how IVNs can be useful in improving the urban mobility and infrastructure of cities. As cities continue growing, smart vehicular networks represent the way towards a smarter and more efficient urban space. Deploying IVNs will bring new ways for better transportation management in terms of managing road networks, eliminating congestion, and raising mobility generally.

But with it come many issues to be tackled, which are network scalability, security in terms of data handling, and interoperability between diverse vehicle manufacturers and infrastructure service providers. Research will be geared toward designing robust algorithms and protocols capable of dealing with such challenges. IVNs must then be capable of supporting large numbers of connected vehicles, and at the same time, provide privacy and security. Further policy frameworks and regulatory standards will have to emerge to help spur the wider adoption of IVNs for a more connected and intelligent transportation future.

Future Directions in VANET-Based Traffic Management Systems

Despite the notable advancements in VANET-based traffic management systems, there are several challenges that need to be addressed for their wide-scale adoption. One of the primary challenges is scalability. As urban areas continue to grow, the density of vehicles on the road increases, putting pressure on communication protocols and traffic management systems. For this, future research is highly focused on the protocol development that would support scalable communication with hundreds of vehicles and reliable, low-latency exchange of data. Such protocols have to adapt dynamically to shifting conditions of traffic and ensure that real-time data can be processed adequately, even in a very dense urban environment. Research in this area likely will focus on advanced routing algorithms, techniques for data aggregation, and decentralized systems that can fairly distribute the workload across multiple nodes of vehicles and infrastructure.

Currently, different vehicle manufacturers, infrastructure providers, and technology vendors use proprietary communication protocols, which can lead to issues regarding interoperability and hinder the seamless operations of VANETs across regions. International communication standards will help ensure that all systems, irrespective of a particular vendor or region to which they belong, work with each other. Thus, the standardization will also support smooth implementation of VANET at a global level since various manufacturing units and different regional agencies can communicate with one another perfectly and thus enhance traffic management and efficiency along with safety and environmental concerns. The economics of the deployment of VANET infrastructure remains a big issue, especially in developing countries.

The installation and maintenance of the necessary infrastructure for VANETs such as roadside units (RSUs), sensors, and communication systems is very expensive. Future research focuses on making the deployment of VANET technologies more economical. Innovations such as low-cost, high-performance communication technologies, energy-efficient devices, and cost-sharing models for infrastructure development are critical to ensuring that VANETs can be widely adopted, even in regions with limited financial resources. This also includes exploring the potential of using existing infrastructure, such as cell towers and streetlights, to deploy VANET components, reducing the overall costs of new installations. Integration of AVs into VANETs is a great opportunity for improving traffic flow and safety on roads. Autonomous vehicles can be integrated into the ecosystem in such a way that contributes to a more coordinated and efficient traffic system, using their advanced sensors and real-time data processing capabilities.

With AVs integrated into the VANET ecosystem, traffic management systems are further optimized, reducing congestion, minimizing accidents, and improving fuel efficiency. The research will focus on algorithm and protocol development that makes seamless communication between autonomous and non-autonomous vehicles possible, so that the entire network functions like one cohesive system. Integration of AVs with VANETs requires collaborative work among researchers, policymakers, and industry leaders in framing regulatory frameworks to ensure that these systems operate safely, efficiently, and securely in urban settings.

CONCLUSION

In summary, VANETs can transform the urban traffic management system. The communication between vehicles and infrastructure will be in real time. The integration of 5G, AI, and edge computing into VANETs can enhance the flow of traffic, reduce congestion, improve road safety, and contribute to environmental sustainability. The reinforcement learning and machine learning algorithms are quite promising in resolving dynamic traffic conditions, optimized routing strategy, and efficient communication in a dense urban area. It is evident from the current development of the VANET-based systems that their actual deployment relies heavily on solving some of the major issues, such as scalability, interoperability, and economic feasibility.

The communication protocols of VANETs face a significant challenge for their future scalability. As the number of people in urban cities increases and more vehicles get connected, the protocols currently being used may not be able to cope with the amount of data traffic, which may lead to communication latency. Future research will be aimed at designing scalable protocols that adaptively manage the increasing mass of data while ensuring good communication between a large population of vehicles and infrastructure units. The development of such protocol designs will make VANETs quite practical for major cities around the globe. Some scalability challenges related to performance are overcome with research on decentralized systems, aggregation techniques of data, and the intelligent routing mechanisms applied within these systems.

One major issue related to interoperability is associated with the multi-vendor, multi-manufacturer, and infrastructure provisioning across various regions and countries. There is also an absence of standardized protocols used in communication between these varied components of VANET. To make VANETs widely available, international communication standards must be developed so that systems from different vendors can work together cohesively. Standardization will also enable more efficient coordination between vehicles, traffic management systems, and even autonomous vehicles, ultimately leading to smarter, safer, and more efficient urban mobility solutions.

The economic viability of VANET deployment in both developed and developing countries is essential for its long-term success. The installation and maintenance of VANET infrastructure, such as roadside units, sensors, and communication equipment, are quite expensive. One of the challenges in future research will be finding cost-effective solutions to decrease the cost of deploying the system while keeping its performance levels. Finally, autonomous vehicle integration into VANET systems offers an opportunity to change traffic management. VANETs will achieve more synchronized and efficient traffic flow by enabling the AVs to communicate with one another and the surrounding infrastructure, enhancing road safety, and reducing congestion. Addressing these challenges, VANETs are expected to play a crucial role in shaping smarter, more sustainable urban transportation systems worldwide.

REFERENCES

Chen, J., Wang, H., & Zhang, L. (2020). Game-theoretic reinforcement learning for congestion control in VANETs. *IEEE Transactions on Vehicular Technology*, *69*(5), 5406–5418.

Chen, L., Zhang, X., & Wang, Y. (2021). Multi-agent reinforcement learning for adaptive routing in VANETs. *IEEE Transactions on Intelligent Transportation Systems*, *22*(4), 2987–2997.

Gao, R., Liu, T., & Wang, Z. (2023). Hierarchical reinforcement learning for scalable congestion control in VANETs. *Computer Communications*, *191*, 24–33.

Gupta, R., Sharma, V., & Malik, A. (2023). RL-driven priority management for mixed vehicular networks. *IEEE Access : Practical Innovations, Open Solutions*, *11*, 49256–49267.

Huang, Y., Zhang, L., & Li, W. (2021). Multi-agent reinforcement learning for dynamic task allocation in VANET-based emergency services. *IEEE Access : Practical Innovations, Open Solutions*, *9*, 89756–89766.

Kihl, M., Linden, C., & Zhou, W. (2012). Adaptive message dissemination for congestion control in VANETs. *Journal of Communications and Networks (Seoul)*, *14*(6), 620–630.

Kumar, P., Singh, A., & Verma, R. (2022). Blockchain-enhanced reinforcement learning framework for secure VANET communication. *IEEE Internet of Things Journal*, *9*(3), 1804–1815.

Li, H., Zhao, J., & Xu, T. (2022). Edge computing-based reinforcement learning for real-time congestion control in VANETs. *Future Generation Computer Systems*, *135*, 122–132.

Li, Y., Zhao, J., & Xu, W. (2022). Exploring 6G for RL-based congestion control in vehicular networks. *IEEE Communications Magazine*, *60*(5), 28–35.

Liu, Y., Wang, H., & Chen, J. (2016). Clustering-based message dissemination for congestion mitigation in VANETs. *International Journal of Distributed Sensor Networks*, *12*(9), 243–254.

Pandey, R., Singh, K., & Kumar, A. (2022). Blockchain-augmented RL frameworks for secure VANETs. *IEEE Transactions on Network and Service Management*, *19*(1), 340–353.

Rahman, T., Liu, P., & Qiu, J. (2018). Q-learning for dynamic congestion control in vehicular networks. *Ad Hoc Networks*, *72*, 13–25.

Singh, P., Sharma, K., & Gupta, S. (2021). 5G-enabled reinforcement learning for adaptive VANET communication. *IEEE Vehicular Technology Magazine*, *16*(1), 28–37.

Sun, Y., Wang, F., & Zhang, Q. (2023). Hybrid reinforcement learning for traffic prediction and control in VANETs. *IEEE Transactions on Intelligent Transportation Systems*, *24*(3), 1654–1665.

Wang, H., Liu, Y., & Zhou, J. (2019). Deep reinforcement learning for spectrum management in VANETs. *Wireless Communications and Mobile Computing*, *2019*, 1–11.

Wang, S., Chen, K., & Zhang, X. (2023). Multi-agent RL for adaptive traffic signal optimization in smart cities. *IEEE Internet of Things Journal*, *10*(2), 1568–1578.

Wang, Y., Qian, J., & Xu, F. (2018). Reinforcement learning-based routing protocol for dynamic vehicular networks. *Wireless Networks*, *24*(3), 723–734.

Xie, L., Chen, M., & Zhao, Y. (2019). Reinforcement learning for priority-based message dissemination in VANETs. *IEEE Transactions on Mobile Computing*, *18*(12), 2907–2918.

Xu, T., Zhang, H., & Li, W. (2021). Edge-assisted RL frameworks for real-time congestion control in VANETs. *Future Generation Computer Systems*, *120*, 15–25.

Ye, L., Wu, F., & Zhang, R. (2020). Deep reinforcement learning for adaptive traffic signal control in urban areas. *IEEE Transactions on Intelligent Transportation Systems*, *21*(2), 665–674.

Zhang, Q., Zhao, L., & Gao, M. (2020). Predictive congestion control in VANETs using reinforcement learning and analytics. *IEEE Systems Journal*, *14*(3), 3058–3067.

Zhao, W., Li, Q., & Huang, S. (2023). Resource allocation optimization in VANETs using deep reinforcement learning. *IEEE Communications Surveys and Tutorials*, *25*(1), 165–182.

Zhou, M., Wu, F., & Chen, T. (2020). Collaborative multi-agent reinforcement learning for efficient message dissemination in VANETs. *IEEE Access : Practical Innovations, Open Solutions*, *8*, 190453–190463.

Chapter 12
Smart Cities on the Move Leveraging IoT Communication in Connected Cars for Enhanced Traffic Management

Manikandan Arunachalam
https://orcid.org/0000-0003-0461-2859
Department of Electronics and Communication Engineering, Amrita Vishwa Vidyapeetham, India

A. Abhiram
Department of Electronics and Communication Engineering, Amrita Vishwa Vidyapeetham, India

Vaishnav Raj
Department of Electronics and Communication Engineering, Amrita Vishwa Vidyapeetham, India

Rakesh Thoppaen Suresh Babu
https://orcid.org/0009-0003-2671-4618
Fiserv Inc, Alpharetta, USA

ABSTRACT

The integration of Connected Cars in Smart Cities marks a significant shift in urban mobility, accelerated by the Internet of Things (IoT) communication infra-

DOI: 10.4018/979-8-3693-6422-2.ch012

structure. Cars on the road collectively function as distributed networks of sensors, continuously gathering diverse data types. Vehicles have long been equipped with various sensors, from multi-purpose cameras to infrared and GPS systems, enabling comprehensive data collection. With the advent of Vehicle-to-Vehicle (V2V) and Vehicle-to-Infrastructure (V2I) communication techniques, data exchange between vehicles becomes feasible, allowing comparative analysis and integration to enhance situational awareness on the road. This paper investigates the potential applications of integrated vehicle data and augmented reality techniques in traffic management, leveraging real-time exchange to mitigate congestion, enhance safety, and optimize transportation systems, also applications of these augmented reality techniques in enhancing traffic management, reducing accidents, and optimizing transportation systems within Smart Cities.

INTRODUCTION

Nowadays, IoT transformation in urban projects has given the solution of smart cities a new perspective. Of all the cases of using IoT in urban environments, the connected car is an innovation that has a huge impact on society(Cvar et al. 2020). These linked vehicles are somewhat more than vehicles for traveling; they are systems composed of several sensing, controlling, and signalizing systems in mutually connected automotive vehicles as well as potentially with other objects in the city(Deng et al. 2023). This integration holds the potential to enhance the safety, effectiveness, and environmentally friendly nature of urban mobility and consequently, the design of sprightlier and better-suited lifestyles. The growth of IoT technology has however been rampant coupled with progression in vehicle-to-everything (V2X) connectivity, making it possible for connected cars to seamlessly integrate with the city environment. These are the essential components of the concept of smart technologies, sensor fusion, and independent analysis of the data obtained in real-time in connected car systems. Automotive connectivity advances, particularly with linked automobiles, have propelled the automobile sector forward to levels of technology and integration(Bučko et al. 2021). Integral to these calls is the use of IoT devices, smart sensors, and enhanced communication systems that hold the capability of enabling cars to actively communicate with other cars, the physical environment, and infrastructure in real-time(Vermesan et al. 2020). V2X communication points to new temporary traffic control and the provision of safe roads with little probability of an accident, on the other hand, sensor fusion means using many sensors to provide a complete view of the driving landscape(Adnan Yusuf, Khan, and Souissi 2024; Teixeira et al. 2023). The concept of connectivity, edge computing, and cloud integration along with the ability to process data in real-time analyze the data instantly and deliver over-the-air updates augment the scalability and the modularity of automotive systems(Cavicchioli, Martoglia, and Verucchi 2022; Vermesan et al. 2021). Innovative attributes like Predictive Maintenance and Personalized Driver Assistance are the primary use cases of OT where advanced AI and Machine learning play a central role, and Cyber Security for a wide range of cyber threats and potential breach attempts is the protection component of the product(Vermesan et al. 2021). The introduction of 5G enhances connection capabilities to enhance the capacities of applications and services by facilitating the actualization of bandwidth-intensive applications and mission-critical services. These technological advancements are making connected vehicles possible, thereby making the future of transportation smarter and safer with more efficiency(Bathla et al. 2022). Intelligent cars have a massive potential to enhance the efficiency of traffic lights and to de-congest urbanization rapidly(Gohar and Nencioni 2021). Real-time traffic information can be readily exchanged through the networks enabling vehicles to alter route plans if there are congested zones. It was also identified that, with linked automobile systems, traffic flow is enhanced and, at the same time, the general travel time for commuters is reduced(E.I. et al. 2023). Technological advancements in Intelligent Transport Systems, specifically in traffic management and optimizing traffic for connected automobiles have revolutionized the nature of transporting people within metropolitan cities(Nguyen et al. 2020). These smart vehicles can connect and other structures in real-time due to the embedded advanced technologies, leading to adjustments to the traffic conditions and environment(Farman et al. 2022). Due to this exchange of information, connected automobiles can reroute the drivers to the less congested roads and help optimize the traffic signals and actively coordinate with fellow vehicles in enhancing the general traffic conditions(Yue et al. 2021). Along with saving commuters time, it also decreases traffic congestion and carbon emissions, making our cities more efficient and environmentally pleasant(Campisi et al. 2021). Finally, traffic management and optimization for connected automobiles constitute a significant step forward in urban mobility, promising safer, smoother, and more sustainable transportation for all(Jaramillo-Alcazar, Govea, and Villegas-Ch 2023). Safety is a top priority in the development and deployment of connected vehicle technology. Connected automobiles can use IoT sensors and V2V connectivity to proactively identify and minimize potential road safety concerns(Hejazi and Bokor 2021). Comprehensive cybersecurity measures are critical for protecting connected automobile systems from cyber-attacks and ensuring data transmission integrity(Mudhivarthi, Thakur, and Singh 2023). The combination of connected cars and IoT infrastructure shows promise for lowering greenhouse gas emissions and reducing environmental pollutants related to urban travel(Zeng, Pang, and Tang 2023). By managing traffic flow and promoting shared mobility services, linked automobile systems can help to create a more sustainable urban mobility ecosystem(Golbabaei, Yigitcanlar, and Bunker 2021). Some of the possible environmental benefits of connected car deployment in smart cities. Effective implementation of connected car technology is dependent not just on technological capabilities, but also on human acceptance and trust. User perceptions and attitudes about connected cars, highlighting the necessity of usability, privacy protection, and open communication in building public trust and acceptance were investigated(Nastjuk et al. 2020). Tying together the discussions in the literature on connected cars in smart cities, it is possible to apprehend the topic as a growing line of academic inquiry that comprises technological advancement, urban management, and social impact domains. Although significant progress has been achieved in assessing the advantages and limitations of applying connected car technology and infrastructures, the spectrum of challenges has not been investigated thoroughly from an interdisciplinary perspective to address the pressing issues of privacy, equality, and policymaking for smart urban mobility. This chapter focuses on the impact of integrating IoT innovation with automobiles, overcoming the traditional models of urban transportation. It starts with such remarkable technological trends paving the evolution of connected cars like V2X communication, sensor fusion, edge computing, and AI. These innovations make the cars intelligent enough to communicate with each other, other vehicles, and road structures, making control and regulation of traffic flow better; enhancing on-board safety and comfort facilities, and optimizing the overall vehicle usage. This chapter also expands the discussion of the benefits of traffic management and optimization from connected cars more specifically stressing the fact that connected cars can help organize traffic flow and reduce traffic densities, hence cutting off time-consuming jams and travel time as well as permitting dynamic re-routing of traffic in concerted driving style. Furthermore, it also allows confronting the emerging issues and possibilities linked to cybersecurity, data protection, and social acceptance of connected car solutions. This chapter also discusses in detail how connected cars will play pivotal roles in achieving smarter, safer, and more efficient transportation systems in the smart city context.

1. INTELLIGENT TRAFFIC MANAGEMENT SYSTEMS

When we are in a large maze of roads with automobiles zooming all around, it may be chaotic, right Well, it's where Traffic Management Systems (TMS) come into play. Think of TMS as a giant mind that allows preserving the whole lot. It uses unique sensors and cameras located on the roads to maintain a watch on all the cars and vans. These sensors watch how speedy automobiles are going, how many there are, or even if there are any troubles like accidents or production. Then, this huge mind analyzes all the facts it collects and makes selections to keep site visitors moving easily. It would possibly alternate visitors lighting to allow more vehicles through, or it can warn drivers approximately a site visitors jam up ahead if they want to take an exclusive route. TMS additionally allows buses and trains to get to where they need to head quicker by giving them special precedence at traffic lighting fixtures.

It even cares about the surroundings, trying to reduce pollutants through maintaining vehicles shifting in preference to sitting in traffic. So, in simple phrases, Traffic Management Systems are like amazing clever helpers that make sure every person gets where they need to go competently and without too much trouble(N. Zhou 2021). Moreover, these systems are adept at swiftly addressing incidents like accidents, roadblocks, or construction, ensuring a seamless flow of traffic.

The (TMS) relies on three main steps:

1.1 Information gathering

where data streams from various sensors capture a kaleidoscope of traffic-related insights— ranging from the density and speed of vehicles to prevailing weather conditions, predicting potential rain, snow or storms on the horizon collection of diverse traffic-related data from multiple origins, including vehicles, in-road and roadside sensors, traffic lights, Roadside Units (RSUs), connectivity sources available to the public, and participation. For example, vehicles contribute to this dataset through their internal sensor modules such as GPS speedometer, and odometer. Once collected, the collected information is integrated into the vehicle's OnBoard Units (OBUs). These OBUs then transmit the appropriate data to the Traffic Management Center (TMC) or neighboring vehicles for analysis and subsequent use This system ensures that the TMS can obtain a complete set of data in real time, thereby improving traffic flow and increasing overall transport efficiency.

1.2 Information processing

Information processing, a cornerstone of Traffic Management Systems (TMS), hinges upon the meticulous aggregation and analysis of obtained traffic information. Through this important step, TMS endeavors to pinpoint capacity visitors' dangers that might undermine the efficiency of visitor's drift. By sifting through massive streams of facts, TMS algorithms figure styles, anomalies, and rising risks, allowing proactive interventions to optimize roadway protection and overall performance.

1.3Service delivery

service delivery in Traffic Management Systems (TMS), where TMS addresses traffic challenges and optimizes overall efficiency. orchestrating a range of solutions to tackle issues such as accidents and congestion. Through timely interventions, TMS ensures smoother traffic flow, making our journeys safer and more efficient. Guiding us through obstacles and helping us reach our destinations with ease.

2. VEHICULAR AD-HOC NETWORK (VANET)

Vehicle ad hoc networks (VANETs) have similarities with mobile ad hoc networks (MANETs), acting as self-configuring networks in which vehicles communicate wirelessly with each other and with infrastructure components. This communication takes place in two main ways: vehicle-to-vehicle (V2V) and vehicle-to-vehicle Infrastructure (V2I).

In VANETs, vehicles act as nodes in a dynamic traffic network, facilitating the exchange of data with each other. This network enables a wide range of applications in a traffic management system. For example, VANETs support collision avoidance techniques by enabling vehicles to share real-time location and routing information. Similarly, they enable the use of traffic monitoring, where congested roads can be flagged, and drivers advised to choose alternative routes to travel more efficiently. In addition, VANETs provide passengers with services including Internet connectivity and updates on nearby vehicles(Upadhyay et al. 2023). By providing real-time traffic updates and navigation assistance, this connectivity improves the overall driving experience for road users. In V2V communication, vehicles exchange data directly without relying on a centralized hub(Álvares, Silva, and Magaia 2021). This form of communication facilitates accident avoidance, road safety, and the sharing of multimedia information such as accident photos or available parking spaces. On the other hand, V2I communication involves exchanging information between vehicles and intelligent infrastructure such as street signs, roadside sensors, and traffic lights. (Masood, Lakew, and Cho 2020; Ta and Dvir 2020) This communication is crucial for effective traffic management and often relies on

Dedicated Short-Range communications (DSRC) technology.

Overall, VANETs play an important role in road safety, traffic efficiency, and passenger convenience to enable seamless communication between vehicles and infrastructure components.

Continued research and development in this field promise to unlock further potential for enhancing transportation systems worldwide.

2.1 Security Concerns for VANETS

Privacy risks: VANETs collect and exchange data, including location information and driving patterns, raising concerns about user privacy. Failure to secure this data could lead to privacy violations and identity theft.(Masood, Lakew, and Cho 2020).

Denial-of-Service (DoS) Attacks: DoS attacks aim to disrupt the normal operation of VANETs by overwhelming network protocols or blocking communication channels. These attacks can prevent vehicles from receiving critical safety alerts or receiving critical services(Malik, Bandi, and Sun 2021).

Example-: An attacker floods the VANET network with bogus messages, causing congestion and preventing legitimate communication between vehicles.

Sybil Attack: In a Sybil attack, a malicious vehicle impersonates multiple users within a network to alter communication and disrupt services. This can cause false data propagation and synchronization failure(Parham and Pouyan 2020).

Addressing the security concerns of vehicular ad hoc networks (VANETs) is critical to passenger safety and network reliability. Failure to do so can result in violence or accidents. Strong security measures, including authentication and encryption, are needed. Continued collaboration and research are needed to ensure the resilience of future transport systems to stay ahead of emerging threats

3. NETWORK INFRASTRUCTURE

The emergence of IoT has come up with the latest way to look at how we interact with technology and is very evident that transport systems nowadays have connected car models as part of them(R and A 2023). These modern cars have been transformed into high-tech mobile devices with sensors, processors, and communication gadgets built into them. This helps them to establish real-time interactions with other vehicles, roadside infrastructure, and cloud platforms. The main feature of this technological revolution is a strong set of network infrastructure that is complex and serves as the basis of effective communication without any disruptions in data exchange. This infrastructure allows connected cars as well as the roadside systems and cloud services to contact directly so that different IoT devices can also communicate. It plays a critical role in ensuring the reliability, security, and efficiency of connected car systems, while simultaneously unlocking a plethora of opportunities for innovation and enhanced user experiences. Thus, ensuring that these systems are reliable, secure, and efficient while opening possibilities for better user experiences and innovations. The modern IoT environment involves many different devices with limited resources spread throughout it to serve different purposes. Inside these networks, various protocols have been put in place and agreed upon to allow the devices to communicate with each other such as Zigbee, Bluetooth Low Energy (BLE), LoRaWAN, and Sigfox(Shilpa, Radha, and Movva 2022). However, IoT network's 'intrinsic limitations like device diversity and limited resources are a great hindrance to their performance. In the IoT networks today, there are prevalent issues such as linking quality degeneration, network congestion, and device failure. Hence leading to poor performance. It becomes crucial therefore that, for example, low end-to-end delay as well as energy efficiency can be best maintained through effective management of such networks just like every other technology. To mitigate these challenges, efficient management of IoT networks is imperative to ensure optimal performance metrics such as low end-to-end delay and energy efficiency. Software applications that are used for handling networks are NMS. These systems have tools and features that make it easy to manage many areas of network infrastructure like devices, connections, services, and performance. When it comes to managing IoT environments, NMS contribute significantly through the installation of various network administrators' software and hardware tools as well. Tools like these allow administrators to watch, set up, guard, and improve network resources and gadgets. Among the key features of NMS within the IoT settings are: Under this category are two areas- Network Configuration Management – which implies setting up the network to achieve defined goals like the required security levels or minimal latency(Aboubakar, Kellil, and Roux 2022). It involves activities such as configuring/reconfiguring parameters on any writeable network device.

3.1 Network management operations in IoT networks

NMS in IoT is like the nervous system of a smart network. It keeps everything running smoothly and securely(Mahesh, Bodhisatwa, and Somnath 2021). Imagine having a central control room where you can see and manage every part of your network. This system helps map out the entire network layout, showing how each device connects and interacts with others. It's flexible too, making it easy to add new devices or adjust for changes without disrupting the flow. Communication is key, and an NMS ensures data moves efficiently between devices, maintaining high quality and speed, which is especially important for tasks that need real-time data. As the network grows, scalability becomes crucial. An NMS makes it simple to expand by adding or removing devices seamlessly and balancing the data load to prevent any one part from getting overwhelmed. Security is another big part of what an NMS does. It acts like a vigilant guard, ensuring only authorized devices and users can access the network, encrypting data to keep it safe, and watching for any signs of trouble. By bringing all these functions together, an NMS keeps your IoT network running like a well-oiled machine, ready to support whatever tasks and services you need. There are many key aspects of Network management operations in IoT networks such as:

Topology Management: The purpose of topology management operations is to make sure that the network is always connected by overseeing network structure preventing chokepoints and creating efficient routes.

Security Management: Prevention of unauthorized access by intruders is the aim of security management. It includes implementing encryption techniques, key distribution mechanisms, threat detection, and recovery measures to safeguard IoT networks from cyber threats.

Quality of Service (QoS) Management: Configuring the network to achieve desirable performance metrics such as low data latency, minimal packet loss, and high throughput is what QoS management is about. This involves prioritizing traffic as well as allocating resources effectively and implementing mechanisms for shaping traffic.

Fault management mechanisms: Detect, isolate, and resolve network problems without disrupting network operations; these consist of proactive monitoring (e.g., fault detection), root cause analysis, and remediation (automated) to guarantee uninterrupted service provision on networks.

Network Maintenance: Network maintenance involves a series of activities that aim to maintain the operational state of the network. Some of the activities involved include bug fixing, software maintenance, firmware updates, and troubleshooting network issues among many others to guarantee that IOT services remain reliable and available

Besides traditional ways of controlling networks, software-defined networking (SDN) has been identified as a business strategy for boosting the functions of a network, cutting costs as well and facilitating innovative research in IoT systems.

3.2 SDN architecture layers

An SDN architecture has three layers(Shamsan and Faridi 2018):

Infrastructure Layer: At this layer, there are network devices like virtual switches, routers, and switches, Network hardware including routers, switches, servers, and so forth, and wireless access points that establish the actual and virtual framework of the network. Control Layer: The control layer contains centralized control points of the network, i.e., SDN controllers. These controllers respond to real-time changes in available resources and application requirements by enforcing routing decisions and policies.

Application Layer: Many applications that use the abilities of SDN to set up and manage the network can be found in the application layer. These applications serve different activities, for example, access control or energy saving, and are meant for inspecting traffic or managing the network. In this way, using the SDN framework combined with network management systems on a large scale, companies can cope effectively with the difficulties encountered due to IoT network intricateness and lack of resources. This makes it possible for them to guarantee the effective functioning of IoT networks, maximize performance, heighten security as well as bring out all the potential benefits of the connected car system to foster Innovation among other advantages like improvement in the daily commutes for people living within groups.

Figure 1. IoT network architecture

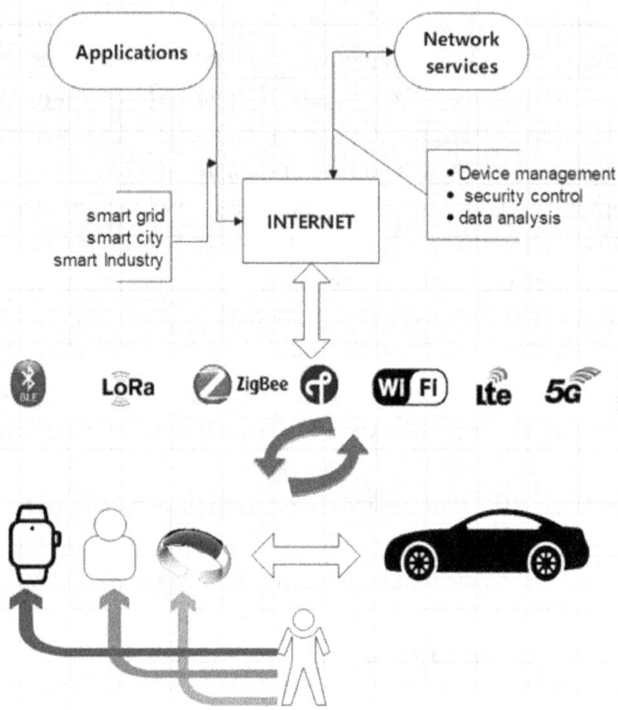

Figure 1 shows a connected ecosystem depicting a smart city and industry; integrated through a well-laid network system. All these are critical in improving operations, survival, and service delivery in urban centers. In managing power production and allocation, technology is key among many other things. All these involve real-time watching of power use, connections to alternative energy sources, automatic demand-response schemes, and sudden grid protection provisions. Smart city pilot schemes are practical as far as making town services and operations perfect. Traffic management cannot be realized without intelligent transport systems(A. Sakr, El-Afifi, and team 2023); on the other hand, efficient energy utilization is guaranteed by smart grids. Connected infrastructure allows us to gather data for good judgment, while citizen services assist in enhancing communication and association among inhabitants and authorities. On the other hand, through automation plus sharing of data smart industry practices are changing manufacturing processes. Production is made more efficient by industrial automation systems whereas machine performance is monitored through data analytics. Systems for managing supply chains and smart warehouses improve efficiency and productivity in industrial environments. The network infrastructure illustrated is composed of various communication technologies that are customized to make it easy to connect these smart systems. Therefore, LoRaWAN (LoRa) technology can enable low-power, long-range types of networks. Zigbee allows low-power, short-range wireless communication for device networking, while Wi-Fi gives high-speed internet access confined locally. Further, LTE and 5G standards offer high-speed mobile communication for diversified connectivity needs. Additionally, smart ecosystems come with such components as connected cars. These cars use sophisticated communication technologies for connection to network infrastructure. Other than that, they enable real-time traffic updates, permit remote diagnostics, and support autonomous driving. Such vehicles support safety, sustainability, and improvement of transport in smart cities. In general, it indicates that smart systems are working together to advance sustainability, resilience, and innovation in cities. By being assimilated and working together these technologies are enabling an efficient, connected, and greener tomorrow, as epitomized by the role that connected cars are playing in determining the smart ecosystem.

4. DATA ANALYTICS AND TRAFFIC PREDICTIONS

As global populations continue to grow, the ought to establish secure and effective road networks becomes increasingly urgent. Historically, expanding nations have trailed behind industrialized counterparts in terms of daily vehicle usage. However, this idea is shifting as many of these nations transition into the ranks of developed economies. With this transition comes an increase in the number of vehicles being utilized daily. The vast number of vehicles on our roads necessitates the utilization of Big Data techniques for acquiring, processing, and leveraging traffic data to optimize traffic systems. Numerous methods have focused on constructing examples and recreations of traffic systems within both microscopic and macroscopic simulation frameworks. Macroscopic models typically adopt constrained network flow models, assuming continuous streams of traffic flow through nodes without considering individual vehicle behavior.

To generate authentic and convincing simulations of traffic systems, it's important to gather and analyze real-world data. Many states worldwide have started data collection efforts on traffic, providing researchers with a great prospect to create precise models of traffic systems. A vast number of drivers and vehicles go through busier roads and routes each day, which leads to the complexity of traffic models that could handle this data. This data will be processed to make simulations. A vast quantity of vehicle count data is accessible from various regions globally, offering the potential to construct mathematical models depicting vehicle arrival rates.

4.1 Big Data and ITS

Numerous algorithms within Intelligent Transportation Systems (ITS) have proven effective in various applications such as traffic flow prediction, travel time estimation, traffic sign recognition, incident detection, and vehicle and road safety optimization(Rahul Anand and Smita Sankhe 2022). These approaches can be broadly categorized into parametric and non-parametric techniques. Among the wide variety of statistical parametric and non-parametric techniques, several algorithms have been applied in ITS.

Former traffic flow prediction methodologies primarily rely on shallow traffic prediction models and often fall short in meeting the requirements of many real-world applications. So deep learning, a variant of machine learning, has come into action for predicting traffic . Deep learning algorithms leverage multi-layer architectures to extract intrinsic features from data, spanning from low-level to high-level representations. They excel at uncovering extensive structures within data. Given the complex nature of traffic flow processes, deep learning algorithms can effectively capture traffic features without relying on prior knowledge, resulting in strong performance for traffic flow prediction. The literature highlights a significant increase in Big Data algorithms and advancements in computational tools that manage modern models with larger datasets as discussed in Table I. As a result, there is a clear trend toward combining different algorithms to overcome the limitations of earlier models and address more complex problems in the ITS domain. This approach leads to more accurate results in shorter timeframes.

Table 1. Applications and their respective tools

Application	Model
Traffic flow prediction	Bayesian Combined Neural Network (BCNN)
Traffic flow prediction	Fuzzy ARTMAP Neural Network (FAMNN)
Traffic flow prediction	Neural Network (NN) and Gaussian Process Regression (GPR)
Traffic sign recognition	Adaptive Resonance Theory Neural Network (ARTNN) and GA
Traffic sign recognition	Convolutional Neural Network and Stochastic Gradient Descent (CNN and SGD)
Travel time prediction	Spectral Basis Artificial Neural Network (SNN)
Travel time prediction	Counter Propagation Neural Network (CPN)
Bus arrival time prediction	Neural Network
Car-following behavior prediction	Neural Network
Incident detection	Neural Network and Data Fusion
Pedestrian detection	RBFN
Incident detection	Convolutional Neural Network and Probabilistic Neural Network (CPNN)
Traffic flow prediction	Stacked Autoencoders (SAES)
Traffic flow prediction	Deep Belief Network (DBN) and Decision-Level Data Fusion
Traffic flow prediction	Deep Belief Network and Multitask Learning (DBN and MTL)

In network-constrained movement, discovered data transformations enabling them to visualize the correlations between two crucial sides of movement: traffic intensity and speed. we can see most of the models prevailing today are considering speed and traffic intensity as their basic go-throughs.

The collaboration between machine learning algorithms and IoT yields enhanced performance in communication and computation, bolstering controllability and decision-making capabilities. This synergy uses advanced monitoring across thousands to billions of ubiquitous sensing devices and enhanced communication capabilities. The bulk data is processed as needed through data analytics.

4.2 Data analytic techniques for bulk data

Analyzing large amounts of data can seem daunting, but modern techniques make it manageable and incredibly insightful. Think of data mining as digging through mountains of information to find valuable nuggets of patterns and trends. Machine learning, on the other hand, is like teaching a computer to learn from past data to predict future outcomes. Statistical analysis helps make sense of the numbers, revealing the story behind the data. Before any of this can happen, data often needs to be cleaned and organized, a process known as ETL (extract, transform, load). Tools like Hadoop and Spark come into play, acting like super-efficient team members that can handle huge amounts of data across many computers. Together, these techniques transform large amounts of data into meaningful insights that help businesses and researchers make smarter decisions, streamline operations, and uncover new opportunities.

Data analytics can be classified into(Tsai et al. 2015):

- Descriptive analytics
- Predictive analytics
- Prescriptive analytics
- Adaptive analytics

Figure 2. Data Analytics classification

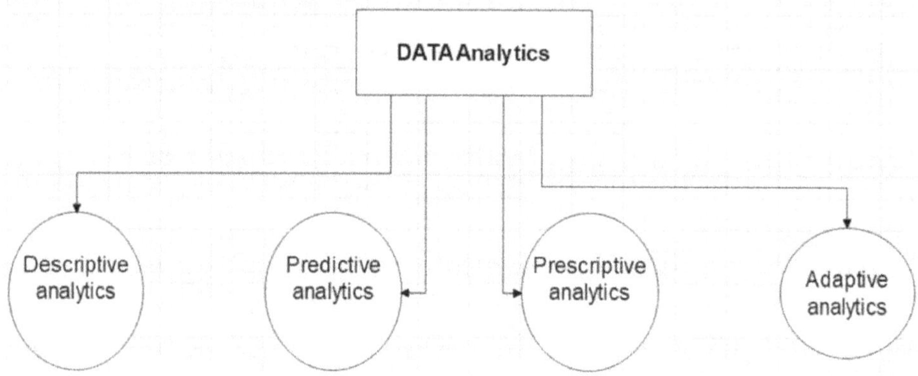

Figure (2) shows the divisions of data analytics into its main four sub-divisions. Descriptive Analytics: Descriptive analytics entails the comprehensive examination of historical data to distill and elucidate past events or trends. Employing methodologies such as data aggregation, visualization, and summary statistics, it tends to offer meaningful insights into the occurrences that have happened.

Predictive analytics: This involves trying to know what will happen in the future by observing previous information. It helps to forecast trends or behaviors that are likely to occur by looking into data patterns and relationships, the techniques in use are based on statistics and machine learning practices. Prescriptive Analytics: Instead of only predicting the future, prescriptive analytics provides actionable advice to achieve desired results. By using complex analytics methods as well as decision models, prescriptive analytics gives input into what would lead to the best results.

With adaptive analytics, analytical insights will surely be useful as well as efficient because they adjust accordingly using dynamic modeling techniques as well as adaptive algorithms that enable them to adapt in real time.

In simple terms, Descriptive analytics looks backward, Predictive analytics looks forward, Prescriptive analytics tells you what to do, and Adaptive analytics recognizes situations instantly and adjusts to the imperatives. Essentially, these forms of analytics converge to enhance intelligent decision-making in multiple sectors as well as industries. Integrating these advanced analytics forms assists in enabling dynamic planning in autonomous cars and unmanned surface vehicles hence facilitating their ability to adjust to changing environments and conditions in real-time.

5. DYNAMIC PLANNING FOR AUTONOMOUS CARS AND UNMANNED SURFACE VEHICLES

Dynamic route planning and navigation refers to the process of capturing real-time data on traffic conditions, weather conditions, road closures, and other relevant factors to determine the most efficient and effective route from a starting point to the point of destination(Khadka, Li, and Wang 2022). In recent years, advances in technology, especially in artificial intelligence, data analytics, and GPS navigation have led to a revolution in road planning and dynamic navigation Mobile apps and GPS devices equipped with dynamic road planning algorithms in to now able to provide users with real-time updates and new options(Okrepilov et al. 2022). There are no accidents, or other obstacles, saving time and improving overall travel efficiency

Dynamic road systems and transportation have applications in a variety of sectors, including transportation, logistics, emergency services, and urban planning. For example, delivery companies can optimize their delivery processes to reduce fuel consumption and shorten delivery times, while manufacturers a emergency can move quickly through vehicles and reach their destination quickly and efficiently

Overall, dynamic road design and navigation play an important role in modern transportation systems, providing users with a comprehensive and flexible journey, and helping to alleviate traffic congestion and reduce and improve the overall travel experience.

5.1 Dynamic course planning for traffic efficiency

Dynamic planning is a critical aspect of the operation and efficiency of autonomous vehicles, including both autonomous cars and unmanned surface vehicles (USVs)(N. Wang et al. 2022). This study explores the methodologies and technologies that enable dynamic planning, focusing on real-time data processing, adaptive algorithms, and the integration of advanced sensor systems. The study addresses the unique challenges faced by autonomous systems in diverse environments, such as urban road networks for cars and maritime conditions for USVs. By leveraging artificial intelligence and machine learning, dynamic planning systems can make real-time decisions to navigate obstacles, optimize routes, and ensure safety. It is must to examine the impact of dynamic planning on overall system performance, highlighting improvements in navigation accuracy, operational efficiency, and response to environmental changes. Through a comprehensive analysis of current technologies and future trends, this research aims to contribute to the development of more robust and reliable autonomous vehicles capable of operating independently in complex and unpredictable environments.

5.2 Dynamic planning for autonomous cars and unmanned surface vehicles

5.2.1 Unmanned Surface Vehicle

Navigating USVs affords demanding situations because of the growing complexity of maritime traffic. These vessels depend on static and dynamic environmental information from diverse detection systems like shipborne radar, electronic charts, and AIS. USVs can strategize sailing routes for green and secure projects of entirety. However, in maritime settings and unfavorable weather conditions, barriers in sensor layout, power, and reception hinder the USV's potential to perceive international environmental information(Briguglio and Crupi 2024). As a result, comprehension of complex environments will become restricted. Research work on USVs' direction planning are classified into two essential processes: global path planning and local path planning. Global path planning involves determining the most appropriate path from the starting point to destination based on available maritime environment information. This information is typically obtained through various environmental data source. Path planning algorithms Dijkstra and A* are generally utilized in global path planning to discover the maximum appropriate path. On the other hand, local path planning is based on sensor statistics accumulated by the USV at some stage in navigation. However, conventional direction-planning algorithms won't be highly efficient in complex maritime environments. To address this drawback, researchers have explored alternative methods along with geometric correlation modeling and potential field methods. These strategies' goal to improve neighborhood route planning by thinking about varying environmental parameters.

Despite efforts to enhance local path planning, it was noted that its average overall performance might also still be suboptimal. This shows that further research is needed to broaden greater effective path-planning techniques for USVs running in dynamic maritime environments. By exploring modern techniques and refining existing algorithms, researchers can try to improve the navigational competencies of USVs, in the long run enhancing their efficiency and protection in challenging maritime conditions.

5.2.2 Autonomous vehicles or connected vehicles

Autonomous vehicles, also known as connected cars, represent a breakthrough in transportation technology, allowing vehicles to sense their surroundings, make decisions, and pick up an environment without human intervention Similar to USV, autonomous vehicles use the same algorithms for modified applications with the terrestrial environment, to develop key technologies.

Dynamic Guidance systems:

Autonomous vehicles use dynamic guidance systems, including V2V, V2I, and V2T technologies, to ensure efficient navigation in urban environments. Through real-time communication with other vehicles, infrastructure elements like traffic lights, and smart terminals, autonomous cars can adaptively plan routes and make informed decisions to optimize traffic flow.

Optimization Algorithms:

Various optimization algorithms play an important role in routing for autonomous vehicles, affecting road traffic and overall efficiency Classic algorithms such as the Dijkstra algorithm, Floyd algorithm, and A* are used to find the most efficient routes between locations. These algorithms consider factors such as distance, travel time, and road conditions to optimize route planning.

Additionally, neural network algorithms utilize machine learning techniques to analyze vast amounts of data and improve navigation decisions over time. By learning from past experiences and adapting to changing conditions, neural networks enhance the accuracy and reliability of autonomous vehicle navigation.

Ant colony optimization stands out as a particularly effective algorithm for route planning in dynamic environments. Inspired by the foraging behavior of ants, this algorithm iteratively optimizes routes based on pheromone trails, effectively navigating through traffic congestion and unforeseen obstacles.

5.2.3 Planning Algorithms

We are using planning algorithms to reach the right destination along with avoiding collision and providing safety for our passengers along with maneuvering capabilities in the presence of obstacles. Here path planning is also divided into two local and global path planning as mentioned in 5.2.1.

Dynamic path planning is divided into four main categories:

Grid-based approach, potential field approach, sample-based approach, and discrete optimization approach. Grid-based approaches involve mapping the environment onto a grid of cells, with each cell indicating the presence or absence of obstacles within the environment. To find the optimal path from the starting point to the goal while avoiding obstacles, optimal search algorithms like A* and D* are commonly employed. These algorithms efficiently navigate the grid to figure out the global best path.

In recent years, advancements have been made to overcome non-holonomic limitations. These enhancements allow for more complex roadways, such as U-turns on roadblocks, parking lots, and unstructured transit routes. These enhancements expand the capabilities of grid-based approaches, making them more versatile and applicable in a wider range of scenarios.

Potential field approaches involve assigning virtual forces to obstacles to repel the vehicle and

Attractive forces to guide it towards the goal position. These forces create a gradient in the potential field, and the vehicle navigates along the steepest gradient to reach the goal. For instance, a path planning method introduced, uses stream functions to plan paths for autonomous vehicles, while daily proposed a harmonic potential field approach specifically for high-speed vehicles in reference.

One major advantage of potential-field-based approaches is their computational efficiency, as

They require minimal computation to generate trajectories. However, there are scenarios were these approaches may meet challenges. For example, in some situations, the vehicle may become trapped in local minima of the potential field, resulting in suboptimal paths or an inability to find a path altogether. These limitations highlight the need for further research and refinement of potential field-based methods to enhance their effectiveness and reliability in various navigation scenarios. Local minima of a potential domain, resulting in poor or completely inaccessible paths.

These limitations emphasize the need for further research and adaptation based on potential areas and ways to enhance their performance and reliability in different shipping conditions. Dynamic route planning transforms transportation by optimizing real-time data and optimizing routes for efficiency. It uses AI, data analytics, and GPS to create personalized navigation, benefiting a variety of industries such as logistics and emergency management.

Dynamic path planning improves travel by adapting to real time data and optimizing routes for efficiency. It uses AI, data analytics and GPS to personalize navigation, benefiting various sectors such as logistics and emergency services.

6. AUTOMATED INCIDENT MANAGEMENT

In order to ensure safety, efficacy, and reliability in today's motor vehicle systems, it is a must to have automatic incident management in connected cars. To effectively deal with emergencies, breakdowns, and road crashes, the trend is toward more and more networked and equipped with cutting-edge technologies. From this, one may deduce that strong event organization schemes are vital. Autonomous and connected cars (ACCs), coupled with Multi-access Edge Computing (MEC) and V2I and I2V communication, enable cars to operate with minimum human intervention whilst facilitating communication among themselves and with infrastructure. Edge computing depicted in Figure 1, is a computing paradigm within cellular networks that provides cloud computing capabilities for rapid processing at the edge of a network. This has been identified as a potential solution for environmentally sustainable smart cities in the field of intelligent transportation systems. A computational tool crucial for looking forward to upcoming site visitors' styles via studying both historical information and real-time information is called a traffic prediction algorithm. These algorithms serve critical roles throughout various packages, ranging from traffic control and route optimization to urban planning initiatives.

The process of developing a traffic prediction algorithm includes several key steps:

- Data collection,
- Data organization,
- Model construction,
- Parameter estimation, and
- Model verification.

Data Collection: Data Collection focuses on collecting a wide range of traffic related data from various sources. These sources include historical levels of road congestion, weather patterns, special occasions in taxable days, or any other factors that affect vehicle movement. To acquire this information, we use sensors located along the highways and cameras posted on road junctions or highways. Also, we rely heavily on GPS devices and traffic control systems to provide most of the relevant data.

Data Organization: Following data collection, data must be organized meticulously and pre-processed to guarantee their integrity and coherence. In this context, data cleansing procedures are used to correct any errors that might have occurred during the process of data entry such as those related to missing values or abnormal observations; rigorous quality control is carried out to determine if the data used in the models are reliable and consistent ensuring that they perform best when trained on or evaluated.

Model Construction: A critical part of the algorithm design process is the selection and employment of the most suitable model architecture. A lot of different approaches are examined including statistical methods or complex machine learning techniques to see how well they do at predicting future road use. Among those could be regression analysis based on historical data about specific routes while others make use of neural networks which have been trained beforehand by feeding them some inputs along with some expected outputs so that they learn patterns from it overtime before making their predictions without any further input otherwise known as unsupervised learning. Neural networks are among the hybrid models that also include statistical time series models. Math models may involve regression analysis models which are also known as statistical models and can further be classified into multiple variations based on their complexity level as well as the types of variables that they use to establish relationships among them such as linear versus nonlinear or univariate/multivariate ones depending upon whether we are focusing just one variable

Parameter Estimation: After selecting a model, the next important step is to determine the parameters, especially when the model has parameters that can be set. During this phase, we calibrate and optimize our model's parameters using advanced methods like cross-validation, grid search, or optimization algorithms. We tune parameters to increase prediction accuracy and make sure that it can deal with different traffic scenarios at the same time.

Model verification: The final phase in which rigorous validation and analysis are conducted to determine how effective and trustworthy the model is. To determine its accuracy and reliability, this model is tested using other data sets with the use of performance metrics like Mean Absolute Error (MAE), Root Mean Square Error (RMSE), accuracy measures, etc. By comparing the anticipated and actual traffic flow patterns, one can determine if the model can detect the underlying dynamics and make a decision on whether to refine algorithm deployment, this is a fact known to many researchers.

It is important to notice that these steps are connected and sometimes proceed in a cycle that requires changes and revisions at all points during the process of developing algorithms as well as carrying out experiments related to them. Moreover, detailed records on approaches used at different points in time, outcomes, as well as conclusions made in each step, serve as a necessary condition for explaining how efficient the algorithm is and adding new information to the field concerned with predictive modeling for transportation.

A new model was developed for predicting short-term traffic flows(S. Zhou et al. 2023). The model comprises two components, which consider the restricted computational power available on MEC servers. It involves a traffic light simulator and a vehicle speed prediction module. With this integration into automatic incident management systems, transportation agencies can now pre-plan for any eventuality ahead of its occurrence. Acting in advance enables improvement in traffic flow, escalating security and reducing road congestion. Traffic lights can be regulated before the occurrence of traffic jams through the application of forecast models. Drivers can also be alerted about expected traffic patterns based on the same data(J. Wang and Chen 2021). In the long run, this combination makes things run better and safer for those who travel every day illustrating how this projection thing can help to better transportation.

Various forms of AV technology are in existence with different levels of automation such as "connected" vehicles, shared vehicles, driverless vehicles, electric vehicles, and tailored vehicles(Sadaf et al. 2023). Traffic signal timing data originates from inductive loop detectors which provide traffic flow count information at given points. Nonetheless, the utilization of floating car data can serve as the best alternative for an adaptable traffic signal control system. This is possible because floating car data offers a lot of detailed information about every vehicle's precise speed, among other things. The proposed system not only makes it cheaper and easier to set up, but it might also be a leap forward in traffic signal optimization due to the use of a dataset. It allows for traffic signal control systems to become more adaptable and efficient through the optimization of comprehensive and detailed data. This offers an approach that can increase their capabilities by including Floating Car Data (FCD) in autonomous incident management systems(Astarita et al. 2020). Awareness of incidents in real-time is delivered by a system that manages autonomous incidents using FCD such that information on when they occur is accurate and prompt. It is possible to detect events such as accidents and traffic jams with a higher level of precision because FCD includes detailed data on the speed and location of vehicles Autonomous incidence resolution systems could respond dynamically having learned from the FCD. when there is an obvious slow down in traffic caused by an accident, for instance, the program can react by directing other vehicles elsewhere or changing lights so that cars do not bunch up at one place creating inconvenience for road users in general. The allocation of resources is better managed by autonomous incident management using exact locations and details on the seriousness of these events. It enables the dispatch of emergency responders to do their job more effectively and thus quickens the arrival time of reaction teams while increasing survival rates for victims involved in accidents through prompt assistance.

Feedback mechanisms are used to continuously improve the performance of incident handling systems that operate independently by comparing their reaction outcomes from their use with data obtained from FCD(ElSahly and Abdelfatah 2022). This feedback loop allows it to keep refining its algorithms and methodologies through time to make the management of traffic incidents more effective and flexible.

Integration with self-driving car networks makes it possible for autonomous incident management systems to rely on FCD provided by these vehicles hence improving incident detection and response rates. Through sensors and communication devices built into them, self-driving cars give out live information about road status and incidents hence enhancing situational awareness and response coordination for incident management systems.

By integrating FCD into autonomous incident management systems, Organizations can thus enhance their capacity to detect, respond to, and mitigate incidents on roadways through incorporating FCD into autonomous incident management systems, which is a major step towards using real-time data and the latest technology to better traffic management and increase road safety(Druml et al. 2021).

7. DATA SECURITY AND PRIVACY CONSIDERATIONS

Vehicles are no longer merely a mode of transportation; it has become a platform for an array of new services that have completely transformed the way we think about automobiles. For one, they are equipped with numerous sensors and software systems that make them ideal choices for entertainment purposes, as well as providing unmatched levels of convenience and safety compared to other types of vehicles available today. With technological advancements like radar-controlled cruise control or autonomous parking systems being incorporated into regular passenger vehicles, there is very little doubt about whether such cars will improve people's lives or not(Velsberg et al. 2021). Moreover, the more our automobiles become sophisticated technically and connected to networks, the more personal data they collect, leading to serious safety and privacy issues. The blend of automotive technology and information technology has brought various as-of-yet-unknown potentials to life, ranging from remote controls for diagnosing problems in vehicles through updates of over-the-air software to modernized navigation devices with the inclusion of real-time updates concerning traffic jams among the rest. On the flip side, however; closely knit in these advanced technologies are risks posed by data confidentiality invasion and compromising vehicle systems' security which would lead to alarming consequences if left unattended. Data travels over a complicated network that can be exploited by hostile actors as it moves between automakers, suppliers, and third-party apps.

At present, it is very important to protect information received from transport vehicles that are connected (for its transfer, storage, or processing purposes). There is a danger to the car's safety and reliability. there is also a danger to the privacy of its driver and passengers. Connected car technologies are capable of collecting a wide range of information- from multimedia interests, vehicle diagnostics, location tracking, or even biometric details this makes them an easy target for hacking purposes or unauthorized access(Rathore et al. 2022). As we embark on our journey to the era of connected mobility, it is important that we cautiously and strategically proceed along the data highway. Understanding the cybersecurity vulnerabilities facing connected vehicles and taking strong measures to mitigate risks while making them part and parcel of the design is crucial. Furthermore, legislators need to carefully balance encouraging innovation with implementing strict data privacy regulations as they struggle to manage this rapidly growing industry. This series delves into the profound domain of data privacy matters in connected cars. We wish to explore the major features of the pivotal intersection between car technology and information safety, ranging from analyzing the ethical implications related to data monetarization strategies to investigating the susceptibilities of car-to-car communication protocols. Come along as we discuss both the prospects and the pitfalls of maintaining our means of movement within an interconnected global world.

Wireless technologies are used by connected cars to enable real-time operation. These technologies allow them to communicate with the network, the cloud, other vehicles (V2V), and infrastructure. Vehicle sensors that regulate airbags, shut down systems, lock doors, and disable the vehicle are all accessible to hackers. They could mislead drivers by giving them false data, bringing down the vehicle network through 'Denial of Service assaults, illegally reprogramming ECUs with malicious software, and even downloading inaccurate navigation maps.

Figure 3. Various attacks

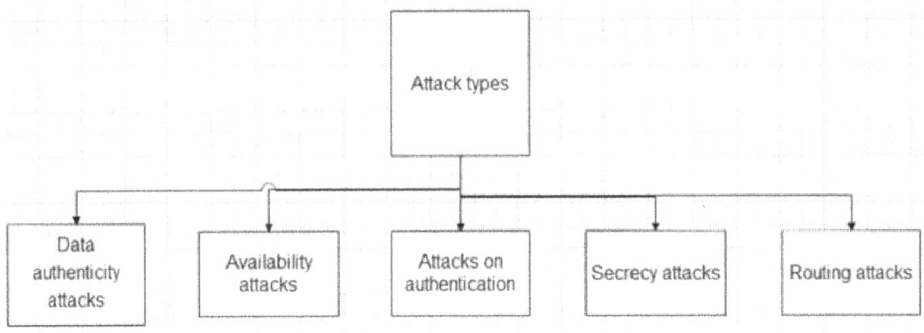

7.1 Attack types

VANETs are susceptible to various types of cyber-attacks due to their decentralized nature and reliance on wireless communication(Dibaei et al. 2020). There are some common types of attacks in VANETs classified in Figure 3.

Data Authenticity Attacks

It's crucial to make sure that the original data doesn't change while data packets go over the network. Attacks on data authenticity take many different shapes.

Availability Attacks

Common availability attacks that take advantage of transmission power and bandwidth restrictions include denial of service (DoS) and channel interference, which have the potential to bring down the Internet of Vehicles (IoV) system.

Attacks on Authentication

- *Sybil Attack*: Because IoV is dynamic, Sybil nodes may take advantage of sporadic and unreliable vehicle connections to interrupt regular services and compromise personal information.
- *GPS Deception*: False evidence and property damage can result from manipulating

GPS data in Internet of Vehicles (IoV) applications such as payment and navigation.

- *Masquerading Attack*: IoV systems are susceptible to confusion and operational disturbances in the absence of unique identification.
- *Wormhole Attack*: Because of its dynamic character and dependence on effective routing algorithms, it is extremely harmful to IoV systems and causes IoV elements' responses to be disrupted.

Secrecy Attacks

Ensuring that only authorized nodes can access critical data and resources is crucial for maintaining system integrity.

Routing Attacks

There are four different sorts of attacks that can occur throughout the routing process: route modification, denial of service, masquerading, and eavesdropping. A user notices several indicators that the server is untrustworthy, hence it is important to protect their privacy. But these methods introduce noise into each user's sensory input separately, then compile the noisy data on the server side. Be aware that some of the current processes may violate users' privacy because they fail to take into account any correlations between user characteristics and data features. The following specifications should be met by any safe IoV(Internet of the vehicle) information-gathering plan to ensure secure Big Data collection:

Data authentication: used to confirm the identities of the cars or vehicular clouds (i.e., vehicles near one another that create vehicular clouds for resource-sharing purposes and have similar aims).

Data integrity: the process of making sure that data has been sent correctly and hasn't been altered or destroyed because it will eventually be included in the Big Data stream.

Programs and Services Access Control: to ensure that every vehicle has access to the services and applications to which it is only authorized

Data confidentiality: necessary to ensure safe communication between cars taking part in a network.

Information non-repudiation: to guarantee that no vehicle could contest the legitimacy of any other vehicle.

Anti-jamming: to stop malevolent cars from attempting to compromise systems, which might block the network entirely or just a portion of it

7.2 Security for AV

Because of the growing attack surface brought about by the inclusion of additional safety and comfort features, automakers continue to encounter difficulties in resolving security and privacy concerns in AVs(Limbasiya et al. 2022). Some situations can be solved with software-based security solutions, but others require tamper-proof security measures. Comprehensive system protection is provided by techniques including digital signatures, intrusion detection systems (IDS), message encryption, authentication, and over-the-air (OTA) firmware updates. In terms of privacy, it's unclear how much personal information AVs gather during testing. But for navigational purposes, GPS position data, route information, speed, and trip duration are least kept, causing strong security measures to protect drivers' and passengers' privacy and safety. But for navigational purposes, GPS position data, route information, speed, and trip duration are at least kept, causing strong security measures to protect drivers' and passengers' privacy and safety. Physical security measures are also required due to the operational characteristics of components, such as electromagnetic emissions and power consumption, which present dangers for advanced side-channel attacks and physical reverse engineering. Software agents that are modified for secure status updates can improve system resilience against cyberattacks. Implementing safeguards to preserve personal data gathered and kept for vehicle operations is part of addressing privacy issues, a topic of discussion for the government and the automotive sector.

7.3 Security for V2V

Establishing a secure Public Key Infrastructure (PKI) and putting in place a Security Credential Management Scheme (SCMS) for V2V) communication is essential to the security and privacy of Vehicle-to-Infrastructure (V2I) communication. Using reputable certificate authorities, a vehicular public key infrastructure (PKI) makes it easier for every car to have a public-private key pair for message signing and verification. For V2V security and privacy, SCMS implementation is essential because it guarantees the integrity of linked vehicles and communication authenticity. It also makes safe Over-The-Air (OTA) software updates possible. To protect vehicle traceability and location privacy, pseudonym-based authentication is suggested. This helps to allay worries about illegal surveillance and protect driver privacy. Pseudonym refreshment, which prevents identity linking, is a difficulty that is handled by algorithms that strike a balance between privacy and expense. Group signatures allow members of a group to sign communications together without showing individual identities, adding an extra degree of security and location privacy.

7.4 Security for V2I

Organizations can use the NIST Cybersecurity Framework for Critical Infrastructure as a framework to help them secure their vital infrastructure. The five main components of the NIST framework's risk management strategy are identification, protection, detection, response, and recovery. Threat categorization is included in the term "identify," whereas "protect, detect, and respond" refers to the standard cybersecurity defensive methods and incident response used in security. In the case of critical infrastructure, recovery is especially crucial since it guarantees the infrastructure's resilience. Numerous cybersecurity rules unique to the information technology domain have adapted and refined the NIST framework.

8. SMART PARKING MANAGEMENT

Smart parking systems represent a significant advancement in urban management, particularly in the context of smart cities where efficiency is paramount(Rocco, Pipino, and Pagano 2023). These systems utilize IoT technology to optimize parking spot utilization and simplify the parking process for drivers by providing real-time information about available spaces. This addresses a crucial challenge faced by urban dwellers—the significant time spent searching for parking spots, which contributes to congestion and frustration.

By leveraging technology to offer drivers visibility into available parking spaces, smart parking systems alleviate this challenge, ultimately enhancing the overall urban mobility experience. The integration of IoT in parking has revolutionized the monitoring process, with sensors or cameras regularly checking the state of parking spots. Upon detecting any change, such as a vehicle occupying a spot, the system captures a photo and updates the Parking Management System (PMS). Additionally, mobile apps allow users to reserve parking spaces in advance, further optimizing the parking experience.

The parking process itself is streamlined through automation. Upon activating the mobile app, the driver signals their intention to park(Amras et al. 2023). As the vehicle enters the parking lot, sensors or cameras detect its arrival, triggering a check for available parking spaces. If a spot is available, the system captures the vehicle's license plate for identification and grants access by controlling the entry gate.

Once a vehicle is parked, the system timestamps the start of the parking period. When the driver leaves, the system detects the exit, records the completion time, calculates the total parking time, and then asks the user of the mobile app based on this length of time the role to pay the bill. Once payment is made, the system completes the parking period and the exit door automatically opens, allowing the vehicle to leave.

This mode of operation not only eliminates the need for manual payment stations but also reduces the waiting time for finding a parking space and completing the payment, thus providing parking management is all very efficient and convenient. And it benefits all employers

In the modern era of intelligent parking systems, the parking system is well-designed and optimized to improve user experience and performance. This begins when the user books a parking space through a dedicated application, facilitating and ensuring a reservation upon arrival.

Upon arrival at a parking lot, advanced sensors or cameras detect that the user's vehicle is approaching the entrance. At the same time, the system checks for parking spaces. When a path is empty, the door

automatically opens, allowing the user to enter. However, the user is notified immediately if the parking spaces are fully occupied, allowing him or her to search for alternative parking options without wasting time.

As the user enters the parking vicinity and parks their vehicle in a to-be-had slot, the machine captures a photograph of the car's license plate for identification and documentation functions. This step ensures duty and protection within the parking facility.

Throughout the parking length, the device diligently statistics the date and time when the automobile enters and leaves the car parking zone. This fact is vital as it should be calculating the parking period, which bureaucracy is the premise for figuring out the parking charge.

When the consumer is ready to depart, they retrieve their vehicle from the parking slot, signaling them to give up in their parking consultation. Once more, the device captures the date and time of departure to complete the parking document. Upon exiting the parking lot, the person makes a parking fee based on the period of their life. This payment method may be facilitated through numerous methods, which include mobile apps, automated price kiosks, or incorporated fee gateways. Ultimately, the vehicle leaves the parking lot, concluding the parking transaction. This seamless and automated procedure no longer enhances user comfort but also optimizes parking management, contributing to smoother visitor go-with-the-flow and improved city mobility.

9. INTEGRATION WITH VEHICLE SYSTEMS

Automating integration with vehicle systems includes several technologies and methods that can improve functionality, safety, efficiency, and user experience. This includes integrating ADAS, infotainment systems, telematics, and V2X communications(Damaj, Yousafzai, and Mouftah 2022). Integrating automation with vehicle systems is a complicated process that pulls advanced technologies and sophisticated communication protocols to enhance functionality, safety, and overall user experience. An Electronic Control Unit (ECU) manages everything from engine control to transmission to more ADAS. Middleware problems like adaptive Automotive Open System Architecture (AUTOSAR) and operating systems such as Android automotive OS offer the required infrastructure for integrating diverse software modules, ensuring interoperability and reducing development phase time.

A critical aspect of automation in-vehicle system is sensor integration and data fusion. Recent and modern vehicles are provided with an array of sensors, including cameras, radar, LiDAR, and ultrasonic sensors, which cooperatively provide a comprehensive understanding of the vehicle's environmen(Raouf et al. 2022)t. Sensor fusion links data from these sensor sources to enhance the accuracy and consistency of ADAS features. For instance, adaptive cruise control relies on radar and camera inputs to maintain a safe distance from other vehicles, while lane-keep assist uses cameras to ensure that the vehicle stays within the lane markings. Automated parking systems force ultrasonic sensors for precise planning in tight spaces. These technologies not only ensure safety but also to contribute a more comfortable and convenient driving experience. V2X communication further improves the resources of automated vehicle systems by allowing real-time data switches between the vehicle, infrastructure, other vehicles, and pedestrians. Dedicated Short Range Communications (DSRC) and Cellular V2X (CV2X) enable this communication, providing key information that can heighten traffic management and safety. For example, V2V communication allows vehicles to share the details of speed, position, and direction to prevent collisions, while V2I communication improves traffic signal timings to reduce congestion. V2P communication can alert drivers and pedestrians about potential hazards, thereby enhancing safety for all road users. Telematics combines telecommunications and informatics to offer services such as GPS navigation, real-time traffic updates, vehicle diagnostics, and remote monitoring systems. Cloud integration supports these utilities by providing storage, processing power, and advanced analytics capabilities. This enables features like predictive maintenance management, where data from vehicle sensors is analyzed to predict and prevent potential failures, thereby reducing the downtime for repairs and costs.

In summary, the integration of automation with vehicle systems signifies a paradigm shift in automotive technology, combining advanced communication protocols, sensor fusion, V2X communications, telematics, and AI to create safer, more efficient, and tightly connected vehicles. V2X systems extend these capabilities by enabling real-time data sharing within infrastructure, other vehicles, and pedestrians further improving traffic management and road safety.

CONCLUSION

In summary, leveraging IoT communications in connected cars presents a robust prospect to transform traffic management and enhance overall transportation productivity. IoT-enabled vehicles can communicate with each other and with infrastructure, creating a network of real-time data sharing that communicates better traffic flow, reduces congestion, and develops safety. Key technologies such as V2V and V2I enable this exchange, allowing for dynamic traffic signal adjustments, real-time navigation updates and coordinated vehicle movements. By directing the power of IoT, traffic management systems can become more pre-emptive and quicker to respond, effectively reducing travel time and fuel consumption, while also minimizing the risk of accidents. As IoT continues to advance in smart city applications, the potential to create smarter, more efficient, and safer urban environments grows, highlighting the crucial role of IoT communication in the future of traffic management and connected transportation systems.

REFERENCES

Aboubakar, M., Kellil, M., & Roux, P. (2022). A Review of IoT Network Management: Current Status and Perspectives. *Journal of King Saud University. Computer and Information Sciences*, *34*(7), 4163–4176. Advance online publication. DOI: 10.1016/j.jksuci.2021.03.006

Álvares, P., Silva, L., & Magaia, N. (2021). Blockchain-Based Solutions for UAV-Assisted Connected Vehicle Networks in Smart Cities: A Review, Open Issues, and Future Perspectives. *Telecom*, *2*(1), 108–140. Advance online publication. DOI: 10.3390/telecom2010008

Amras, M., Yulianto, E., Dewi, D. A., & Setiawan, A. (2023). Implementation of the REST API Model Using QR Codes on Mobile Devices to Order Parking Tickets. *International Journal of Advanced Computer Science and Applications*, *14*(10). Advance online publication. DOI: 10.14569/IJACSA.2023.0141043

Anand, R., & Sankhe, S. (2022). Traffic Prediction for Intelligent Transportation Systems Using Machine Learning. *International Journal for Modern Trends in Science and Technology*, *8*(7), 276–280. Advance online publication. DOI: 10.46501/IJMTST0807041

Astarita, V., Giofré, V. P., Festa, D. C., Guido, G., & Vitale, A. (2020). Floating Car Data Adaptive Traffic Signals: A Description of the First Real-Time Experiment with 'Connected' Vehicles. *Electronics (Basel)*, *9*(1), 114. Advance online publication. DOI: 10.3390/electronics9010114

Bathla, G., Bhadane, K., Singh, R. K., Kumar, R., Aluvalu, R., Krishnamurthi, R., Kumar, A., Thakur, R. N., & Basheer, S. (2022). R. N. Thakur, and Shakila Basheer. 2022. "Autonomous Vehicles and Intelligent Automation: Applications, Challenges, and Opportunities.". *Mobile Information Systems*, *2022*, 1–36. Advance online publication. DOI: 10.1155/2022/7632892

Briguglio, G., & Crupi, V. (2024). Review on Sensors for Sustainable and Safe Maritime Mobility. *Journal of Marine Science and Engineering*, *12*(2), 353. Advance online publication. DOI: 10.3390/jmse12020353

Bučko, B., Michálek, M., Papierniková, K., & Zábovská, K. (2021). Smart Mobility and Aspects of Vehicle-to-Infrastructure: A Data Viewpoint. *Applied Sciences (Basel, Switzerland)*, *11*(22), 10514. Advance online publication. DOI: 10.3390/app112210514

Campisi, T., Severino, A., Al-rashid, M. A., & Pau, G. (2021). The Development of the Smart Cities in the Connected and Autonomous Vehicles (Cavs) Era: From Mobility Patterns to Scaling in Cities. *Infrastructures*, *6*(7), 100. Advance online publication. DOI: 10.3390/infrastructures6070100

Cavicchioli, R., Martoglia, R., & Verucchi, M. (2022). A Novel Real-Time Edge-Cloud Big Data Management and Analytics Framework for Smart Cities. *Journal of Universal Computer Science*, *28*(1), 3–26. Advance online publication. DOI: 10.3897/jucs.71645

Cvar, N., Trilar, J., Kos, A., Volk, M., & Duh, E. S. (2020). The Use of Iot Technology in Smart Cities and Smart Villages: Similarities, Differences, and Future Prospects. *Sensors (Basel)*, *20*(14), 3897. Advance online publication. DOI: 10.3390/s20143897 PMID: 32668714

Damaj, I. W., Yousafzai, J. K., & Mouftah, H. T. (2022). Future Trends in Connected and Autonomous Vehicles: Enabling Communications and Processing Technologies. *IEEE Access : Practical Innovations, Open Solutions*, *10*, 42334–42345. Advance online publication. DOI: 10.1109/ACCESS.2022.3168320

Deng, B., Nan, J., Cao, W., & Wang, W. (2023). A Survey on Integration of Network Communication into Vehicle Real-Time Motion Control. *IEEE Communications Surveys and Tutorials*, *25*(4), 2755–2790. Advance online publication. DOI: 10.1109/COMST.2023.3295384

Dhinesh Kumar, R., & Rammohan, A. (2023). Revolutionizing Intelligent Transportation Systems with Cellular Vehicle-to-Everything (C-V2X) Technology: Current Trends, Use Cases, Emerging Technologies, Standardization Bodies, Industry Analytics and Future Directions. *Vehicular Communications*, *43*, 100638. Advance online publication. DOI: 10.1016/j.vehcom.2023.100638

Dibaei, M., Zheng, X., Jiang, K., Abbas, R., Liu, S., Zhang, Y., Xiang, Y., & Yu, S. (2020). Attacks and Defences on Intelligent Connected Vehicles: A Survey. *Digital Communications and Networks*, *6*(4), 399–421. Advance online publication. DOI: 10.1016/j.dcan.2020.04.007

Druml, N., Ryabokon, A., Schorn, R., Koszescha, J., Ozols, K., Levinskis, A., Novickis, R., ... (2021). "Programmable Systems for Intelligence in Automobiles (PRYSTINE): Final Results after Year 3." In *Proceedings - 2021 24th Euromicro Conference on Digital System Design, DSD 2021*, DOI: 10.1109/DSD53832.2021.00049

ElSahly, O., & Abdelfatah, A. (2022). A Systematic Review of Traffic Incident Detection Algorithms. *Sustainability (Basel)*, *14*(22), 14859. Advance online publication. DOI: 10.3390/su142214859

Farman, H., Khan, Z., Jan, B., Boulila, W., Habib, S., & Koubaa, A. (2022). Smart Transportation in Developing Countries: An Internet-of-Things-Based Conceptual Framework for Traffic Control. *Wireless Communications and Mobile Computing*, *2022*, 1–11. Advance online publication. DOI: 10.1155/2022/8219377

Gohar, A., & Nencioni, G. (2021). The Role of 5g Technologies in a Smart City: The Case for Intelligent Transportation System. *Sustainability (Basel)*, *13*(9), 5188. Advance online publication. DOI: 10.3390/su13095188

Golbabaei, F., Yigitcanlar, T., & Bunker, J. (2021). The Role of Shared Autonomous Vehicle Systems in Delivering Smart Urban Mobility: A Systematic Review of the Literature. *International Journal of Sustainable Transportation*, *15*(10), 731–748. Advance online publication. DOI: 10.1080/15568318.2020.1798571

Hejazi, H., & Bokor, L. (2021). A Survey on the Use-Cases and Deployment Efforts Toward Converged Internet of Things (IoT) and Vehicle-to-Everything (V2X) Environments. *Acta Technica Jaurinensis*, *15*(2), 58–73. Advance online publication. DOI: 10.14513/actatechjaur.00627

Ihama, E. I., Akazue, M. I., Omede, E., & Ojie, D. (2023). A Framework for Smart City Model Enabled by Internet of Things (IoT). *International Journal of Computer Applications*, *185*(6), 6–11.

Jaramillo-Alcazar, A., Govea, J., & Villegas-Ch, W. (2023). Advances in the Optimization of Vehicular Traffic in Smart Cities: Integration of Blockchain and Computer Vision for Sustainable Mobility. *Sustainability (Basel)*, *15*(22), 15736. Advance online publication. DOI: 10.3390/su152215736

Khadka, S., Li, P. T., & Wang, Q. (2022). Developing Novel Performance Measures for Traffic Congestion Management and Operational Planning Based on Connected Vehicle Data. *Journal of Urban Planning and Development*, *148*(2), 04022016. Advance online publication. DOI: 10.1061/(ASCE)UP.1943-5444.0000835

Limbasiya, T., Teng, K. Z., Chattopadhyay, S., & Zhou, J. (2022). A Systematic Survey of Attack Detection and Prevention in Connected and Autonomous Vehicles. *Vehicular Communications*, *37*, 100515. Advance online publication. DOI: 10.1016/j.vehcom.2022.100515

Mahesh, J., Bodhisatwa, M., & Somnath, D. 2021. "Biometric-Based Secure Authentication for IoT-Enabled Devices and Applications." In *Security of Internet of Things Nodes*, DOI: 10.1201/9781003127598-4-4

Malik, S., Bandi, P., & Sun, W. (2021). "An Experimental Study of Denial of Service Attack against Platoon of Smart Vehicles." In *Proceedings - 2021 4th International Conference on Connected and Autonomous Driving, MetroCAD 2021*, DOI: 10.1109/MetroCAD51599.2021.00013

Masood, A., Lakew, D. S., & Cho, S. (2020). Security and Privacy Challenges in Connected Vehicular Cloud Computing. *IEEE Communications Surveys and Tutorials*, *22*(4), 2725–2764. Advance online publication. DOI: 10.1109/COMST.2020.3012961

Mudhivarthi, B. R., Thakur, P., & Singh, G. (2023). Aspects of Cyber Security in Autonomous and Connected Vehicles. *Applied Sciences (Basel, Switzerland)*, *13*(5), 3014. Advance online publication. DOI: 10.3390/app13053014

Nastjuk, I., Herrenkind, B., Marrone, M., Brendel, A. B., & Kolbe, L. M. (2020). What Drives the Acceptance of Autonomous Driving? An Investigation of Acceptance Factors from an End-User's Perspective. *Technological Forecasting and Social Change*, *161*, 120319. Advance online publication. DOI: 10.1016/j.techfore.2020.120319

Nguyen, D. D., Rohács, J., Rohács, D., & Boros, A. (2020). Intelligent Total Transportation Management System for Future Smart Cities. *Applied Sciences (Basel, Switzerland)*, *10*(24), 8933. Advance online publication. DOI: 10.3390/app10248933

Okrepilov, V. V., Kovalenko, B. B., Getmanova, G. V., & Turovskaj, M. S. 2022. "Modern Trends in Artificial Intelligence in the Transport System." In *Transportation Research Procedia*, DOI: 10.1016/j.trpro.2022.01.038

Parham, M., & Pouyan, A. A. (2020). An Effective Privacy-Aware Sybil Attack Detection Scheme for Secure Communication in Vehicular Ad Hoc Network. *Wireless Personal Communications*, *113*(2), 1149–1182. Advance online publication. DOI: 10.1007/s11277-020-07272-8

Raouf, I., Khan, A., Khalid, S., Sohail, M., Azad, M. M., & Kim, H. S. (2022). Sensor-Based Prognostic Health Management of Advanced Driver Assistance System for Autonomous Vehicles: A Recent Survey. *Mathematics*, *10*(18), 3233. Advance online publication. DOI: 10.3390/math10183233

Rathore, R. S., Hewage, C., Kaiwartya, O., & Lloret, J. (2022). In-Vehicle Communication Cyber Security: Challenges and Solutions. *Sensors (Basel)*, *22*(17), 6679. Advance online publication. DOI: 10.3390/s22176679 PMID: 36081138

Rocco, G., Pipino, C., & Pagano, C. (2023). An Overview of Urban Mobility: Revolutionizing with Innovative Smart Parking Systems. *Sustainability (Basel)*, *15*(17), 13174. Advance online publication. DOI: 10.3390/su151713174

Sadaf, M., Iqbal, Z., Javed, A. R., Saba, I., Krichen, M., Majeed, S., & Raza, A. (2023). Connected and Automated Vehicles: Infrastructure, Applications, Security, Critical Challenges, and Future Aspects. *Technologies*, *11*(5), 117. Advance online publication. DOI: 10.3390/technologies11050117

Sakr, A., El-Afifi, M. I., & Team, P. (2023). Intelligent traffic management systems: a review. Nile journal of communication and computer science, 5(1), 42-56.

Shamsan, A. H., & Faridi, A. R. (2018). "SDN-Assisted Iot Architecture: A Review." In *2018 4th International Conference on Computing Communication and Automation, ICCCA 2018*, DOI: 10.1109/CCAA.2018.8777339

Shilpa, B., Radha, R., & Movva, P. (2021). Comparative analysis of wireless communication technologies for IoT applications. In Artificial Intelligence and Technologies: Select Proceedings of ICRTAC-AIT 2020 (pp. 383-394). Singapore: Springer Singapore.

Ta, V. T., & Dvir, A. (2020). A Secure Road Traffic Congestion Detection and Notification Concept Based on V2I Communications. *Vehicular Communications*, *25*, 100283. Advance online publication. DOI: 10.1016/j.vehcom.2020.100283

Teixeira, P., Sargento, S., Rito, P., Luis, M., & Castro, F. (2023). A Sensing, Communication and Computing Approach for Vulnerable Road Users Safety. *IEEE Access : Practical Innovations, Open Solutions*, *11*, 4914–4930. Advance online publication. DOI: 10.1109/ACCESS.2023.3235863

Tsai, C. W., Chin, F. L., Chao, H. C., & Vasilakos, A. V. (2015). Big Data Analytics: A Survey. *Journal of Big Data*, *2*(1), 21. Advance online publication. DOI: 10.1186/s40537-015-0030-3 PMID: 26191487

Upadhyay, P., Marriboina, V., Goyal, S. J., Kumar, S., El Sayed, M. E.-K., Ibrahim, A., Alhussan, A. A., & Khafaga, D. S. (2023). An Improved Deep Reinforcement Learning Routing Technique for Collision-Free VANET. *Scientific Reports*, *13*(1), 21796. Advance online publication. DOI: 10.1038/s41598-023-48956-y PMID: 38066104

Velsberg, O., Jonsson, K., Westergren, U. H., & Saarikko, T. (2021). IoT Triggers. How Municipalities Are Transforming to Smarter Cities through IoT Use. *Scandinavian Journal of Information Systems*, *33*(1).

Vermesan, O., Bahr, R., Falcitelli, M., Brevi, D., Bosi, I., Dekusar, A., Velizhev, A., ... 2020. "IoT Technologies for Connected and Automated Driving Applications." In *Internet of Things - The Call of the Edge: Everything Intelligent Everywhere*, DOI: 10.1201/9781003338611-6

Vermesan, O., John, R., Pype, P., Daalderop, G., Kriegel, K., Mitic, G., Lorentz, V., Bahr, R., Sand, H. E., Bockrath, S., & Waldhör, S. (2021). Automotive Intelligence Embedded in Electric Connected Autonomous and Shared Vehicles Technology for Sustainable Green Mobility. *Frontiers in Future Transportation*, *2*, 688482. Advance online publication. DOI: 10.3389/ffutr.2021.688482

Wang, J., & Chen, Q. (2021). A Traffic Prediction Model Based on Multiple Factors. *The Journal of Supercomputing*, *77*(3), 2928–2960. Advance online publication. DOI: 10.1007/s11227-020-03373-0

Wang, N., Zhang, Y., Ahn, C. K., & Xu, Q. (2022). Autonomous Pilot of Unmanned Surface Vehicles: Bridging Path Planning and Tracking. *IEEE Transactions on Vehicular Technology*, *71*(3), 2358–2374. Advance online publication. DOI: 10.1109/TVT.2021.3136670

Yue, W., Li, C., Mao, G., Cheng, N., & Zhou, D. (2021). Evolution of Road Traffic Congestion Control: A Survey from Perspective of Sensing, Communication, and Computation. *China Communications*, *18*(12), 151–177. Advance online publication. DOI: 10.23919/JCC.2021.12.010

Yusuf, A., Syed, A. K., & Souissi, R. (2024). Vehicle-to-Everything (V2X) in the Autonomous Vehicles Domain – A Technical Review of Communication, Sensor, and AI Technologies for Road User Safety. *Transportation Research Interdisciplinary Perspectives*, *23*, 100980. Advance online publication. DOI: 10.1016/j.trip.2023.100980

Zeng, F., Pang, C., & Tang, H. (2023). Sensors on the Internet of Things Systems for Urban Disaster Management: A Systematic Literature Review. *Sensors (Basel)*, *23*(17), 7475. Advance online publication. DOI: 10.3390/s23177475 PMID: 37687929

Zhou, N. (2021). "The Future Integrated Traffic Management System to Optimize the Automobile Traffic." In *Journal of Physics: Conference Series*, DOI: 10.1088/1742-6596/1972/1/012097

Zhou, S., Wei, C., Song, C., Pan, X., Chang, W., & Yang, L. (2023). Short-Term Traffic Flow Prediction of the Smart City Using 5G Internet of Vehicles Based on Edge Computing. *IEEE Transactions on Intelligent Transportation Systems*, *24*(2), 1–10. Advance online publication. DOI: 10.1109/TITS.2022.3147845

Chapter 13
Smart Cities, Smart Mobility:
How Mobile Intelligence Is Transforming Urban Transportation

Pradeep Pandian
https://orcid.org/0000-0002-2417-7228
Bengaluru City University, India

Chandan Chavadi
Presidency Business School, India

Ravindran Kandasamy
Presidency Business School, India

Monika Sirothiya
Research Rescue, India

ABSTRACT

The rapid urbanisation of cities around the world has increased the need for more efficient, intelligent transportation solutions. In this chapter, the mobile intelligence solution that is reshaping city driving leads to the next evolution of cities as smart cities accelerate smart transportation by incorporating V2X communication, Intelligent Transportation Systems (ITS), telematics, and autonomous vehicle technologies. Fundamentally, this means they are good at gathering, sifting through, and using real-time data to minimise traffic congestion and maximise people's safe movement. The chapter opens by defining what is meant by the term smart cities and how smart mobility can be an essential ingredient in making cities efficient. This

DOI: 10.4018/979-8-3693-6422-2.ch013

will cover V2X communications, the integrated mode of vehicles, infrastructure and pedestrians, and 5G by building on current solutions to better road safety through major communication systems.

I. INTRODUCTION

Overview of Smart Cities and Mobility

Definition of Smart Cities and the Role of Smart Mobility in Urban Development

A smart city is an urban area where multiple services and facilities are offered based on an interconnected framework of digital technology, data-driven solutions, and the Internet of Things (IoT) to optimize infrastructure and overall civic management (Musa, 2016). According to Kuchta (2014), cities have turned into "smart cities" when they aim to use information technologies to solve daily problems in the lives of citizens. A smart city consistently collects information in order to get a clear overview of the situation or place, analyse the data and use it wisely to upgrade decision-making processes to make an environment work more efficiently for sustainability.

Within this, smart mobility is central to offering residents a range of seamless, efficient and sustainable transport choices. It is a method of incorporating cutting-edge technologies, including IoT, AI and big data analytics, to create smarter public transportation systems, manage traffic flow better, promoting environment-friendly transportation options such as electric vehicles (EVs) and shared mobility services (Vakula & Raviteja, 2017). Smart mobility offers a way to help a city meet its sustainability goals by reducing congestion and emissions while increasing connectivity.

Brief Introduction to Vehicle-to-Everything (V2X) Communication

At the core of smart mobility is *Vehicle-to-Everything (V2X) communication*, which enables vehicles to communicate with each other and their surroundings. This includes:

- **Vehicle-to-Vehicle (V2V)**: Vehicles communicate with one another to share data such as speed, location, and direction to avoid collisions and optimize traffic flow.

- **Vehicle-to-Infrastructure (V2I)**: Vehicles connect with traffic signals, road signs, and other infrastructure to receive information that helps manage traffic and improve road safety.
- **Vehicle-to-Pedestrian (V2P)**: Communication between vehicles and pedestrians helps prevent accidents, particularly in busy urban areas.
- **Vehicle-to-Network (V2N)**: This allows vehicles to connect to broader networks, such as the internet, for real-time updates on traffic conditions, weather alerts, or road hazards.

By facilitating this continuous exchange of information, V2X communication enhances safety, reduces traffic congestion, and enables autonomous vehicles to navigate efficiently (Abboud et al., 2016).

Importance of Mobile Intelligence in Transforming Urban Transportation

Mobile intelligence is the integration of real-time data analytics, connectivity, and AI-powered systems in order to streamline urban mobility. In the context of smart mobility, mobile intelligence powers:

Dynamic Traffic Management: Mobile intelligence can analyse data collected from traffic signals and strategically redirect vehicles to alleviate congestion (Cassandras, 2017).

Enhanced safety systems: Through V2X data, mobile intelligence assists in prompt signalling to drivers about potential hazards, thus reducing crash possibilities (Y. Zhang & Letaief, 2019).

Enhanced public transportation: Mobile intelligence, with its real-time tracking of public transit vehicles, helps manage routes and schedules more efficiently to provide seamless operations and reduce waiting times (Singh & Gupta, 2015).

Data-driven urban planning: City planners could use mobile intelligence to better understand and plan transportation usage (Conway-Beaulieu et al., 2015).

One key aspect needed to build the smart mobility ecosystems is mobile intelligence to ensure the seamless exchange of data between vehicles, infrastructure and public systems. This shifts traditional, reactive urban transportation design to proactive, predictive systems that contribute to the general purpose of improved overall efficiency and sustainability in a city (Tan & Leon-Garcia, 2018). With the urban population rising, the way forward is in Mobile Intelligence to create future-proof urban spaces that are not only intelligent but also responsive.

II. VEHICLE-TO-EVERYTHING (V2X) COMMUNICATION

Definition and Core Concepts of V2X Communication

Vehicle-to-Everything (V2X) communication is a general term that encompasses the information exchange between every vehicle on the road and any entity in the area. That includes vehicular communication between vehicles, infrastructure, pedestrians and networks. The goal of V2X communication is the improvement of road safety, traffic efficiency and autonomous driving technology operations (Wang et al., 2019).

V2X is a critical enabler of connected and autonomous vehicles since it supplies a constant stream of data that informs on-vehicle decision-making (Mavromatis et al., 2020). This kind of communication allows drivers and autonomous systems to know not only what the vehicle can see by its sensors but also what is farther from it, which allows safer and more efficient driving (Sakaguchi et al., 2020).

Explanation of V2X

1. Vehicle-to-Infrastructure (V2I):

These services include the ability for vehicles to communicate with traffic lights, toll booths, road signs, parking meters, and other forms of roadside equipment via V2I communication. This helps vehicles receive information on traffic signals, road conditions, and construction zones via V2I, which assists drivers in making informed decisions and hence lessens the chances of accidents (Daddanala et al., 2021). For instance, it can tell approaching vehicles that the light is about to turn red so they can slow down gently.

2. Vehicle-to-Vehicle (V2V):

Vehicles can communicate with one another, sharing details like their speed, GPS location, and heading through V2V communication. This is useful in protecting from collisions, especially in busy places or when weather visibility is low (Sewalkar & Seitz, 2019). Situational awareness via V2V communications allows vehicles to communicate with one another, assisting in smoother lane changes, merging, and emergency braking.

3. Vehicle-to-Pedestrian (V2P):

V2P communication focuses on the interactions between vehicles and pedestrians or cyclists. The technology can help vehicles see pedestrians in complex urban environments and dark areas or low-light situations, such as night-time or lousy weather. Another example of how V2P can be used is in the case when warning a driver that a pedestrian is walking across the road and is not at a designated crosswalk or if there is a bike nearby (Sewalkar & Seitz, 2019).

4. Vehicle-to-Network (V2N):

V2N is the capability of the vehicle to connect to networks that are larger than itself—say, probably a cloud—to ascertain data in real-time about traffic, weather or road conditions. V2N has an important part to play in ensuring vehicles have the data they need to navigate more intelligently, update their route recommendations, and react to environmental conditions (Amadeo et al 2019). V2N could, for example, give before-time warnings of road blockages, accidents or severe weather to allow vehicles to reroute.

Technological Advancements in V2X Communication

Several key technologies have emerged to support V2X communication, making it faster, more reliable, and more efficient.

1. 5G Networks:

The deployment of 5G has been a game-changer for V2X, enabling ultra-low latency, higher bandwidth, and faster data delivery. These capabilities are critical for the safety and operation of autonomous vehicles, as they allow vehicles to communicate with each other and the infrastructure in real-time (Noor-A-Rahim et al., 2022). The improved connectivity of 5G also allows for more accurate positioning and a real-time response to changing traffic conditions.

2. Dedicated Short-Range Communication (DSRC)

DSRC is a dedicated short-range wireless communication technology designed explicitly for automotive safety applications. It operates in the 5.9 GHz band and provides low-latency, high-speed communications between vehicles and their environment. It has been widely tested and deployed in pilot programs for connected vehicles. DSRC is extremely effective for short-range communication and holds considerable importance in collision avoidance systems (Dasanayaka et al., 2020).

3. Cellular V2X (C-V2X):

C-V2X is an updated version of V2X communication that uses cellular networks to allow cars to communicate at longer ranges with infrastructure. This technology will work in concert with 5G and DSRC, allowing vehicles to communicate not only with each other but also with a limitless number of external systems. C-V2X works in both direct (V2V) and network-based (V2C) models, allowing it to be used for a wide range of use cases (Nguyen et al., 2022).

The Role of V2X in Enhancing Traffic Safety and Efficiency

Automobiles can detect their environment better and be able to decide faster which in turn results in greater traffic safety being facilitated by V2X communication. This two-way communication with other vehicles and infrastructure can prevent accidents and lessen the severity of collisions. For example:

Collision avoidance: V2X can warn drivers that are in a collision with other vehicles or pedestrians so they can prevent it.

Less congestion: V2X communication allows vehicles to utilize more optimal lane changes and exits/merging, leading to better traffic patterns as a whole, which helps lower the overall congestion level.

Emergency vehicle alignment: In an emergency, V2X can guide vehicles and facilitate communication with traffic signals so that they respond faster and pass through congested streets.

Furthermore, V2X communication aids in fuel efficiency and reducing pollution in the atmosphere. It helps the environment by reducing unnecessary idling and stopping, improving traffic flow.

Applications and Use Cases

1. Traffic Management:

V2X communication systems are also being rolled out to help manage traffic flow in cities. Connected traffic lights and smart road signs in cities like Hamburg, Germany, can also communicate with vehicles to help optimize the flow of traffic. As a consequence, making fewer stops and travelling at higher speeds from source to destination would accelerate emissions (Schutera et al., 2018).

2. Low-Latency Emergency Vehicle Communication:

In the US, this V2X technology is being deployed to increase the response time of emergency vehicles. Relating to the graphic above, if an ambulance or fire truck is approaching a junction, V2I can change the traffic lights ahead to allow that vehicle through, which decreases delays and allows emergency services to reach them faster (Kavas-Torris et al., 2023).

3. Pedestrian Safety:

In heavy urban traffic, pilot projects for pedestrian safety use V2P. For example, in China, pedestrian-detection systems are used near crosswalks to alert approaching vehicles if there is a pedestrian present and help reduce accidents and fatalities (Sewalkar & Seitz, 2019).

4. Autonomous Driving:

In South Korea, Hyundai has included V2X communication technology in its autonomous driving programs. In complex urban environments, vehicles utilize V2V and V2I communication to change routes according to traffic signals, road conditions, and pedestrians' locations. This has increased the safety and reliability of autonomous systems in the real world (Savić et al., 2017).

V2X communication, powered by next-gen technologies such as 5G and DSRC, is a game-changer in the way vehicles communicate with one another and their environment. Driving the building blocks of safer, more efficient and intelligent transportation systems, we get closer to the potential of fully autonomous smart cities.

III. INTELLIGENT TRANSPORTATION SYSTEMS (ITS)

Definition and Components of ITS

Intelligent Transportation Systems (ITS) refer to the application of advanced information and communication technologies to improve the efficiency, safety, and sustainability of transportation systems (Meneguette et al., 2018). ITS encompasses a wide range of tools and technologies designed to enhance the coordination, management, and flow of traffic, reduce accidents, and optimise the use of transportation infrastructure (Singh & Gupta, 2015).

Components of ITS include real-time data collection, mobile intelligence, Vehicle-to-Everything (V2X), Intelligent Transportation Systems (ITS), telematics, autonomous vehicles and decision-making to offer an uninterrupted transportation experience. Key components of ITS include

1. Transportation Management System (TMS)

These systems are designed to track and control the flow of traffic in real-time based on data collected from sensors, cameras and GPS-equipped vehicles. Traffic signal control systems, incident detection systems and congestion monitoring technologies are among the tools used in TMS (Malik et al., 2018). Optimised traffic signals that respond to real-time data can help reduce congestion, improve travel times and decrease the environmental consequences of transportation.

2. Advanced driver assistance systems (ADAS)

ADAS is used to make safety-related decisions such as braking, steering, and vehicle navigation. Some of the more common ADAS features are lane-keeping assistance, automatic emergency braking, adaptive cruise control, and blind-spot monitoring. ADAS is a crucial part of AV technology to help drivers conduct safer driving and also reduce human factors (Zhou et al., 2020).

3. Dynamic Routing Algorithms:

Dynamic Routing Algorithms use real-time traffic data to provide drivers with the fastest and most optimal directions. They incorporate variables like existing traffic, road closures, and weather conditions to route around bottlenecks. Navigation systems or mobile apps often contain dynamic routing to provide real-time directions with the latest traffic data and the best routes available (Jahn et al., 2005).

4. Connected Traffic Signals:

IoT-enabled traffic lights send signals by communicating with vehicles and central traffic management centres. These systems optimise the operation of a network of traffic signals in coordination with real-time traffic conditions, resulting in reduced stop-and-go driving patterns that normally impose unnecessary fuel waste and improve overall flow (Wang et al. 2021). Smart cities, for example, can let priority traffic lights sail emergency vehicles or public transportation to guarantee effective transit through intersections.

Impact of ITS on Urban Mobility

ITS technologies have a profound impact on urban mobility by optimizing traffic flow, improving road safety, and making transportation systems more responsive to changing conditions. The key benefits include:

1. **Improving Traffic Flow:** Roving monitoring and management systems can accomplish dynamic changes to traffic signals, routing, and lane usage. This reduces gridlock and bottlenecks, making vehicles operate more efficiently within urban areas (Souza et al. 2017). Some adaptive traffic control systems, like those deployed in many major cities, use cameras and road sensors to adjust signal timings based on the number of vehicles present. This results in a more even flow of vehicles during rush hours.
2. **Reducing Congestion:** By using dynamic routing and traffic management tools, intelligent transportation systems can redirect vehicles off of main thoroughfares, reducing congestion. This not only shortens driver travel times but also reduces total traffic congestion to some extent, especially in dense urban areas (Guidoni et al., 2020).
3. **Enhancing Safety:** ITS technologies like ADAS and connected traffic signals drastically enhance safety on the street. ADAS provides drivers with helpful hints automatically based on what their cars react to, and traffic signals can telegraph changes in the status of a light up ahead. Second, real-time data collected from ITS can discover accidents and hazards to avoid or reduce potential vehicle collisions (Li et al., 2021).
4. **Environmental Benefits:** ITS has the additional dimension of supporting sustainability by providing more efficient use of fuel and reducing emissions. Less fuel is wasted due to smoother traffic flow and fewer stops at traffic signals, leading to less congestion and reduced CO_2 emissions from urban transportation (Barth et al. 2015).

Case Studies on Cities Implementing ITS for Smarter Transportation

1. Singapore:

A world leader in ITS, Singapore uses cutting-edge traffic management systems that are uniquely suited to its highly-density urban setting. As an example, we can cite Singapore, which implemented the Electronic Road Pricing (ERP) system, charging drivers automatically according to distance and time of day and thereby reducing congestion (Goh, 2002). Singapore also leverages real-time traffic data from its smart traffic signals to help manage the movement of vehicles, increasing system-wide efficiency. There are also fully integrated modes of trains, buses and taxis across the board for a seamless journey throughout the city (Boyle et al., 2013).

2. Copenhagen, Denmark:

Copenhagen has turned to ITS to improve efficiency for traffic management and environmental purposes. The Green Wave of the city, in which traffic light phase adjustment preferences are given to cyclists, thus guarantees a safer and more efficient bicycle tour (Gössling, 2013). The initiative is part of Copenhagen's ambition to become carbon neutral by 2025. The city also enables the use of real-time data coming from connected traffic lights, and sensors that help authorities to monitor traffic patterns and respond by reducing waiting time at intersections, which in turn improves flow and results in less emissions, to some extent necessary for public mode (public transport/ biking) use too (Massink et al., 2011).

3. Los Angeles, USA:

The Adaptive Traffic Control System (ATCS) used in LA is one of the largest Intelligent Transport Systems deployments on Earth. The innovative traffic system also collects real-time data based on the actual condition of traffic through a transponder sensor and virtual sensors alongside video cameras, hence optimising the timing of signals (Penerapan, 2019; R & Rajendran, 2021). It has managed to cut travel times and emissions, particularly in peak traffic hours. A different type of dynamic pricing in LA is through the use of toll pricing on some highways, where higher costs can be paid at peak hours to incentivise drivers to choose less congested routes (Ghiasi et al., 2018).

Challenges and Opportunities

1. Infrastructure Challenges:

To implement ITS, besides the infrastructure necessary to have sensors/cameras installed and a communication network available. Funding and technical expertise for large-scale ITS deployment are sorely lacking in many cities, especially those in the developing world. Furthermore, MALIK (2018) says that the technology is evolving, and traditional ITS Infrastructure has various issues as you have to maintain and upgrade, which is a continuously ongoing project.

2. Data Privacy Concerns:

ITS providers must work together to ensure data is collected and stored safely within guidelines as to how it can be utilised. Trust, particularly public trust, is as essential and concerning as the dissemination of data-sharing practices, mainly when this involves personal data collected from vehicles and mobile devices (Prevost & Kettani, 2019).

3. The Role of Public and Private Partnership

Many successful ITS implementation require cooperation between the public authority and private companies. Governments set the regulatory framework and infrastructure, whereas private companies deliver innovative technology and know-how (Ogunleye & Belle 2014). Tech companies like Google and IBM have worked with cities to design traffic solutions that manage routes dynamically. Nevertheless, the usual public-private alignment is difficult to achieve when profit competes with public goods (Lazarus et al. 2018), or one goes into grey territory as far as profitability goes.

4. Future Paths for Innovation:

New areas such as AI, machine learning, and edge computing can also be put into use to add capabilities for predictive analytics in traffic management and autonomous vehicle functions. Moreover, the fusion of electric vehicles (EVs) and autonomous vehicles (AVs) with ITS is expected to yield more intelligent and green transportation networks, which would mould the forthcoming years of urban mobility (Mahrez et al., 2021).

To sum up, ITS is changing the urban mobility equation by optimizing traffic control, creating safer streets, and delivering sustainable cities. Infrastructure, data privacy, and partners are challenges to developing ITS, but the rewards are much larger in the field of smart cities.

IV. TELEMATICS AND AUTONOMOUS VEHICLES

Understanding Telematics

Telematics is a telecommunications and informatics-based system for structured vehicle data storage, fleet management, and driver inspection in one place. Telematics systems utilize wireless communication capabilities to record information on vehicle operations, location data, driving behavioural patterns of a driver, and other surrounding conditions in real-time. The data is sent to a central system, which can be used for analysis, and the information gathered can help optimise vehicle systems, safety, and fleet management (Haque & Hossain, 2014).

It is increasingly beneficial for private drivers as well as companies, specifically in fleet management, to monitor vehicle condition and efficacy and pinpoint when a service is required. This is a linchpin for the way forward in mobility, especially with respect to autonomous cars and intelligent transportation systems (Jadaan et al., 2017)

Primary features of Telematics:

1. Remote Vehicle Diagnostics:

Telematics helps track engine live conditions, including fuel usage and service requirements. Early diagnosis of problems can help avoid expensive repairs and breakdowns so that vehicles run at their best efficiency (Huang et al., 2012). A simple example is alerting fleet managers when a vehicle service is due, helping reduce downtime and enhancing operational efficiency.

2. Fleet Management:

Telematics are critical in fleet management as they allow the real-time monitoring of exactly where vehicles are, how much fuel is being consumed and driver behaviour. With real-time tracking, fleet managers can drive the most optimised routes to minimise fuel consumption and ensure that their vehicles are being used efficiently. It also improves the process of vehicle scheduling and coordination, which makes it possible for businesses to manage large fleets (Gopal & Prakash, 2012).

3. Driver Behavior Monitoring:

These systems monitor such driver behaviours as speed, acceleration, braking and compliance with traffic rules. This data is significant for enhancing road safety since it enables fleet managers and insurance companies to pinpoint dangerous driving behaviours and address them (Horrey et al., 2011). These insights allow the app to learn how individual drivers are using their vehicle/customising it, which can result in them receiving prompts to make driving adjustments for safer operations (i.e. excessive hard braking or speeding leads to feedback via an alert warning and recommendations, allowing for potential avoiding of an on-road collision/accident).

Telematics in Autonomous Vehicles

Autonomous vehicles rely heavily on telematics, as it offers the necessary insight to allow their operational systems to think and make smarter decisions in real-time. In autonomous vehicles, telematics is part of an automated vehicle system that links with the sensors and onboard computer and has cameras to collect, process and transmit data in your surrounding environment while you are driving safely.

1. Enhancing Autonomous Driving:

Autonomous vehicles depend on telematics for road conditions, traffic, and obstacles. This data is critical for making decisions, since for the autonomous systems it is very important to know about the environment in order to do a safe navigation (Khalil, 2017). A technology termed telematics allows these vehicles to analyze large quantities of real-time data to drive in a much better manner backed by making instantaneous responses including moving lanes, avoiding obstacles or adjusting speed (Okumura et al., 2016).

2. Enhancing Car Performance:

Telematics helps to analyse these data for fueling, engine health, and driving patterns thereby optimising the performance of autonomous vehicles. Geared towards increasing performance and efficiency, telematics can optimise operations by tracking vehicle health via continuous monitoring (Jambotkar et al., 2021). As an example, the development of autonomous vehicles allows for a design where your car itself alerts you or may even adjust its driving style (also considering that fuel efficiency is decreasing) to be more demanding with maintenance checks.

3. User Experience:

Telematics is also part of what the driver experiences as an autonomous vehicle, with real-time traffic, time of arrival, and other route adjustments made in the car. Autonomous vehicles can provide smoother rides to passengers by saving them time and traffic-related fatigue. They travel in optimized routes enabled by telematics systems, which can unburden traditional GPS and improve safety (Mittal et al., 2019).

Data Analytics in Telematics

Data analytics is at the heart of telematics, enabling the extraction of actionable insights from the vast amounts of data generated by vehicles. This data is analysed to predict maintenance needs, optimise fuel consumption, and monitor driver behaviour.

1. **Predictive Maintenance:** Telematics data can predict when to service a vehicle. Telematics reads and logs a lot of data to analyze patterns based on thousands or sometimes millions of kilometres worth of historical data, predicting future problems ahead of time. This reduces maintenance costs and vehicle downtime since recharge/replacement repairs can be planned rather than performed due to a failure occurrence (Meng et al., 2013).
2. **Fuel Optimization:** Fleet managers and vehicle owners can track fuel consumption and recognise inefficiencies via data analytics. In other words, telematics systems can analyse driving behaviour, route selection, and vehicle load to recommend methods for reducing fuel consumption. This includes suggesting detours with decreased traffic or informing motorists how to alter driving behaviour to decrease fuel usage (Barth et al. 2004).
3. **Enhanced Safety:** Telematics systems use data analytics to monitor driving behaviour and assess risk factors. By analysing patterns such as hard braking, rapid acceleration, or speeding, telematics can provide feedback to drivers to help them improve their behaviour. This not only reduces the likelihood of accidents but also helps businesses lower insurance premiums by demonstrating a commitment to safety (Soleymanian et al., 2019).

Challenges in Telematics Implementation

While telematics offers many benefits, its widespread adoption faces several challenges that need to be addressed:

1. Data Security:

Location and driver behaviour data from cars are sensitive in nature; here, the concern is regarding how they get collected and transmitted. The hackability of telematics will render personal and corporate data susceptible to cyberattacks (Singh & Kim, 2017). Protecting the privacy of vehicle owners Automotive cybersecurity includes safeguarding and ensuring high standards in encrypting communication transmitting between vehicles and other devices (infotainment, Wi-Fi, etc.), data storage on your domains, security among fleets and ultimately secure transferring to you big-data analytic tools.

2. Cost of Telematics Systems:

Telematics systems can be costly to implement and are often prohibitively so for small businesses or fleets owned by individuals. Installing and maintaining telematics hardware, software, and the data it generates can be expensive. On the other hand, the high costs could be off-putting for organisations since companies' return on investment is mainly acquired from the fuel savings and efficiency gains for large businesses, whereas if you are a small or medium business, it might not outweigh the value added to justify an upfront investment (Said et al., 2014).

3. Legal and Regulatory Issues:

There are a myriad of data privacy, liability, and insurance-related regulations related to telematics systems. There is also the issue of who controls the data stemming from telematics systems. From Roger et al. (2000), questions include whether it should be owned by the vehicle owner or Manufacturers and third-party organisations that have data services built into their business models. Autonomous vehicles using telematics to navigate the world also deal with regulatory issues, such as safety standards, liability in accidents and rules around data sharing.

Nonetheless, as autonomous driving technology continues to progress, telematics is becoming more fundamental than ever for contemporary cars. Telematics can significantly disrupt fleet management, and if the security concerns, cost benefits, and regulatory landscape are adequately addressed, so are personal and commercial transportation locations in general.

V. ROLE OF MOBILE INTELLIGENCE IN URBAN TRANSPORTATION

Mobile Intelligence and Data-Driven Decision Making

Mobile intelligence, powered by smartphones, IoT devices, and cloud-based analytics platforms, plays a crucial role in transforming urban transportation. By enabling real-time data collection and analysis, mobile intelligence allows city planners, transportation authorities, and private companies to make informed, data-driven decisions that improve traffic flow, safety, and efficiency (Munhoz et al., 2020).

1. **Data Collection and Traffic Control:**

Mobile intelligence systems continuously Collect data from multiple sources like GPS, mobile apps, sensors, and traffic cameras. Data is collected in real-time by the mobile intelligence systems throughout all the mobile network participants. The data is analysed after the fact to get a sense of how traffic patterns are moving over time, where congestion points might exist, or if an accident or road has been closed (Dabiri & Heaslip 2018). Using this information, traffic management systems can dynamically control traffic light timings and provide alternative routes that would reduce the time of delays in the networks. Mobile intelligence makes networks of transportation more responsive and adaptive to the conditions which characterise mobile systems leading directly to efficient mobility.

2. Predictive Analytics in Urban Mobility Planning

One of the most essential parts of mobile intelligence is that it can predict traffic based on past and current information to work efficiently for city planning. Models predict future congestion in real-time based on past traffic data, weather reports and social events, suggesting proactive measures to avoid such congestion (Chavhan & Venkataram, 2019). This could mean, for example, city workers changing the timing of traffic lights during busy times or sending more public buses to accommodate a big occasion. By taking this data-driven approach, urban mobility will be greatly enhanced as congestion can be prevented before it happens, and resources will be utilised more effectively.

3. Congestion Management:

When it comes to managing congestion, mobile intelligence provides direction and rerouting suggestions to drivers and fleet owners in real time. Navigation apps like Google Maps and Waze are able to recommend the fastest paths to dodge traffic (Souza et al., 2015) by analysing real-time traffic conditions. In high-capacity cities, mobile data is used to dynamically reorganise routes in order to better service peak demand while reducing the load on road networks and providing more reliable options for commuters (Çolak et al., 2016).

Impact on Public and Private Transportation Systems

Mobile intelligence significantly impacts both public and private transportation systems by improving efficiency, reducing costs, and enhancing the overall experience for commuters. Its integration into various platforms, from public buses to ride-sharing services, allows for smarter, more responsive mobility solutions.

1. Incorporating Mobile Intelligence Into Public Transit Systems:

Mobile intelligence is also helping public transportation systems, like buses and trains, route more efficiently, arrive on schedule, and provide better passenger experiences. GPS-tracked smart buses provide real-time information on location, delay of the bus etc. (Vakula & Raviteja, 2017). Similar data is accessible for passengers on their mobile apps, which could help them to pre-plan their journeys more effectively. Transport authorities are implementing mobile intelligence to have a real-time view of passenger flow and adjust services using AI-based demand prediction models as needed, reducing overcrowding and providing guaranteed seamless commuting (Xie, 2020).

Case in Point: In cities such as London and Singapore, smart buses, along with real-time public transportation apps, are able to deliver new schedules and routes immediately, which ensures Swiss-perfection timing in everyday commutes and perhaps a whole lot more.

2. Autonomous taxis and ride-sharing incentives:

Even ride-sharing platforms like Uber and Lyft are using mobile intelligence to match passengers and nearby drivers dynamically. These platforms are operated by companies that use smartphone data and GPS to better plan rides, reduce waiting times, and update prices according to the demand situation in real time (Lee et al. 2021). Meanwhile, autonomous taxis with mobile intelligence will eradicate the need for private cars, improving traffic flow and, in turn, easing parking requirements on city land.

Example: Phoenix, Arizona and Singapore, where trials are set for autonomous taxis, showed how the power of mobility intelligence can help guide self-driving cars through urban labyrinths, giving commuters easy access to a more efficient solution to traditional Taxis.

3. Benefits for Commuters:

Introducing mobile intelligence into urban mobility will deliver many benefits to the end-users, which enhance their overall journey experience:

Reduced Travel Time:

With mobile intelligence, routes can be adjusted in real-time based on the traffic conditions at that very moment, providing a break from stop-and-go frustration and resulting in faster arrivals. Predictive analytics help control city traffic lights or the route optimisation of ride-sharing platforms, which can reduce commute times (Yang et al., 2021).

Enhanced Safety:

Transportation systems can identify safety hazards such as road accidents or dangerous weather with real-time data sensing and processing. These data are transmitted to drivers and riders through mobile apps or navigation systems so that they can change their routes and driving behaviour in order to steer clear of accidents. We can also include mobile intelligence in autonomous vehicles and ride-sharing platforms for safe routing, decreasing human driving errors (Ioniţă, 2017).

Sustainability:

This enhances sustainability by minimising the waste of fuel and lowering emissions. Intelligent routing in public transit systems helps to reduce idle time and save fuel, and ride-sharing platforms encourage shared rides, decreasing the number of single-occupancy vehicles (Razakova et al., 2023). In the future, there is scope also to extend this environmental savings for urban transportation with autonomous electric vehicles which are integrated with mobile intelligence.

Conclusion

To summarise, mobile intelligence revolutionises urban transportation by collecting real-time data, predicting behaviours and improving how congestion is managed. These advantages directly benefit commuters either as part of public or private transportation systems (smart buses, autonomous taxis, ride-sharing applications), reducing travel times and increasing safety and sustainability. Mobile intelligence will play a crucial role in defining the future of intelligent, interconnected urban mobility as cities expand and the availability of efficient transportation solutions climbs.

VI. FUTURE PROSPECTS AND CHALLENGES

Advancements in Smart Mobility

Technologies will continue to be developed as intelligent mobility becomes a mainstay in the evolution of smart cities. These breakthroughs are designed to make urban transportation systems more efficient, safe, and less harmful to the climate—a more connected mobility experience, in essence.

1. AI-Powered Traffic Lights:

One of the most notable technological revolutions in traffic management involves the use of artificial intelligence, which allows traffic lights to adjust to actual traffic in real-time. For example, utilising AI capabilities to monitor vehicle and pedestrian flow and adjust signal timings accordingly on a fly basis will optimise traffic flow and reduce congestion (Gandhi et al., 2020). That infrastructure will also include bright traffic lights that give priority to emergency vehicles or public transportation, enabling faster and more efficient safety responses and the flow of all road users.

2. Traffic Monitoring using Drones:

Drones can observe traffic patterns, accidents, and road infrastructure via cameras and sensors (Khan et al., 2017). From this data, educated decisions are made regarding traffic routing and congestion control. They are also capable of monitoring construction zones and providing up-to-the-minute information to drivers about potential road closures or detour activities (Coifman et al., 2006).

3. Smart Parking Systems:

Parking is the most common cause of congestion and air pollution in cities. Smart parking systems integrate sensors and mobile apps to help drivers locate the nearest available parking spots in real-time. These systems cut the time drivers spend hunting for parking, which in turn reduces fuel use and emissions. In some cities, smart parking is combined with payment systems that provide the facility to book and pay for parking spots from a distance as an additional convenience (Sadhukhan 2017).

4. The Future of EVs: Electric Vehicles and Autonomous Transport

The future of intelligent mobility is being shaped around electric vehicles (EVs) and the use of mobility sharing as well as autonomous transport systems. In regards to emissions, EVs cause greenhouse gas reduction compared to traditional combustion-engine vehicles when they are powered using clean energy (Vaidya and Mouftah 2019). Before long, AI-driven offerings like self-driving cars and buses will provide a fantastic alternative to urban transportation. However, in upcoming smart cities, the combination of EVs and autonomous vehicles is considered to provide zero-emission and driverless transportation methods (Nikitas et al., 2017).

Policy, Regulation and Ethical Matters

Innovative mobility technologies are evolving rapidly, and they pose a number of policy, regulatory, and ethical questions that require exploring and answering if these systems are to be widely adopted.

1. **Data Privacy Issues:** Smart mobility relies heavily on mobile intelligence, telematics and autonomous systems, which in turn generate massive data, including sensitive private passenger-driver information. This raises a data privacy concern, as well as who and where this data is collected from/ stored/ used .There is a clear need for governments and transportation authorities to enforce stringent privacy protection regulations while still allowing data-driven mobility solutions (Mahmood et al., 2018).
2. **Regulatory Frameworks:** The rising rate of smart mobility technologies like self-driving cars and the fast-expanding AI-controlled traffic systems necessitates fresh regulatory frameworks. Regulators have to provide clear frameworks that specify the safety requirements, accident liability rules and protocols for sharing data. In this regard, the integration of electric vehicles in urban transportation will necessitate policy measures to promote EV charging infrastructure deployment and to drive demand for clean energy technologies (Brown et al., 2010)

3. **Public Acceptance of Smart Mobility Technologies** Innovative mobility technologies like autonomous vehicles and AI-driven systems have little hope of succeeding without public trust and compliance. There are doubts concerning safety, potential job loss (taxi drivers, etc.), and the level of reliability these technologies must reach in order to gain mass acceptance. Trust and adoption come hand in hand; however, for this trust to be accomplished, public awareness campaigns, transparent technology development, and government backing will be essential (Kaur & Rampersad, 2018).

Sustainable Urban Mobility and Smart City Technology for a Better Tomorrow

Smart mobility is also able to positively influence urban sustainability by decreasing carbon emissions and increasing green cities.

1. Reducing Carbon Emissions:

Smart Technologies and mobility, such as electric vehicles or AI traffic management and smart parking systems, optimise energy consumption and reduce carbon emissions and road congestion (Barth et al., 2015). As an example, they are a zero-tailpipe emission vehicle, and this means cleaner air in cities. Meanwhile, intelligent traffic lights and dynamic routing systems can reduce idling times and fuel consumption to decrease the average carbon footprint of corresponding transportation networks (Wang et al., 2020)

2. Creating Eco-Friendly Cities:

Mobility as a solution is, first of all, an instrument to stitch together intelligent cities. A city today, in order to be eco-friendly and focused on sustainability, must have an intelligent mobility system. In this capacity, by supporting the adoption of e-mobility within urban areas and engaging in clean energy infrastructure construction coupled with reducing dependency on fossil fuels, smart cities can foster more resilient low-carbon urban environments (Bibri & Krogstie, 2017). Smart mobility development programs such as bike-sharing, carpooling and eco-friendly urban planning strategies for pedestrians further reduce the air pollution level brought on by urban transportation (Li, 2016).

3. Circular economy

The creation of sustainable, intelligent mobility systems also abets the advent and growth of a circular economy where resources benefit from being used more efficiently while waste is minimised (Taleb & Farooque, 2020). For instance, the demand for new vehicle production and resource extraction can be less if fewer cars are required due to autonomous vehicles or hate-sharing platforms. Equally, intelligent logistics and fleet management systems optimise the use of resources, which offer a very greener perspective in relation to goods transportation and delivery services.

CONCLUSION

The scope for the future of smart mobility would be to assist in changing urban transportation structures through AI-informed traffic control, drone-based surveillance, and the incorporation of electric and autonomous vehicles. Although these technologies open up new possibilities, they also raise concerns about privacy and regulation as well as acceptance in the community. Smart mobility, with its focus on sustainability and eco-friendly solutions, can be a boon on two fronts — it helps build smarter cities that are green and efficient whilst giving due importance to keeping balance in the environment on one side and the well-being of humans on the other. That vision can be achieved but only with intelligent and sustainable planning, working together in a public-private partnership.

REFERENCES

Abboud, K., Omar, H. A., & Zhuang, W. (2016, July 14). Interworking of DSRC and Cellular Network Technologies for V2X Communications: A Survey. *IEEE Transactions on Vehicular Technology*, *65*(12), 9457–9470. DOI: 10.1109/TVT.2016.2591558

Amadeo, M., Campolo, C., Molinaro, A., Härri, J., Rothenberg, C. E., & Vinel, A. (2019, September 18). Enhancing the 3GPP V2X Architecture with Information-Centric Networking. *Future Internet*, *11*(9), 199–199. DOI: 10.3390/fi11090199

Barth, M., Scora, G., & Younglove, T. (2004, April 23). Intelligent off-board management of vehicle operating parameters. DOI: 10.1109/ITSC.2003.1251976

Barth, M., Wu, G., & Boriboonsomsin, K. (2015, July 25). Intelligent Transportation Systems and Greenhouse Gas Reductions. Springer Science+Business Media, 2(3), 90-97. DOI: 10.1007/s40518-015-0032-y

Bibri, S E., & Krogstie, J. (2017, March 6). Smart sustainable cities of the future: An extensive interdisciplinary literature review. Elsevier BV, 31, 183-212. DOI: 10.1016/j.scs.2017.02.016

Boyle, L N., Chin, H C., & Debnath, A K. (2013, March 1). Sustainable, safe, smart—three key elements of Singapore's evolving transport policies. Elsevier BV, 27, 20-31. DOI: 10.1016/j.tranpol.2012.11.017

Brown, S., Pyke, D F., & Steenhof, P A. (2010, March 17). Electric vehicles: The role and importance of standards in an emerging market. Elsevier BV, 38(7), 3797-3806. DOI: 10.1016/j.enpol.2010.02.059

Cassandras, C G. (2017, January 1). Automating mobility in smart cities. Elsevier BV, 44, 1-8. DOI: 10.1016/j.arcontrol.2017.10.001

Chavhan, S., & Venkataram, P. (2019, July 4). Prediction based traffic management in a metropolitan area. Elsevier BV, 7(4), 447-466. DOI: 10.1016/j.jtte.2018.05.003

Coifman, B., McCord, M R., Mishalani, R G., Iswalt, M., & Ji, Y. (2006, January 1). Roadway traffic monitoring from an unmanned aerial vehicle., 153(1), 11-11. DOI: 10.1049/ip-its:20055014

Çolak, S., Lima, A., & González, M. C. (2016, March 15). Understanding congested travel in urban areas. *Nature Communications*, *7*(1), 10793. Advance online publication. DOI: 10.1038/ncomms10793 PMID: 26978719

Conway-Beaulieu, J., Athaide, A., Jalali, R., & El-Khatib, K. (2015, November 2).. . *Smartphone-based Architecture for Smart Cities.*, *55*, 79–83. DOI: 10.1145/2815347.2826698

Dabiri, S., & Heaslip, K. (2018, January 1). Transport-domain applications of widely used data sources in the smart transportation: A survey. Cornell University. https://doi.org//arxiv.1803.10902DOI: 10.48550

Daddanala, R., Mannava, V., Tawlbeh, L., & Al-Ramahi, M. (2021, January 1). Vehicle to Vehicle (V2V) Communication Protocol: Components, Benefits, Challenges, Safety and Machine Learning Applications. Cornell University. https://doi.org//arxiv.2102.07306DOI: 10.48550

Dasanayaka, N., Hasan, K. F., Wang, C., & Feng, Y. (2020, January 1). Enhancing Vulnerable Road User Safety: A Survey of Existing Practices and Consideration for Using Mobile Devices for V2X Connections. Cornell University. https://doi.org/DOI: 10.48550/arXiv.2010

Gandhi, M., Solanki, D S., Daptardar, R S., & Baloorkar, N S. (2020, December 1). Smart Control of Traffic Light Using Artificial Intelligence. DOI: 10.1109/ICRAIE51050.2020.9358334

Ghiasi, A., Hale, D A., Bared, J., Kondyli, A., & Ma, J. (2018, November 1). A Dynamic Signal Control Approach for Integrated Ramp and Mainline Metering. DOI: 10.1109/ITSC.2018.8569567

Goh, M. (2002, March 1). Congestion management and electronic road pricing in Singapore. Elsevier BV, 10(1), 29-38. DOI: 10.1016/S0966-6923(01)00036-9

Gopal, K., & Prakash, R. (2012, September 28). Building the Business Case for Telematics Based Diagnostics at Mahindra Reva. *Multidisciplinary Digital Publishing Institute*, *5*(3), 722–729. DOI: 10.3390/wevj5030722

Guidoni, D. L., Maia, G., Souza, F. S. H. D., Villas, L. A., & Loureiro, A A F. (2020, January 1). Vehicular Traffic Management Based on Traffic Engineering for Vehicular Ad Hoc Networks. *IEEE Access : Practical Innovations, Open Solutions*, *8*, 45167–45183. DOI: 10.1109/ACCESS.2020.2978700

Haque, M A., & Hossain, M D. (2014, December 1). Technology survey of wireless communication for in-vehicle applications. DOI: 10.1109/SKIMA.2014.7083523

Horrey, W J., Lesch, M F., Dainoff, M J., Robertson, M M., & Noy, Y I. (2011, December 24). On-Board Safety Monitoring Systems for Driving: Review, Knowledge Gaps, and Framework. Elsevier BV, 43(1), 49-58. DOI: 10.1016/j.jsr.2011.11.004

Huang, A., Bao, J., Zhang, J., & Zhang, T. (2012, August 1). Vehicle information monitoring system based on mobile phone. DOI: 10.1109/ICAL.2012.6308216

Ioniţă, S. (2017, October 1). Autonomous vehicles: From paradigms to technology. *IOP Conference Series. Materials Science and Engineering*, *252*, 012098–012098. DOI: 10.1088/1757-899X/252/1/012098

Jadaan, K., Zeater, S., & Abukhalil, Y. (2017, January 1). Connected Vehicles: An Innovative Transport Technology. Elsevier BV, 187, 641-648. DOI: 10.1016/j.proeng.2017.04.425

Jahn, O., Möhring, R. H., Schulz, A. S., & Stier-Moses, N. E. (2005, August 1). System-Optimal Routing of Traffic Flows with User Constraints in Networks with Congestion. *Operations Research*, *53*(4), 600–616. DOI: 10.1287/opre.1040.0197

Jambotkar, S., Guo, L., & Jia, Y. (2021, September 27). Adaptive Optimization of Autonomous Vehicle Computational Resources for Performance and Energy Improvement. DOI: 10.1109/IROS51168.2021.9635828

Kaur, K., & Rampersad, G. (2018, April 1). Trust in driverless cars: Investigating key factors influencing the adoption of driverless cars. Elsevier BV, 48, 87-96. DOI: 10.1016/j.jengtecman.2018.04.006

Kavas-Torris, O., Cantas, M. R., Gelbal, Ş. Y., & Güvenç, L. (2023, January 1). Hardware-in-the-Loop and Road Testing of RLVW and GLOSA Connected Vehicle Applications. Cornell University. https://doi.org/DOI: 10.48550/arXiv.2306

Khalil, O K. (2017, June 1). A Study on Road Accidents in Abu Dhabi Implementing a Vehicle Telematics System to Reduce Cost, Risk and Improve Safety. DOI: 10.1109/DeSE.2017.41

Khan, M. A., Ectors, W., Bellemans, T., Janssens, D., & Wets, G. (2017). UAV-based traffic analysis: A universal guiding framework based on literature survey. *Transportation Research Procedia*, *22*, 541–550.

Kuchta, R. (2014, January 1). Smart City Concept, Applications and Services. *OMICS Publishing Group*, *03*(02). Advance online publication. DOI: 10.4172/2167-0919.1000117

Lazarus, J., Ugirumurera, J., Hinardi, S., Zhao, M., Shyu, F., Wang, Y., Yao, S., & Bayen, A M. (2018, November 1). A Decision Support System for Evaluating the Impacts of Routing Applications on Urban Mobility. DOI: 10.1109/ITSC.2018.8569622

Lee, E., Cen, X., Lo, H. K., & Ng, K. F. (2021, January 1). Designing zonal-based flexible bus services under stochastic demand. Cornell University. https://doi.org/ DOI: 10.48550/arXiv.2102

Li, H. (2016, January 1). Study on Green Transportation System of International Metropolises. Elsevier BV, 137, 762-771. DOI: 10.1016/j.proeng.2016.01.314

Li, X., Lin, K., Meng, M., Li, X., Li, L., & Hong, Y. (2021, January 1). Composition and Application of Current Advanced Driving Assistance System: A Review. Cornell University. https://doi.org//arxiv.2105.12348DOI: 10.48550

Mahmood, A., Zen, H., & Hilles, S M S. (2018, January 1). Big Data and Privacy Issues for Connected Vehicles in Intelligent Transportation Systems. Springer Nature, 1-7. DOI: 10.1007/978-3-319-63962-8_234-1

Mahrez, Z., Sabir, E., Badidi, E., Saad, W., & Sadik, M. (2021, June 11). Smart Urban Mobility: When Mobility Systems Meet Smart Data. *IEEE Transactions on Intelligent Transportation Systems*, *23*(7), 6222–6239. DOI: 10.1109/TITS.2021.3084907

Malik, F., Shah, M A., & Biomechanics, A B A. (2018, September 1). Intelligent Transport System: An Important Aspect of Emergency Management in Smart Cities. DOI: 10.23919/IConAC.2018.8749062

Mavromatis, I., Tassi, A., Piechocki, R J., & Sooriyabandara, M. (2020, May 1). On Urban Traffic Flow Benefits of Connected and Automated Vehicles. DOI: 10.1109/VTC2020-Spring48590.2020.9128758

Meneguette, R I., Grande, R E D., & Loureiro, A A F. (2018, January 1). Intelligent Transport System in Smart Cities. Springer International Publishing. DOI: 10.1007/978-3-319-93332-0

Meng, L., Luo, J., & Wei, X. (2013, June 11). A Telematics Service System for Vehicle Safety Based on Vehicular Data Acquisition Module. DOI: 10.1061/9780784413036.233

Mittal, G., Sharma, M., Khatri, S K., & Johri, P. (2019, February 1). Optimizing Autonomous Vehicles to Prevent On-Road Congestion. DOI: 10.1109/AICAI.2019.8701385

Munhoz, P A M. S. A., Dias, F. D. C., Chinelli, C. K., Guedes, A. L. A., Santos, J A N. D., Silva, W. D. S. E., & Soares, C A P. (2020, December 21). Smart Mobility: The Main Drivers for Increasing the Intelligence of Urban Mobility. *Sustainability (Basel)*, *12*(24), 10675–10675. DOI: 10.3390/su122410675

Musa, S. (2016, January 1). Smart Cities - A Roadmap for Development. *OMICS Publishing Group, 05*(03). Advance online publication. DOI: 10.4172/2167-0919.1000144

Nguyen, V., Hwang, R., Lin, P., Vyas, A., & Nguyen, V. (2022, September 12). Toward the Age of Intelligent Vehicular Networks for Connected and Autonomous Vehicles in 6G. *IEEE Network, 37*(3), 44–51. DOI: 10.1109/MNET.010.2100509

Nikitas, A., Kougias, I., Alyavina, E., & Tchouamou, E. N. (2017, November 30). How Can Autonomous and Connected Vehicles, Electromobility, BRT, Hyperloop, Shared Use Mobility and Mobility-As-A-Service Shape Transport Futures for the Context of Smart Cities? *Urban Science (Basel, Switzerland), 1*(4), 36–36. DOI: 10.3390/urbansci1040036

Noor-A-Rahim, M., Liu, Z., Lee, H., Khyam, M. O., He, J., Pesch, D., Moessner, K., Saad, W., & Poor, H. V. (2022, May 20). 6G for Vehicle-to-Everything (V2X) Communications: Enabling Technologies, Challenges, and Opportunities. *Proceedings of the IEEE, 110*(6), 712–734. DOI: 10.1109/JPROC.2022.3173031

Ogunleye, O S., & Belle, J P V. (2014, May 1). Exploring the success, failure and factors influencing m-Government implementation in developing countries. DOI: 10.1109/ISTAFRICA.2014.6880656

Okumura, B., James, M., Kanzawa, Y., Derry, M., Sakai, K., Nishi, T., & Prokhorov, D. (2016, March 1). Challenges in Perception and Decision Making for Intelligent Automotive Vehicles: A Case Study. *IEEE Transactions on Intelligent Vehicles, 1*(1), 20–32. DOI: 10.1109/TIV.2016.2551545

Prevost, S., & Kettani, H. (2019, October 23). On Data Privacy in Modern Personal Vehicles. DOI: 10.1145/3372938.3372940

R, S S., & Rajendran, L. (2021, January 27). Real-Time Adaptive Traffic Control System For Smart Cities. DOI: 10.1109/ICCCI50826.2021.9402597

Razakova, D., Shmatko, L., Shatila, K., & Davtyan, A. (2023, January 1). Analysis for the sustainability of intelligent transport systems in the United States. *EDP Sciences, 389*, 05002–05002. DOI: 10.1051/e3sconf/202338905002

Rogers, D., Wallace, S., Emrich, J., Chaput, S., & Noens, R. (2000, March 6). Evolving Telematics Systems and Standards. DOI: 10.4271/2000-01-0814

Sadhukhan, P. (2017, September 1). An IoT-based E-parking system for smart cities. https://doi.org/DOI: 10.1109/ICACCI.2017.8125982

Said, H., Nicoletti, T., & Perez-Hernandez, P. (2014, November 18). Utilizing Telematics Data to Support Effective Equipment Fleet-Management Decisions: Utilization Rate and Hazard Functions. American Society of Civil Engineers, 30(1). DOI: 10.1061/(ASCE)CP.1943-5487.0000444

Sakaguchi, K., Fukatsu, R., Yu, T., Fukuda, E., Mahler, K., Heath, R. W., Fujii, T., Takahashi, K., Khoryaev, A., Nagata, S., & Shimizu, T. (2020, January 1). Towards mmWave V2X in 5G and Beyond to Support Automated Driving. Cornell University. https://doi.org/DOI: 10.48550/arXiv.2011

Savić, V., Schiller, E. M., & Papatriantafilou, M. (2017, January 1). Aiding Autonomous Vehicles with Fault-tolerant V2V Communication. Cornell University. https://doi.org//arxiv.1710.00692DOI: 10.48550

Schutera, M., Goby, N., Smolarek, S., & Reischl, M. (2018, January 1). Distributed traffic light control at uncoupled intersections with real-world topology by deep reinforcement learning. Cornell University. https://doi.org//arxiv.1811.11233DOI: 10.48550

Sewalkar, P., & Seitz, J. (2019, January 17). Vehicle-to-Pedestrian Communication for Vulnerable Road Users: Survey, Design Considerations, and Challenges. *Sensors (Basel), 19*(2), 358–358. DOI: 10.3390/s19020358 PMID: 30658392

Singh, B., & Gupta, A. (2015, April 1). Recent trends in intelligent transportation systems: a review. SciELO, 9(2), 30-34. https://doi.org/DOI: 10.1590/2238-1031.jtl.v9n2a6

Singh, M., & Kim, S. (2017, November 1). Security analysis of intelligent vehicles: Challenges and scope., 8, 13-14. DOI: 10.1109/ISOCC.2017.8368805

Soleymanian, M., Weinberg, C. B., & Zhu, T. (2019, January 1). Sensor Data and Behavioral Tracking: Does Usage-Based Auto Insurance Benefit Drivers? *Marketing Science, 38*(1), 21–43. DOI: 10.1287/mksc.2018.1126

Souza, A. M. D., Brennand, C A R. L., Yokoyama, R. S., Donato, E., Madeira, E. R. M., & Villas, L. A. (2017, April 1). Traffic management systems: A classification, review, challenges, and future perspectives. *International Journal of Distributed Sensor Networks, 13*(4), 155014771668361–155014771668361. DOI: 10.1177/1550147716683612

Souza, A M D., Yokoyama, R S., Botega, L C., Meneguette, R I., & Villas, L A. (2015, October 1). Scorpion: A Solution Using Cooperative Rerouting to Prevent Congestion and Improve Traffic Condition. DOI: 10.1109/CIT/IUCC/DASC/PICOM.2015.71

Taleb, M A., & Farooque, O A. (2020, September 6). Towards a circular economy for sustainable development: An application of full cost accounting to municipal waste recyclables. Elsevier BV, 280, 124047-124047. DOI: 10.1016/j.jclepro.2020.124047

Tan, X., & Leon-Garcia, A. (2018, October 1). Autonomous Mobility and Energy Service Management in Future Smart Cities: An Overview. DOI: 10.1109/UV.2018.8642141

Vaidya, B., & Mouftah, H T. (2019, March 28). Connected Autonomous Electric Vehicles as Enablers for Low-Carbon Future. IntechOpen. DOI: 10.5772/intechopen.84287

Vakula, D., & Raviteja, B. (2017, December 1). Smart public transport for smart cities. https://doi.org/DOI: 10.1109/ISS1.2017.8389288

Wang, J., Shao, Y., Ge, Y., & Yu, R. (2019, January 15). A Survey of Vehicle to Everything (V2X) Testing. *Sensors (Basel)*, *19*(2), 334–334. DOI: 10.3390/s19020334 PMID: 30650658

Wang, Q., Yuan, Y., Yang, X., & Huang, Z. (2021, January 20). Adaptive and multipath progression signal control under connected vehicle environment. Elsevier BV, 124, 102965-102965. DOI: 10.1016/j.trc.2021.102965

Wang, T., Yang, Y., Tang, T., & Qu, X. (2020, January 1). An unnoticed side effect of electric vehicles. Cornell University. https://doi.org//arxiv.2009.10543DOI: 10.48550

Xie, Z. (2020, February 1). Research on Bus Passenger Flow Analysis Method Based on Intelligent Transportation. *IOP Conference Series. Earth and Environmental Science*, *440*(4), 042005–042005. DOI: 10.1088/1755-1315/440/4/042005

Yang, F., Vereshchaka, A., Lepri, B., & Dong, W. (2021, August 24). Optimizing City-Scale Traffic Through Modeling Observations of Vehicle Movements. *IEEE Transactions on Intelligent Transportation Systems*, *23*(8), 10492–10503. DOI: 10.1109/TITS.2021.3094758

Zhang, J., & Letaief, K. B. (2019, October 28). Mobile Edge Intelligence and Computing for the Internet of Vehicles. *Proceedings of the IEEE*, *108*(2), 246–261. DOI: 10.1109/JPROC.2019.2947490

Zhou, T., Ma, S., Yu, D., Li, M., & Hang, T. (2020, July 22). Development of Reliable, High Performance WLCSP for BSI CMOS Image Sensor for Automotive Application. *Sensors (Basel)*, *20*(15), 4077–4077. DOI: 10.3390/s20154077 PMID: 32707858

Chapter 14
Future of Mobility:
Potential Challenges and Ethical Considerations

Sarvesh Sawant
https://orcid.org/0000-0002-6183-4653
National Institute of Technology, Karnataka, India

Bhawana Rudra
National Institute of Technology, Karnataka, India

ABSTRACT

Quantum Computers with its extraordinary computing power, have marked the beginning of a new era of computers. Their presence is not just revolutionising the scientific community, but also transforming the world. However, such a power in the hands of an adversary can be fatal. They will empower adversaries to have unauthorized access to user's data. Technologies based on lightweight cryptographic methods such as Vehicular Communication, will be adversely affected. Thus, there is a need to develop mechanisms that are lightweight and at the same time uncompromisable by Quantum-Powered Adversaries (QPA) before powerful Quantum Computers are developed. In this chapter, we first focus on the current scenario of secure communication and how QPA can exploit it. Later, we focus on how Quantum-Safe Cryptography and Artificial Intelligence can contribute in the development of defensive mechanisms against QPA. Finally, we conclude the chapter by highlighting the challenges in making the solutions feasible for adoption.

DOI: 10.4018/979-8-3693-6422-2.ch014

1. INTRODUCTION

Vehicular Communication have transformed transportation. Particularly, they have enhanced security, reliability and traceability of vehicles. This Vehicular Network when was further connected with Internet, a plethora of possibilities opened in Vehicular Communication. These opportunities revolutionized Vehicular Communication bringing them into the umbrella of Internet-of-Vehicles (IoV). IoV has connected Vehicular Communication with cutting-edge technologies such as Cloud Computing, Edge Computing, Artificial Intelligence to name a few (Partovi et al. 2023).

Figure 1 depicts the architecture of IoV (El-Rewini et al. 2020). Here, we can observe the multi-layered architecture which consists of the perception layer, network layer, and application layer. The perception layer includes sensors and devices embedded in vehicles and infrastructure to collect real-time data such as on road conditions, vehicle health, environmental factors etc. Technologies like LIDAR, radar, cameras, and GPS are integral to this layer. The network layer ensures seamless data transmission between vehicles and infrastructure using advanced communication protocols such as Dedicated Short-Range Communication (DSRC), Cellular V2X (C-V2X), and 5G. Cloud and edge computing systems are employed to manage and process the vast volumes of data that is generated. Finally, in the application layer critical processes such as traffic optimization, predictive maintenance, autonomous driving support, infotainment systems etc. (Taslimasa et al. 2023) are addressed. However, to improve the efficiency of these layers, effective management of Data Communication is of paramount importance (Bucaioni et al. 2020). Figure 2 depicts the communication framework that is used in an IoV Network. Here, we can observe the four types of communication namely, Trusted Authority to Infrastructure, Vehicle to Trusted Authority, Vehicle to Vehicle (V2V) and Vehicle to Infrastructure (V2I). When a vehicle is travelling in an IoV network, it can communicate with each other (V2V) through Dedicated Short-Range Communication (DSRC) such as Bluetooth, Lidar etc. (Azam et al. 2021). It can also communicate to the Road-Side Units (V2R) which are deployed besides the road. They provide the vehicle with necessary information using DSRC. Moreover, these RSU's also receive data through Internet from other infrastructural units such as Trusted Authority, other RSUs, etc. The vehicle can also receive

Figure 1. Architecture of Internet-of-Vehicles (El-Rewini et al. 2020).

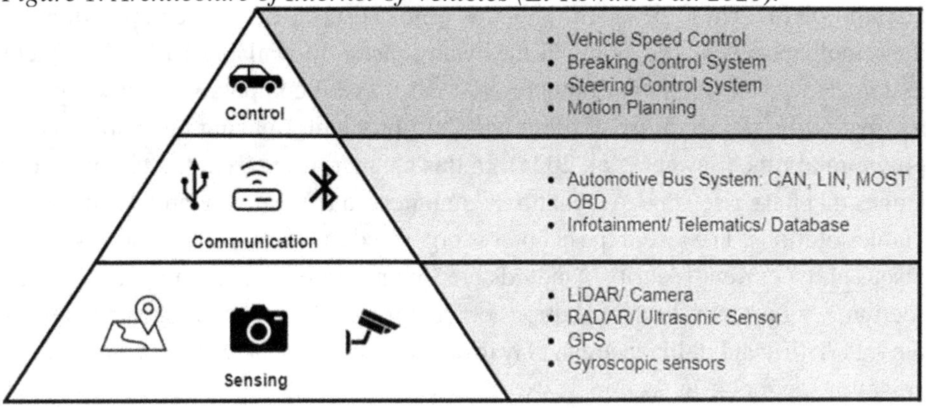

data from other nearby devices such as sensors, and pedestrians through Internet (V2X). However, this connectivity with the Internet has also brought the Vehicular Networks in closer proximity with the adversaries. By exploiting the technological loopholes of IoV, the adversaries have performed attacks such as Impersonation Attack, Modification Attack, Replay Attack, DDoS Attack etc. (Alawany et al. 2024). Thus, to protect the IoV network, researchers

Figure 2. Communication Framework of Internet-of-Vehicles (Sawant et al., 2023).

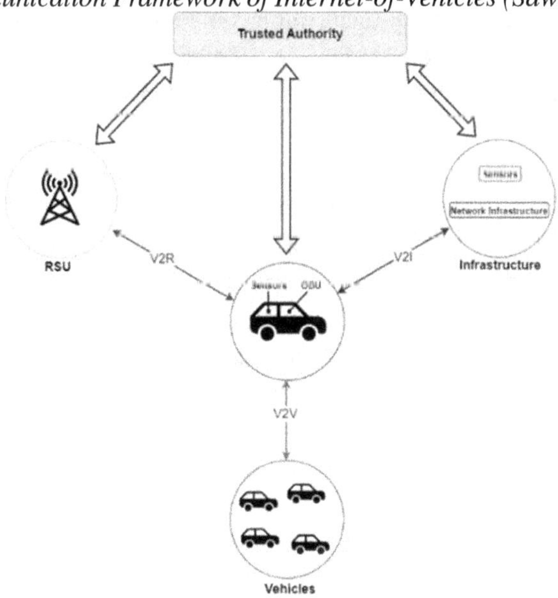

developed various schemes (Kumar et al. 2024) which were mainly based on NP-Hard problems. These NP-Hard problems were chosen as they are computationally intractable to solve in polynomial time by any currently available Turing Complete system. However, Quantum Computers with its massive computational power poses a threat to the intractability of these NP-Hard problems, making them solvable in polynomial time (Sawant et al. 2023). In this chapter, we present the various challenges that have risen due to Quantum Computers to IoV network and discuss their viable solutions. The rest of the chapter is organised as follows, in section 2, we will discuss how current security standards are threatened by Quantum Computers. In section 3, we present the contributions of Quantum-Safe Cryptography and present the role Artificial Intelligence can play in developing efficient Quantum-Safe mechanism in section 4. In section 5, we present the challenges involved in deploying such solutions at a global stage. Finally, in section 6, we conclude our chapter by presenting our conclusion.

2. THE SCARE OF QUANTUM COMPUTERS

2.1 Current Security Standards

Data Communication across the Internet is carried out using ciphertext. These ciphertexts are encrypted data that is derived using encryption keys over the message that needs to be communicated. Encryption is broadly of two types, the symmetric encryption and asymmetric encryption. In the symmetric encryption, the same key is used to decrypt and encrypt the data. In asymmetric encryption, one publicly available key (public key) is used to encrypt the data and a private key is used to retrieve data from the generated ciphertext (Rana et al. 2023).

Symmetric Encryption such as AES are used for data encryption over the Internet. Here, the encrypted text is generated by first dividing the plaintext (message) into blocks of fixed sizes and then performing multiple "rounds" over them. In each round, the plaintext (message) is subjected to a series of transformations such as substitution, permutation and mixing. In order to decrypt data from this encrypted text, reverse of the encryption process is performed (Hamza et al. 2020). The AES encryption scheme is considered to be one of the most secure encryption schemes upon which no attack can be performed (Aouedi et al., 2022). In IoV, this AES can be used for secure vehicle-to-vehicle (V2V) communication and also secure vehicle-to-infrastructure (V2I) communication (Saleem et al., 2024). Since, vehicles exchange real-time critical information such as speed, location etc., these symmetric schemes turn out to be extremely useful in such time-critical applications due to their lightweight nature (Mustafa et al., 2024). However, these schemes also face challenges in the IoV environment such as key management. Specifically, secure distribution of the key among the vast IoV network of devices is a difficult task. Compromising a single key could expose the entire communication system to potential attacks. To address this limitation, IoV systems often integrate symmetric cryptography with asymmetric cryptography, leveraging the strengths of both approaches (Sawant et al., 2023).

Asymmetric Encryption such as Public Key Infrastructure, uses NP-Hard problems to generate public – private key pairs. These key pair ensures that even if the public key is shared widely, only the intended recipient, who holds the corresponding private key, can decrypt the message (Tang et al., 2024). For example, RSA algorithm uses Integer Factorization Problem as its Hardness problem. Integer Factorization Problem states that given a large composite number, finding its prime factors are not feasible in Polynomial Time using the existing Turing Complete systems. Thus, encryption developed using RSA algorithm will enable anyone to encrypt a message using the public key. However, decryption will only be possible by the system that holds the private key with it (Zhou et al., 2011).

Due to the use of hardness problems, asymmetric cryptography is computationally more intensive than symmetric cryptography, however, it offers robust mechanisms, making it indispensable in IoV (Karim et al., 2022). This can be seen in the authentication and key sharing process of asymmetric cryptography. As we know, it is very important that vehicles are properly authenticated before any communication happens. With the help of digital certificates, this identification of vehicles can be efficiently performed (Bojjagani et al. 2022). Asymmetric Cryptography also plays a crucial role in key exchange protocols. They facilitate secure sharing of symmetric keys over the unsecure channel using the protocols such as Diffie-Hellman and Elliptic Curve Diffie-Hellman (ECDH) (Nguyen et al., 2016). Thus, we can observe that digital signatures ensure integrity and non repudiation of messages while key exchange ensures safe sharing of symmetric key for secure communication (Guo et al., 2022). This highlights the crucial role asymmetric cryptography can play in an IoV network. Thus, both the schemes have their own significance in the Data Communication over the Internet. The asymmetric encryption ensures secure symmetric key exchange and once the key is exchanged, the system switches to symmetric encryption for subsequent data transmission. This approach minimizes computational overhead while maintaining security.

However, as highlighted by (Sawant et al. 2023), a Quantum Powered Adversary (QPA) can breach the asymmetric encryption in polynomial time. Since sharing of symmetric key also uses asymmetric key encryption mechanisms, the QPA can easily receive the symmetric key that is being shared. Thus, in the next section we will analyse the working of Quantum Computers and understand how it can break the currently used standard NP-Hard problems.

2.2 The Leverage of Quantum Computers

The smallest unit in a Quantum Computer is the qubit and they work on the principles of Superposition and Entanglement which enhances the reach of Quantum Computers. The possibility of qubits to exists in both 0 and 1 state simultaneously unlike classical bits is called Superposition. This allows a single qubit to explore multiple possibilities at once. The property of entanglement states that the change in the state of one qubit instantly affects the change in states of other qubits regardless of their physical separation. Due to this high correlation, Quantum Computers possess faster information processing capabilities. Thus, the probabilistic nature of quantum computation allows for the exploration of multiple possibilities simultaneously, leading to potential speedups for certain types of problems. With these properties together, multiple scenarios of a given NP Complete problem instance can be solved by powerful Quantum Computers easily (Yang et al., 2023). For example, Google demonstrated the Quantum Supremacy by performing a "certain computation" in 200 seconds which normally would take 10,000 years on the currently available supercomputers (Rinott et al,, 2022). In Figure 2, we present how Shor's Algorithm can be used by Quantum Computers to crack RSA algorithm by solving the Integer Factorization Problem. Here, it can be observed that first a composite number is given as an input to the algorithm. Then, using the Quantum Subroutine call, Shor's algorithm is connected to quantum computing environment to compute the prime numbers (Ugwuishiwu et al., 2020).

Figure 3. Quantum variant of Shor's Algorithm (Ugwuishiwu et al., 2020).

Algorithm 1 Shor's Algorithm

```
Input: N ← Composite Number.
Output: (u, v) ← Prime numbers such that N = u × v.
while True do
    Choose randomly x ∈ {2, N − 1}
    if d = gcd(x, N) ≥ 2 then
        return(u = d, v = N/u)
    end if
    Find order r such that x^r ≡ 1 mod N using Quantum Subroutine.    ▷ Order Finding using phase estimation.
    if r is even && d = gcd(x^(r/2) − 1, N) ≥ 2 then
        return(u = d, v = N/u)
    end if
end while
```

There is another powerful algorithm that poses threat to symmetric key encryption mechanism known as Grover's algorithm (Grassl et al., 2016). It can significantly speed up the process of searching an unsorted database compared to classical computers. With the properties of qubits to function in superposition and entanglement, multiple possibility of solutions gets explored simultaneously. Moreover, with every iteration of Grover's algorithm, a significant amplification of the possibility of finding correct solution is observed. This makes the bruteforce exploitation of symmetric key possible by QPA.

However, the technology is still under development. This is so because, quantum computers face a lot of internal challenges. The first challenge is that they are built using entanglement and superposition. This architecture, though gives multiple outputs at a given time, they are not discrete like the classical computers. They rely on the probability of the expectation of results to be accurate. This inherently leads to lot of errors in finding the correct answer. The second challenge is the task of data communication. Quantum qubits are highly sensitive to external disturbances leading to errors in quantum computations. As a result, many researchers are focussed in developing efficient error correction techniques as well. Finally, the biggest challenge for Quantum Computers is the existential one (Almudever et al., 2017). Quantum Computers must be maintained in a very secure and extremely cool environment so that the qubits function as desired. However, the procedures involved are cost-intensive. Therefore, substantial investments and research are being channelled towards developing lightweight, stable, and efficient quantum computers. Thus, it is only a matter of time for powerful quantum computers to see the light of day making it all the more important to develop quantum-safe mechanisms before the arrival of doomsday.

2.3 Measures Against the Quantum Powered Adversaries

2.3.1 Increasing key size

In order to counter the attacks of Quantum Powered Adversaries, the size of the keys can be increased. Thus, key size of symmetric key encryption and digital signatures can be increased to make the ciphertext secure against these adversaries. However, large key sizes are infeasible to use in time-critical applications like IoV networks (Sawant et al., 2023).

2.3.2 Hybridization of Existing Schemes

Existing asymmetric schemes are not secure under a Quantum Powered Adversary. Also, new schemes that are developed has large key sizes that find themselves unusable for the existing low energy networks and devices rendering them incompatible. Thus, to resolve these issues, researchers started the process of hybridization of the network. One of the best illustrations is the TLS handshake protocol. Below figures depict the classical and hybrid TLS handshakes.

Figure 4. TLS handshake protocol (Schwabe et al., 2020).

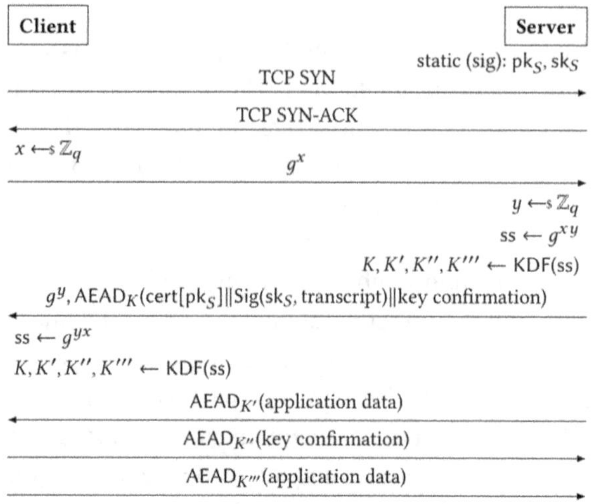

Figure 3 depicts the classical TLS handshake. Here, we can observe that the process begins with a TCP handshake to establish a connection (TCP SYN, TCP SYN-ACK). Subsequently, a Diffie-Hellman key exchange is initiated, where both the client and server generate random values and exchange them to derive a shared secret key (ss). This shared secret key is then used to derive multiple cryptographic keys for encryption and authentication purposes (K, K′, K′′, K′′′). The server authenticates itself by sending a signed message containing its certificate and a key confirmation message(cert[pk], Sig(sk, transcript), key confirmation). The client verifies the signature and the key confirmation to ensure the authenticity of the server. Once authentication is established, both parties can exchange application data securely using the derived encryption keys and an authenticated encryption with associated data (AEAD) scheme.

Figure 5. Hybrid - TLS handshake protocol (Schwabe et al., 2020).

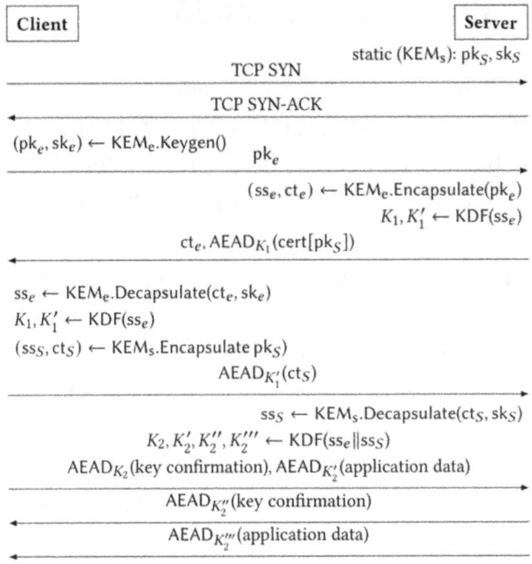

Figure 4 depicts the hybrid TLS handshake. Here, we can observe that the protocol begins with a traditional TCP handshake, followed by a hybrid key exchange phase. The public key (pk) from the public-private key pair that is derived by the client through the derivation function (KEM.KeyGen()) is shared to the server. The server first encapsulates public key (pk) using the KEM.Encapsulate(pk) function and derives the shared secret (ss) and ciphertext (ct). Later, it uses the Key Derivation Function (KDF) over (ss) to produce keys K1 and K1′. The server then sends (ct) and AEAD_K1(cert[pk_s]) to authenticate its public key pk_s. The client decapsulates (ct) using KEM.Decapsulate(ct,sk) to retrieve (ss) and then deriving K1 and K1′. Later, the client performs encapsulation with KEM_s.Encapsulate(pk_s) and generates (sss) and ciphertext (cts_s). This encapsulation ciphertext is sent back to server (AEAD_K′ (ct_s)). The server decapsulates ct_s (KEM_s.Decapsulate(ct_s,sk_s)) and recover (sss) and keys K2,K2′,K2″,K2‴. The difference between the two types of TLS discussed is the use of additional key structure called the Key Encapsulation Mechanism (KEM) in order to initiate communication. However, these schemes have higher computational cost and is challenging to be implemented in an IoV network.

Thus, to make IoV network secure against QPA, there is a need to develop lightweight Quantum-Safe mechanisms. In the next section, we will discuss the Quantum-safe cryptography also known as Post Quantum Cryptography (PQC) and the significant role it can play in the defending against a Quantum Powered Adversary.

3. POST QUANTUM CRYPTOGRAPHY AGAINST QUANTUM POWERED ADVERSARIES

Post-Quantum Cryptography (PQC) or Quantum-Safe Cryptography refers to public-key encryption schemes designed to resist attacks from quantum computers, leveraging mathematical problems that are intractable for Shor's algorithm and its generalizations. There are broadly five categories of PQC systems (Hasija et al., 2022).

3.1 Super singular elliptic curve isogeny crypto systems

These systems are developed over the features of super-singular isogeny graphs and super-singular elliptic curves. Super-singular elliptic curves follow the properties of abelian group over the endomorphism ring, that is a ring that maps from the curve to itself. This endomorphism ring is larger and more structured than that of ordinary elliptic curves, leading to more complex mathematical structures. When used in combination with super-singular isogeny graphs they formulate a strong foundation for the development of PQC schemes. A super-singular isogeny graph is a graph where vertices represent super-singular elliptic curves over a finite field, and edges represent isogenies between these curves (Roy et al., 2019). One of the most widely referred example is the *Supersingular Isogeny Diffie-Hellman (SIDH)* key exchange protocol (Jalali et al., 2017). Here, we see that two parties (Alice and Bob), exchange secret information by executing a random walk over a super-singular isogeny graph. Finally, the shared secret is derived from the final point reached by both the parties.

3.2 Hash-based Cryptography

The Hash-based cryptography finds its application in developing secure digital signatures for PQC (Zeydan et al., 2022). One of the highly regarded schemes in this category is the Lamport's one-time signature scheme (Abdullah et al., 2018). Here, they use multiple secret and public key pairs with a constraint that a single secret-public key pair can be used only once. Moreover, to increase the efficiency and reduce the size of public key, researchers considered to use the Merkle tree to store the keys. This involves arranging multiple Lamport key pairs in a tree-like structure, where each node represents the hash of its child nodes. By unravelling the secret keys over a defined path from the leaf node to the root node, we can efficiently perform signature verification. The hardness of the scheme is reverse hashing. Which means that, an adversary cannot get any information from the hash of a value and this is considered to be resistant to quantum attacks. However, we observe rapid growth in the public key size which is a cause of major concern in these schemes.

3.3 Multivariate Cryptography

Multivariate cryptography is a public-key cryptography that relies on the difficulty of solving systems of multivariate polynomial equations over finite fields (Roy et al., 2019). They create simple mathematical puzzles that can be solved easily with secret key but hard to solve without it. One of the prominent examples is the HFE (Dey et al., 2023). Here, the core idea is to create a complex mathematical puzzle that's easy to solve for the person who knows the secret solution but incredibly difficult for anyone else. The public key consists of a set of multivariate polynomial equations, while the private key is a secret polynomial. To sign a message, the signer uses their secret polynomial to solve a system of equations. The resulting solution is the signature. The security of HFE relies on the assumption that solving these multivariate equations is computationally infeasible, even for quantum computers. However, the major drawback is the size of the public key which HFE algorithm has attempted to address.

3.4 Post-quantum code-based cryptography:

In code-based cryptography, a message is encrypted by adding errors to a codeword (Hasija et al., 2022). The public key is a matrix that defines the code, while the private key is a decoding algorithm that can efficiently correct errors. To decrypt a message, the receiver uses the private key to decode the received codeword and recover the original message. The security of code-based cryptography relies on the difficulty of decoding a random linear code. One of the prominent examples is public-key cryptography introduced by McEliece where the code is secretly generated as "Goppa code" (Janwa et al. 1996). McEliece scheme with time, has undergone many adaptations such as that of Niederreiter (Wang et al., 2018). Such transformations have made the scheme secure against chosen-ciphertext attacks, relevant even against a Quantum Powered Adversary.

3.5 Lattice based Cryptography:

Public-key Cryptography based on mathematical lattice structures formed by repeating patterns of points in space fall under this category (Nejatollahi et al., 2019). Here, the security relies on the hardness of solving intractable lattice-base problems such as finding shortest vector or learning with errors problem. With these fundamental problems as base, different Public-Key Encryption schemes are developed that are reduceable to these hard problems. Lattice-based cryptography presents numerous advantages such as, efficiency, flexibility, and strong security. The 'NTRU' algorithm is one of the most discussed algorithms in this category

(Hoffstein et al. 2003). Here, they generate a pair of polynomials, one public and one private. To perform encryption, the message is multiplied with the public key and then reduced to modulo of a large prime number. For decryption, the private key is used to recover the original message by solving a series of polynomial equations.

Though Post-Quantum Cryptography solutions promise robust security against quantum adversaries, real-world deployment of these schemes is hindered by performance and efficiency limitations. This is due to their high computational complexity, large key sizes and compatibility with devices. Thus, many researchers are also focussing on combining different types of PQC schemes so that they can benefit from each other's strengths. For example, authors (Andreas et al. 2021) proposed the PQ-Wireguard Key Exchange Mechanism (KEM) based handshake protocol using the Classical McEliece and Saber (Anvers et al. 2018) Key Exchange Mechanisms (KEMs). They were observed to be 1000 times faster than the Quantum Resistant variant of OpenVPN handshake protocol. Another example is the Crystal Kyber (Bos et al. 2018) where they used Module-Learning With Error to develop a chosen-plaintext attack secure scheme and then used the Fujisaki–Okamoto transform to create a CCA-Secure KEM (Hofheinz et al., 2017). However, developing highly efficient and lightweight cryptographic schemes that can withstand quantum attacks still remains an active area of research.

4. ARTIFICIAL INTELLIGENCE WITH POST-QUANTUM CRYPTOGRAPHY

Artificial Intelligence is a branch of computational science where machines are trained to mimic human-like intelligence using mathematical models. In simple words, these are systems that have human-like analysing, processing and decision-making capabilities. Artificial Intelligence includes many sub-domains such as Machine Learning, Deep Learning, etc. (Zhang et al., 2021). The below Figure 4 depicts various areas that are under Artificial Intelligence.

Figure 6. Artificial Intelligence in a nutshell.

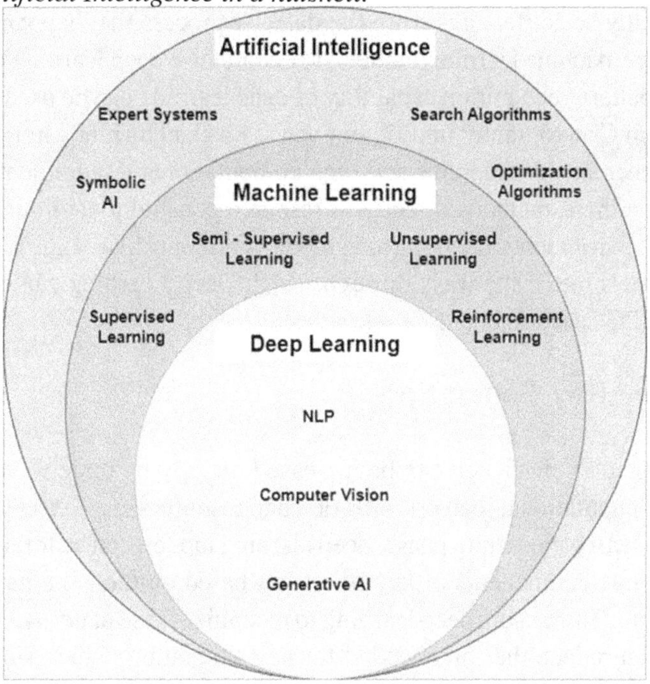

Symbolic Artificial Intelligence (AI) and Expert Systems work with domain knowledge and rules to solve complex problems. They use techniques such as forward chaining and backward chaining to get inferences from knowledge base (Santoro et al., 2021). Search Algorithms like Dijktra (Fang et al., 2020) and Optimization Algorithms like Genetic (Li et al., 2021) helps find out the best way to perform a task such as finding shortest path etc.

Machine Learning models employ statistical and mathematical algorithms to discern underlying patterns and relationships within data. These learned patterns are then utilized to make predictions. However, the specific prediction techniques and outcomes can vary significantly based on the learning paradigm. For instance, in supervised learning, which refers to learning using labels, prediction is of a real value (Abhishek et al., 2021). While, in unsupervised learning, which refers to learning in absence of a label, the models learn hidden patterns and performs prediction such as recommendation (Naeem et al., 2023). However, prediction strategy used in reinforcement learning is completely different. Here, the models are learnt to perform correctly using rewards and punishment (Singh et al., 2022).

On the other hand, deep learning is a subset of machine learning algorithms that use multiple hidden layers made of neurons. These neurons are trained over data, which can be labelled dataset or unlabelled. These layers formulate their own methods for recognizing patterns in the input. This allows deep learning models to automatically select features from the data—a process that happens implicitly, unlike in other machine learning models that do not use deep learning (Dong et al., 2021). This pattern recognition capability of deep learning can be used to empower Post-Quantum Cryptography in different ways. Since algorithms in PQC involves complex mathematical structures and high computational requirements, it is challenging to use these methods directly in resource-constraint environment such as IoV. In such environments, deep learning models can contribute significantly. Below mentioned are some of the ways through which Deep Learning can contribute in empowering PQC (Darzi et al., 2024).

4.1 Efficient Key Generation

Key Generation efficiency can be increased majorly by three ways, predictive modelling, computational load optimization and resource-adaptive key generation.

In the predictive modelling phase, deep learning models can be trained to predict and pre-generate components of key structures based on the patterns observed in PQC algorithm. Thus, using deep learning to recognize these patterns, cryptographic systems can reduce the time needed for key generation, which supports faster cryptographic operations. For computational load optimization, these models can help determine the most efficient path such as selection of specific key parameters. These selections can minimize computation without compromising the security standards. Moreover, in the resource constraint environments, deep learning methods can contribute by adjusting the complexity of key generation based on the available processing power and memory without any compromise on the security standards (Chaeikar et al., 2022).

4.2 Optimizing Cryptography

When we mention of optimizing cryptography, we must mention about parameter tuning, adaptable security levels, key size reduction and compression, and accelerated cryptographic computations to enhance the viability of PQC schemes. Automated parameter tuning such as key size, polynomial degree, and error distribution needs to be rightly balanced. Deep learning models can learn from a large dataset of cryptographic parameters and outcomes, then apply this knowledge to recommend optimal configurations for new scenarios. Deep learning can also be used to dynamically adjust parameters based on real-time security needs.

Deep learning can help optimize the size of PQC keys by identifying redundancies and patterns that allow for compression without sacrificing security. This is similar to how deep learning optimizes data compression in other fields, where it retains critical information while minimizing data size. Techniques like neural network-based autoencoders can be adapted to compress PQC keys. For PQC, this could mean reducing the storage requirements of keys, making the cryptography more scalable and efficient for constrained environments. Moreover, these models can be trained to approximate complex cryptographic functions by learning patterns within the encryption and decryption processes. For example, schemes based on lattice problems, machine learning models can identify shortcuts or simplifications that reduce computational load without compromising security. By doing so, these models speed up calculations involved in encryption and decryption. As Deep Learning models use accelerated devices such as GPU/FGPA's, certain process of encryption and decryption operations can be performed in parallel. This can eventually reduce the time required for computationally heavy cryptographic tasks like key-based transformations or modular arithmetic. Also, deep learning can take advantage of hardware-specific optimizations, such as quantization and model pruning. These techniques reduce the computation load and speed up encryption and decryption on devices with limited resources. As Deep learning excels at recognizing complex patterns within high-dimensional data, it can be applied to analyze mathematical structures and identify new problems that are resistant to quantum attacks. For instance, deep learning could explore novel constructions within lattices or other algebraic frameworks to discover cryptographic problems that are difficult for quantum algorithms to solve. Neural architecture search has been used to automatically generate efficient deep learning models, and similar techniques could be adapted to explore the space of cryptographic algorithms. By defining rules for algorithm construction, deep learning could systematically test and refine new PQC schemes, identifying those with strong security properties and efficient performance. Moreover, Deep learning can aid in analyzing and testing classical cryptographic schemes against potential quantum attacks, helping cryptographers understand vulnerabilities and develop more secure PQC algorithms. Thus, by simulating the behavior of quantum algorithms, deep learning can assist in evaluating the robustness of PQC algorithms under various attack scenarios, even before large-scale quantum computers become available (Mundra et al., 2020).

4.3 Key Management and Distribution

The task of key distribution is challenging in PQC due to its large size, especially when the network is highly dense. Deep Learning can contribute towards key distribution by analysing the network and then suggesting efficient routes and predicting potential vulnerabilities in the transmission. They can also contribute by analysing the possible vulnerabilities in the key management practices ensuring the keys remain secure in the network. Additionally, when these models are trained to detect security incidents with key distribution in the network, they can efficiently identify and mitigate such risks before a catastrophe (Chaeikar et al., 2022).

4.4 Error Correction and Robustness

When data is exposed to noise, error-correcting codes is one of the major techniques used in data communication to maintain its integrity. Even PQC use these techniques to maintain its data integrity when posed against noisy channel. However, PQC algorithms such as lattice-based algorithms, requires a significantly high computational power to handle error correction during decryption. When deep learning models are trained on real-world noise and erroneous data, they can contribute significantly to the error-correction process. This can eventually reduce the computational cost incurred during error correction. As a result, it helps in speeding up the decryption process by reducing the processing time needed for error handling. Moreover, in certain cryptographic schemes such as lattice-based encryption, noise accumulates during encryption and can slow down decryption. Deep learning models can be trained to recognize and filter out unnecessary noise, which makes decryption faster and more accurate (Choukroun et al. 2022).

From the above discussion, we can infer that the synergy of PQC and Artificial Intelligence can contribute greatly in developing efficient PQC algorithms. However, standardization of these schemes is crucial for efficient communication of data over the Internet. Thus, we can say that the challenges of adoption of PQC schemes are both technical and administrative in nature. In the next section, we will discuss upon the policy changes that needs to be addressed so that global adoption of PQC schemes is a tangible reality.

5. POLICY CHANGES FOR GLOBAL ADOPTION OF POST-QUANTUM CRYPTOGRAPHY

With the rise of powerful Quantum Computers, the existing cryptographic protocols such as RSA and ECC face a threat of being compromised. To uphold the security of digital infrastructure, the PQC attempts to develop cryptographic mechanisms which are resistant to quantum attacks. However, there is a need to develop a global consensus and standardization of PQC schemes. In this section, we explore the challenges and the necessary policy and infrastructural changes that needs to be addressed for a world-wide adoption of PQC schemes.

The most crucial step in global adoption of PQC is the standardization of PQC algorithms. National Institute of Standards and Technology (NIST) is working towards the finalization of the PQC algorithms. However, the process can gain significant momentum if there is a collaborative effort from other international standardization institutes such as International Electrotechnical Commission (IEC) and International Standardization Organization (ISO). Moreover, individual countries, their governments and research institutes should also collaborate in the standardization process as it can help reduce the compatibility issues. Additionally, every country should allocate dedicated resources towards the standardization process as this can significantly contribute towards reducing development, testing and other delays that affect the overall timeline.

The process of transitioning to PQC requires heavy investment in technology, training and infrastructure. This can make the transition process cumbersome and challenging to government and businesses alike. Thus, incentivizing the process with regulatory and economic support such as tax reduction, grants, subsidies, etc. can motivate the businesses to implement PQC schemes. Moreover, funded certification and training programmes by government and other regulatory bodies can further support the process of transition.

Awareness and educational drives such as workshops and conferences on the threat of Quantum Powered Adversaries must be conducted. These programmes will help in educating the policymakers about the threat Quantum computers can pose to data security. As a consequence, they will actively participate in enforcing for a compliance in standardization of the quantum-safe cryptography protocols. Also, they will help to contribute in updating the data privacy laws and mauver the legal complexities involved in updating the framework of data security and privacy. Moreover, they can mandate organizations to demonstrate their progress towards implementation of quantum-resistant measures.

To achieve rapid advancements in the PQC technologies, every country should be provided with equal opportunities to participate in the Research and Development (R&D) of PQC. To realise this, institutions and governments should provide the necessary financial and logistical support in their countries to support development of PQC. They should also consider to perform collaboration between academia, industry and government bodies. Such public-private partnership and international research collaborations can help developing countries to actively participate and contribute equally to the R&D of the PQC algorithms.

Global digital infrastructure is interconnected, and PQC algorithms must be compatible across diverse platforms and regions. Policymakers must promote policies that ensure the interoperability of PQC solutions, allowing systems across borders to operate securely and efficiently together. However, ensuring compatibility with complex legacy systems and accounting for the diverse technological landscapes of different countries are key challenges. During the transition period, hybrid cryptographic systems that combine traditional and quantum-safe algorithms can maintain backward compatibility with legacy systems, reducing disruption. Governments should work towards establishing compatibility requirements for PQC algorithms to support the development of cross-border infrastructures for finance, trade, and data sharing, ensuring that systems can securely communicate internationally.

To make the leap towards global adoption of PQC, skilled workforce with the knowledge of quantum-safe cryptography is of paramount importance. Thus, training programs in cryptography, quantum computing and Post Quantum Cryptography to prepare the next generation of cybersecurity professionals must be untaken. These courses should not just include the mathematical analysis of the algorithms, but also practical lessons. Universities, technical institutions and industries should support the upskill of their students and employees through funded trainings and upgradation of curriculum. Also, they should ensure that sufficient infrastructure and conducive environment are available to the participants to learn the technology. These measures will help in building a larger pool of qualified professionals.

Thus, the challenges for the global adoption of Post-Quantum Cryptography are a multi-fold one. Though the shift to PQC will ensure the security of digital data, however, more intense research, awareness and development of skilled workforce for PQC is very important. As the technology is relatively new, incentivization and research funding along with collaborative efforts from all the countries is the need of the hour. Also, implementation of these policy changes will require collective involvement of policymakers, government and international organizations. However, this is a pressing issue that requires an urgent attention from the global community.

6. CONCLUSION

There has been tremendous imputes given for the development of efficient Quantum Computers. Thus, the day is not far when a new chapter for humankind will be unveiled. Quantum Computers will open up many new possibilities for us. However, these systems in the hands of adversaries can also create havoc particularly, for data security of users.

In this chapter, we began by examining IoV network and the role classical encryption system plays in data protection. We analysed different encryption techniques and evaluated their effectiveness within resource-intensive applications, like the Internet of Vehicles (IoV). Later, we explored the prowess that makes a quantum computer so invincible and how they can breach the classical security schemes. We then presented the concepts of Post Quantum Cryptography and how they can be used on classical computers to make them secure against a Quantum Powered Adversary. We also analyzed the challenges these schemes have and the potential role Artificial Intelligence can play in developing efficient Post Quantum Cryptography solutions that can also be used in resource constraint environments such as IoV.

However, global adoption of Post Quantum Cryptography is challenging due to the lack of established standardization policies. Hence, we concluded our discussion by presenting the policy challenges and solutions that needs to be adhered to make global adoption of Post Quantum Cryptography schemes a viable reality.

REFERENCES

Abdullah, G. M., Mehmood, Q., & Khan, C. B. A. (2018). Adoption of Lamport signature scheme to implement digital signatures in IoT. In *2018 International conference on computing, mathematics and engineering technologies (iCoMET)*. IEEE. DOI: 10.1109/ICOMET.2018.8346359

Abhishek, B., & Tyagi, A. K. (2021). An Useful Survey on Supervised Machine Learning Algorithms: Comparisons and Classifications. In: *International Conference on 0Advances in Electrical and Computer Technologies*. Springer.

Alalwany, E., & Mahgoub, I. (2024). Security and trust management in the internet of vehicles (iov): Challenges and machine learning solutions. *Sensors (Basel)*, *24*(2), 368. DOI: 10.3390/s24020368 PMID: 38257461

Almudever, C. G., Lao, L., Fu, X., Khammassi, N., Ashraf, I., Iorga, D., & Bertels, K. (2017, March). The engineering challenges in quantum computing. In *Design, Automation & Test in Europe Conference & Exhibition*. IEEE. DOI: 10.23919/DATE.2017.7927104

Aouedi, O., Piamrat, K., & Parrein, B. (2022). Intelligent traffic management in next-generation networks. *Future Internet*, *14*(2), 44. DOI: 10.3390/fi14020044

Azam, F., Yadav, S. K., Priyadarshi, N., Padmanaban, S., & Bansal, R. C. (2021). A comprehensive review of authentication schemes in vehicular ad-hoc network. *IEEE Access : Practical Innovations, Open Solutions*, *9*, 31309–31321. DOI: 10.1109/ACCESS.2021.3060046

Bojjagani, S., Reddy, Y. P., Anuradha, T., Rao, P. V., Reddy, B. R., & Khan, M. K. (2022). Secure authentication and key management protocol for deployment of internet of vehicles (IoV) concerning intelligent transport systems. *IEEE Transactions on Intelligent Transportation Systems*, *23*(12), 24698–24713. DOI: 10.1109/TITS.2022.3207593

Bos, J., Ducas, L., Kiltz, E., Lepoint, T., Lyubashevsky, V., Schanck, J. M., & Stehlé, D. (2018, April). CRYSTALS-Kyber: a CCA-secure module-lattice-based KEM. In: *IEEE European Symposium on Security and Privacy (EuroS&P)*. IEEE. DOI: 10.1109/EuroSP.2018.00032

Bucaioni, A., & Pelliccione, P. (2020). Technical architectures for automotive systems, in: *IEEE International Conference on Software Architecture, ICSA, IEEE.*

Chaeikar, S. S., Jolfaei, A., & Mohammad, N. (2022). AI-enabled cryptographic key management model for secure communications in the internet of vehicles. *IEEE Transactions on Intelligent Transportation Systems.*

Choukroun, Y., & Wolf, L. (2022). Error correction code transformer. *Advances in Neural Information Processing Systems.*

D'Anvers, J. P., Karmakar, A., Sinha Roy, S., & Vercauteren, F. (2018). Saber: Module-LWR based key exchange, CPA-secure encryption and CCA-secure KEM. In *Progress in Cryptology–AFRICACRYPT 2018:10th International Conference on Cryptology in Africa,Marrakesh, Morocco,May 7–9, 2018, Proceedings.* Springer International Publishing.

Darzi, S., & Yavuz, A. A. (2024). PQC meets ML or AI: Exploring the Synergy of Machine Learning and Post-quantum Cryptography. *Authorea Preprints.*

Dey, J., & Dutta, R. (2023). Progress in multivariate cryptography: Systematic review, challenges, and research directions. *ACM Computing Surveys, 55*(12), 1–34. DOI: 10.1145/3571071

Dong, S., Wang, P., & Abbas, K. (2021). A survey on deep learning and its applications. *Computer Science Review, 40*, 100379. DOI: 10.1016/j.cosrev.2021.100379

El-Rewini, Z., Sadatsharan, K., Selvaraj, D. F., Plathottam, S. J., & Ranganathan, P. (2020). Cybersecurity challenges in vehicular communications. *Vehicular Communications, 23*, 100214. DOI: 10.1016/j.vehcom.2019.100214

Fang, Y., Huang, X., Qin, L., Zhang, Y., Zhang, W., Cheng, R., & Lin, X. (2020). A survey of community search over big graphs. *The VLDB Journal.*

Grassl, M., Langenberg, B., Roetteler, M., & Steinwandt, R. (2016, February). Applying Grover's algorithm to AES: quantum resource estimates. In *International Workshop on Post-Quantum Cryptography.* Cham: Springer International Publishing. DOI: 10.1007/978-3-319-29360-8_3

Guo, H., & Yu, X. (2022). A survey on blockchain technology and its security. Blockchain: research and applications, 3(2), 100067.

Hamza, A., & Kumar, B. (2020). A review paper on DES, AES, RSA encryption standards. *In:9th International Conference System Modeling and Advancement in Research Trends (SMART),* IEEE. DOI: 10.1109/SMART50582.2020.9336800

Hasija, T., Ramkumar, K., Kaur, A., Mittal, S., & Singh, B. (2022). A survey on nist selected third round candidates for post quantum cryptography, in: *7th International Conference on Communication and Electronics Systems (ICCES)*, IEEE. DOI: 10.1109/ICCES54183.2022.9835864

Hoffstein, J., Howgrave-Graham, N., Pipher, J., Silverman, J. H., & Whyte, W. (2003). NTRUSIGN: Digital signatures using the NTRU lattice. In *Cryptographers' track at the RSA conference*. Springer Berlin Heidelberg. DOI: 10.1007/3-540-36563-X_9

Hofheinz, D., Hövelmanns, K., & Kiltz, E. (2017). A modular analysis of the Fujisaki-Okamoto transformation. In: *Theory of Cryptography Conference*. Cham: Springer International Publishing. DOI: 10.1007/978-3-319-70500-2_12

Hülsing, A., Ning, K. C., Schwabe, P., Weber, F. J., & Zimmermann, P. R. (2021). Post-quantum wireguard. In: *IEEE Symposium on Security and Privacy (SP)*. IEEE. DOI: 10.1109/SP40001.2021.00030

Jalali, A., Azarderakhsh, R., Kermani, M. M., & Jao, D. (2017). Supersingular isogeny Diffie-Hellman key exchange on 64-bit ARM. *IEEE Transactions on Dependable and Secure Computing*.

Janwa, H., & Moreno, O. (1996). McEliece public key cryptosystems using algebraic-geometric codes. *Designs, Codes and Cryptography*, *8*(3), 293–307. DOI: 10.1023/A:1027351723034

Karim, S. M., Habbal, A., Chaudhry, S. A., & Irshad, A. (2022). Architecture, protocols, and security in IoV: Taxonomy, analysis, challenges, and solutions. *Security and Communication Networks*, *2022*, 1–19. DOI: 10.1155/2022/1131479

Kumar, S., Velliangiri, S., Karthikeyan, P., Kumari, S., & Khan, M. K. (2024). A survey on the blockchain techniques for the internet of vehicles security. *Transactions on Emerging Telecommunications Technologies*, *35*(4), e4317. DOI: 10.1002/ett.4317

Li, W., Wang, G. G., & Gandomi, A. H. (2021). A survey of learning-based intelligent optimization algorithms. *Archives of Computational Methods in Engineering*, *28*(5), 3781–3799. DOI: 10.1007/s11831-021-09562-1

Mundra, A., Mundra, S., Srivastava, J. S., & Gupta, P. (2020). Optimized deep neural network for cryptanalysis of DES. *Journal of Intelligent & Fuzzy Systems*, *38*(5), 5921–5931. DOI: 10.3233/JIFS-179679

Mustafa, I., & Huang, C. Y. (2020). Lightweight cryptographic URLLC for 5G-V2X. *In:International Symposium on Networks, Computers and Communications (ISNCC)*, IEEE.

Naeem, S., Ali, A., Anam, S., & Ahmed, M. M. (2023). An unsupervised machine learning algorithms: Comprehensive review. *International Journal of Computing and Digital Systems*.

Nejatollahi, H., Dutt, N., Ray, S., Regazzoni, F., Banerjee, I., & Cammarota, R. (2019). Post-quantum lattice-based cryptography implementations: A survey. *ACM Computing Surveys (CSUR)*.

Nguyen, K. T., Laurent, M., & Oualha, N. (2015). Survey on secure communication protocols for the Internet of Things. *Ad Hoc Networks*, *32*, 17–31. DOI: 10.1016/j.adhoc.2015.01.006

Partovi, Z., Zarei, M., & Rahmani, A. M. (2023). Data-centric approaches in the internet of vehicles: A systematic review on techniques, open issues, and future directions. *International Journal of Communication Systems*, *36*(3), e5383. DOI: 10.1002/dac.5383

Rana, S., Parast, F. K., Kelly, B., Wang, Y., & Kent, K. B. (2023). A comprehensive survey of cryptography key management systems. *Journal of Information Security and Applications*, *78*, 103607. DOI: 10.1016/j.jisa.2023.103607

Rinott, Y., Shoham, T., & Kalai, G. (2022). Statistical aspects of the quantum supremacy demonstration. *Statistical Science*, *37*(3). Advance online publication. DOI: 10.1214/21-STS836

Roy, K. S., & Kalita, H. K. (2019). A survey on post-quantum cryptography for constrained devices. *International Journal of Applied Engineering Research: IJAER*, *14*(11), 2608–2615.

Saleem, M. A., Li, X., Mahmood, K., Shamshad, S., Alenazi, M. J., Das, A. K. (2024). A Cost-Efficient Anonymous Authenticated and Key Agreement Scheme for V2I-Based Vehicular Ad-Hoc Networks. *IEEE Transactions on Intelligent Transportation Systems.*

Santoro, A., Lampinen, A., Mathewson, K., Lillicrap, T., & Raposo, D. (2021). Symbolic behaviour in artificial intelligence. *arXiv preprint arXiv:2102.03406.*

Sawant, S. V., & Rudra, B. (2023). Analysis of magnitude of threats for v2x authentication schemes under quantum powered adversary, in: *International Conference on Information and Communication Technology for Intelligent Systems,* Springer. DOI: 10.1007/978-981-99-3761-5_43

Schwabe, P., Stebila, D., & Wiggers, T. (2020, October). Post-quantum TLS without handshake signatures. In *Proceedings of the 2020 ACM SIGSAC Conference on Computer and Communications Security.*

Singh, B., Kumar, R., & Singh, V. P. (2022). Reinforcement learning in robotic applications: A comprehensive survey. *Artificial Intelligence Review*, *55*(2), 945–990. DOI: 10.1007/s10462-021-09997-9

Tang, Q., & Teague, V. (2024). Public-Key Cryptography–PKC 2024. In: *Proceedings of the 27th IACR International Conference on Practice and Theory of Public-Key Cryptography,* Sydney, NSW, Australia.

Taslimasa, H., Dadkhah, S., Neto, E. C. P., Xiong, P., Ray, S., & Ghorbani, A. A. (2023). Security issues in Internet of Vehicles (IoV): A comprehensive survey. *Internet of Things : Engineering Cyber Physical Human Systems*, *22*, 100809. DOI: 10.1016/j.iot.2023.100809

Ugwuishiwu, C. H., Orji, U. E., Ugwu, C. I., & Asogwa, C. N. (2020). An overview of quantum cryptography and shor's algorithm. *International Journal of Advanced Trends in Computer Science and Engineering*.

Wang, W., Szefer, J., & Niederhagen, R. (2018). FPGA-based Niederreiter cryptosystem using binary Goppa codes. In *Post-Quantum Cryptography: 9th International Conference, PQCrypto 2018, Fort Lauderdale, FL, USA, April 9-11, 2018* [Springer International Publishing.]. *Proceedings*, *9*, 77–98.

Yang, Z., Zolanvari, M., & Jain, R. (2023). A survey of important issues in quantum computing and communications. *IEEE Communications Surveys and Tutorials*, *25*(2), 1059–1094. DOI: 10.1109/COMST.2023.3254481

Zeydan, E., Turk, Y., Aksoy, B., & Ozturk, S. B. (2022). Recent advances in post-quantum cryptography for networks: A survey. In: *Seventh International Conference OnMobileAnd Secure Services(MobiSecServ)*. IEEE. DOI: 10.1109/MobiSecServ50855.2022.9727214

Zhang, C., & Lu, Y. (2021). Study on artificial intelligence: The state of the art and future prospects. *Journal of Industrial Information Integration*, *23*, 100224. DOI: 10.1016/j.jii.2021.100224

Zhou, X., & Tang, X. (2011). Research and implementation of RSA algorithm for encryption and decryption. In: *Proceedings of 2011 6th international forum on strategic*. IEEE. DOI: 10.1109/IFOST.2011.6021216

Conclusion

The rapid advancements in vehicular networking, intelligent transportation, and communication technologies have reshaped the landscape of modern mobility. This book, Networking, Transport, and Quality of Service in Vehicular Networks, has explored various aspects of these advancements, providing insights into cutting-edge research, methodologies, and applications that enhance transportation efficiency, safety, and connectivity.

Through the chapters presented, we have delved into key topics such as artificial intelligence, machine learning, quantum security, and multi-criteria decision models—each contributing to the broader vision of intelligent vehicular networks. These technologies hold the potential to revolutionize urban transportation, address congestion, improve emergency response mechanisms, and foster the development of smart cities.

The discussions and analyses within this book underscore the importance of interdisciplinary collaboration, as the challenges in vehicular networking require expertise across multiple domains, including network engineering, transportation planning, cybersecurity, and data science. By bridging these areas, we can develop innovative solutions that enhance the quality of service, optimize network performance, and ensure robust and secure vehicular communication.

Looking ahead, the future of mobility will be shaped by the integration of emerging technologies such as 5G/6G communication, blockchain for secure data sharing, and AI-driven predictive analytics. As connected and autonomous vehicles become more prevalent, researchers and industry professionals must continue to push the boundaries of innovation to overcome existing challenges and improve the efficiency and reliability of vehicular networks.

We hope that this book serves as a foundational resource for researchers, professionals, students, and policymakers, inspiring further exploration and advancements in this ever-evolving field. The collective efforts of academia and industry will be instrumental in shaping the next generation of smart transportation systems, ensuring a safer, more efficient, and sustainable future for all.

Compilation of References

Abboud, K., Omar, H. A., & Zhuang, W. (2016, July 14). Interworking of DSRC and Cellular Network Technologies for V2X Communications: A Survey. *IEEE Transactions on Vehicular Technology*, *65*(12), 9457–9470. DOI: 10.1109/TVT.2016.2591558

Abdullah, G. M., Mehmood, Q., & Khan, C. B. A. (2018). Adoption of Lamport signature scheme to implement digital signatures in IoT. In *2018 International conference on computing, mathematics and engineering technologies (iCoMET)*. IEEE. DOI: 10.1109/ICOMET.2018.8346359

Abhishek, B., & Tyagi, A. K. (2021). An Useful Survey on Supervised Machine Learning Algorithms: Comparisons and Classifications. In: *International Conference on 0Advances in Electrical and Computer Technologies*. Springer.

Abizar, F., Farman, H., Jan, B., Khan, Z., & Koubaa, A. (2022). A smart energy-based source location privacy preservation model for Internet of Things-based vehicular ad hoc networks. *Transactions on Emerging Telecommunications Technologies*, *33*(2), e3973. DOI: 10.1002/ett.3973

Aboubakar, M., Kellil, M., & Roux, P. (2022). A Review of IoT Network Management: Current Status and Perspectives. *Journal of King Saud University. Computer and Information Sciences*, *34*(7), 4163–4176. Advance online publication. DOI: 10.1016/j.jksuci.2021.03.006

Ackels, S. (2023). Deep Learning Techniques for Optimization of Roadside Unit Placement (Doctoral dissertation, The University of Texas at San Antonio).

Aguado, A., Lopez, V., & Lopez, D.. (2019). The engineering of software-defined quantum key distribution. networks. *IEEE Communications Magazine*, *57*(7), 20–26.

Alalwany, E., & Mahgoub, I. (2024). Security and trust management in the internet of vehicles (iov): Challenges and machine learning solutions. *Sensors (Basel)*, *24*(2), 368. DOI: 10.3390/s24020368 PMID: 38257461

Alawi, M., Alsaqour, R., Abdalla, A., Abdelhaq, M., & Uddin, M. (2021). Multi-criteria prediction mechanism for vehicular Wi-fi offloading. *Computers, Materials & Continua, 69*(2), 2313–2337. DOI: 10.32604/cmc.2021.018282

Alawi, M., Alsaqour, R., Abdelhaq, M., Alkanhel, R., Sharef, B., Sundararajan, E., & Ismail, M. (2022). Adaptive QoS-Aware Multi-Metrics Gateway Selection Scheme for Heterogenous Vehicular Network. *Systems, 10*(5), 142. DOI: 10.3390/systems10050142

Alhabib, R., Ranna, M. M., Farah, H., & Salama, A. A. (2018). Some neutrosophic probability distributions. Infinite Study.

Ali, W. A., Roccotelli, M., Boggia, G., & Fanti, M. P. (2024). Intrusion detection system for vehicular ad hoc network attacks based on machine learning techniques. Information Security Journal: A Global Perspective, 1-19.

Ali, Z., Lagen, S., Giupponi, L., & Rouil, R. (2021). 3GPP NR V2X Mode 2: Overview, Models and System-Level Evaluation. *IEEE Access : Practical Innovations, Open Solutions, 9*, 89554–89579. DOI: 10.1109/ACCESS.2021.3090855 PMID: 37461405

Aljeri, N., & Boukerche, A. (2023, October). Performance Evaluation of Communication Lifetime Prediction Model for Autonomous Vehicular Networks. In *Proceedings of the Int'l ACM Symposium on Performance Evaluation of Wireless Ad Hoc, Sensor, & Ubiquitous Networks* (pp. 17-25).

Alladi, T., Chamola, V., Sahu, N., Venkatesh, V., Goyal, A., & Guizani, M. (2022). A comprehensive survey on the applications of blockchain for securing vehicular networks. *IEEE Communications Surveys and Tutorials, 24*(2), 1212–1239.

Almeida, L. G. D. A. (2023). Driver's behavior classification in vehicular communication networks for commercial vehicles.

Almudever, C. G., Lao, L., Fu, X., Khammassi, N., Ashraf, I., Iorga, D., & Bertels, K. (2017, March). The engineering challenges in quantum computing. In *Design, Automation & Test in Europe Conference & Exhibition*. IEEE. DOI: 10.23919/DATE.2017.7927104

Altahrawi, M., Abdullah, N. F., & Nordin, R. (2022). Service-oriented lstm multi-criteria rat selection scheme for vehicle-to-infrastructure communication. *IEEE Access : Practical Innovations, Open Solutions, 10*, 110261–110284. DOI: 10.1109/ACCESS.2022.3214852

Álvares, P., Silva, L., & Magaia, N. (2021). Blockchain-Based Solutions for UAV-Assisted Connected Vehicle Networks in Smart Cities: A Review, Open Issues, and Future Perspectives. *Telecom, 2*(1), 108–140. Advance online publication. DOI: 10.3390/telecom2010008

Amadeo, M., Campolo, C., Molinaro, A., Härri, J., Rothenberg, C. E., & Vinel, A. (2019, September 18). Enhancing the 3GPP V2X Architecture with Information-Centric Networking. *Future Internet, 11*(9), 199–199. DOI: 10.3390/fi11090199

Amras, M., Yulianto, E., Dewi, D. A., & Setiawan, A. (2023). Implementation of the REST API Model Using QR Codes on Mobile Devices to Order Parking Tickets. *International Journal of Advanced Computer Science and Applications, 14*(10). Advance online publication. DOI: 10.14569/IJACSA.2023.0141043

Anand, M. C. J., Kalaiarasi, K., Martin, N., Ranjitha, B., Priyadharshini, S. S., & Tiwari, M. (2023, September). Fuzzy C-Means Clustering with MAIRCA-MCDM Method in Classifying Feasible Logistic Suppliers of Electrical Products. In *2023 First International Conference on Cyber Physical Systems, Power Electronics and Electric Vehicles (ICPEEV)* (pp. 1-7). IEEE.

Anand, M. C. J., Martin, N., Clementking, A., Rani, S., Priyadharshini, S., & Siva, S. (2023, March). Decision making on optimal selection of advertising agencies using machine learning. In *2023 International Conference on Information Management (ICIM)* (pp. 121-125). IEEE.

Anand, M. C. J., Moorthy, C. B., Sivamani, S., Indrakumar, S., Kalaiarasi, K., & Barhoi, A. (2023). *Fuzzy intelligence inventory decision optimization model of sustainability and green technologies for mixed uncertainties of carbon emission. In: 2023 International Conference on Information Management (ICIM)*. IEEE., https://ieeexplore.ieee.org/document/10145085 DOI: 10.1109/ICIM58774.2023.00020

Anand, M. C. J., Moorthy, C. B., Sivamani, S., Indrakumar, S., Kalaiarasi, K., & Barhoi, A. (2023, March). Fuzzy intelligence inventory decision optimization model of sustainability and green technologies for mixed uncertainties of carbon emission. In *2023 International Conference on Information Management (ICIM)* (pp. 78-82). IEEE.

Anand, M. C. J., Priyadharshini, S. S., Martin, N., Anand, M. E., Poorani, B., & Mohiddin, S. K. (2023, September). Decision Trees in the Selection of Electric Vehicles Based on Various Parameters. In *2023 First International Conference on Cyber Physical Systems, Power Electronics and Electric Vehicles (ICPEEV)* (pp. 1-6). IEEE.

Anand, R., & Sankhe, S. (2022). Traffic Prediction for Intelligent Transportation Systems Using Machine Learning. *International Journal for Modern Trends in Science and Technology*, *8*(7), 276–280. Advance online publication. DOI: 10.46501/IJMTST0807041

Ansari, K. (2020, December 31). Joint use of DSRC and C-V2X for V2X communications in the 5.9 GHz ITS band. *IET Intelligent Transport Systems*, *15*(2), 213–224. DOI: 10.1049/itr2.12015

Aouedi, O., Piamrat, K., & Parrein, B. (2022). Intelligent traffic management in next-generation networks. *Future Internet*, *14*(2), 44. DOI: 10.3390/fi14020044

Appaji, I., & Raviraj, P. (2023). Framework for simulation of vehicular communication using LSTM-based graph attention networks. *Indian Journal of Science and Technology*, *16*(16), 1230–1240. DOI: 10.17485/IJST/v16i16.1777

Arakelyan, D. Machine Learning-Powered Accident Prediction for the Automotive Insurance Industry.

Arif, M., Kumar, V. D., Jayakumar, L., Ungurean, I., Izdrui, D., & Geman, O. (2021). DAHP–TOPSIS-Based Channel Decision Model for Co-Operative CR-Enabled Internet on Vehicle (CR-IoV). *Sustainability (Basel)*, *13*(24), 13966. DOI: 10.3390/su132413966

Arrieta, A. B., Díaz-Rodríguez, N., Del Ser, J., Bennetot, A., Tabik, S., Barbado, A., & Herrera, F. (2020). Explainable Artificial Intelligence (XAI): Concepts, taxonomies, opportunities and challenges toward responsible AI. *Information Fusion*, *58*, 82–115. DOI: 10.1016/j.inffus.2019.12.012

Aslam, B., Amjad, F., & Zou, C. C. (2012, July). Optimal roadside units placement in urban areas for vehicular networks. In 2012 IEEE symposium on computers and communications (ISCC) (pp. 000423-000429). IEEE.

Aslam, B., & Zou, C. C. (2011, January). Optimal roadside units placement along highways. In *2011 IEEE Consumer Communications and Networking Conference (CCNC)* (pp. 814-815). IEEE.

Astarita, V., Giofré, V. P., Festa, D. C., Guido, G., & Vitale, A. (2020). Floating Car Data Adaptive Traffic Signals: A Description of the First Real-Time Experiment with 'Connected' Vehicles. *Electronics (Basel)*, *9*(1), 114. Advance online publication. DOI: 10.3390/electronics9010114

Azam, F., Yadav, S. K., Priyadarshi, N., Padmanaban, S., & Bansal, R. C. (2021). A comprehensive review of authentication schemes in vehicular ad-hoc network. *IEEE Access : Practical Innovations, Open Solutions*, *9*, 31309–31321. DOI: 10.1109/ACCESS.2021.3060046

Azouaghe, Y. (2021). Prediction of engine failures in the automotive industry using machine learning techniques (Doctoral dissertation, Universitat Politècnica de València).

Barth, M., Scora, G., & Younglove, T. (2004, April 23). Intelligent off-board management of vehicle operating parameters. https://doi.org/DOI: 10.1109/ITSC.2003.1251976

Barth, M., Wu, G., & Boriboonsomsin, K. (2015, July 25). Intelligent Transportation Systems and Greenhouse Gas Reductions. Springer Science+Business Media, 2(3), 90-97. https://doi.org/DOI: 10.1007/s40518-015-0032-y

Bathla, G., Bhadane, K., Singh, R. K., Kumar, R., Aluvalu, R., Krishnamurthi, R., Kumar, A., Thakur, R. N., & Basheer, S. (2022). R. N. Thakur, and Shakila Basheer. 2022. "Autonomous Vehicles and Intelligent Automation: Applications, Challenges, and Opportunities.". *Mobile Information Systems*, *2022*, 1–36. Advance online publication. DOI: 10.1155/2022/7632892

Bellman, R. E., & Zadeh, L. A. (1970). Decision Making in a Fuzzy Environment. *Management Science*, *17*(4), B141–B164. DOI: 10.1287/mnsc.17.4.B141

Bergès, C., Bird, J., Shroff, M. D., Lumanauw, E., Raghunathan, S., & Smith, C. (2022, July). Inspection methodologies and machine-learning approaches for defectivity data in semiconductor industry for automotive applications: case study for field-failure prevention. In *2022 IEEE International Symposium on the Physical and Failure Analysis of Integrated Circuits (IPFA)* (pp. 1-7). IEEE. DOI: 10.1109/IPFA55383.2022.9915769

Bergès, C., Bird, J., Shroff, M. D., Rongen, R., & Smith, C. (2021, September). Data analytics and machine learning: root-cause problem-solving approach to prevent yield loss and quality issues in semiconductor industry for automotive applications. In *2021 IEEE International Symposium on the Physical and Failure Analysis of Integrated Circuits (IPFA)* (pp. 1-10). IEEE.

Bergès, C., Céneray, C., Albermann, G., Bird, J., Balakrishnan, B., Mustafa, S. M., & Smith, C. (2020, July). Detection and prevention of assembly defects, by machine learning algorithms, in semiconductor industry for automotive. In *2020 IEEE International Symposium on the Physical and Failure Analysis of Integrated Circuits (IPFA)* (pp. 1-9). IEEE. DOI: 10.1109/IPFA49335.2020.9260582

Bergès, C., Zhang, S., Létévé, A., Anilturk, O., Bird, J., Lumanauw, E., & Smith, C. (2023, July). Test And Reliability Improvement With Defect-Image Classification And Machine-Learning Algorithms In Semiconductor Industry For Automotive Applications. In *2023 IEEE International Symposium on the Physical and Failure Analysis of Integrated Circuits (IPFA)* (pp. 1-9). IEEE.

Bersali, M., Rachedi, A., Bouarfa, H., & Badjara, M. E. A. (2022). A novel cooperative clustering approach based on multi-criteria decision-making for IoV. *International Journal of High Performance Systems Architecture, 11*(1), 36–46. DOI: 10.1504/IJHPSA.2022.121923

Bhabani, B., & Mahapatro, J. (2023). Performance Evaluation of Priority-based Scheduling in Hybrid VANETs for Different Criteria Weights Using AHP-AHP and AHP-TOPSIS. *Journal of the Institution of Electronics and Telecommunication Engineers*, •••, 1–12.

Bharatraj, J., & Anand, M. C. J. (2019). Power harmonic weighted aggregation operator on single-valued trapezoidal neutrosophic numbers and interval-valued neutrosophic sets. Fuzzy Multi-criteria Decision-Making Using Neutrosophic Sets, 45-62.

Bharatraj, J., & Clement Joe Anand, M. (2019). Power harmonic weighted aggregation operator on single-valued trapezoidal neutrosophic numbers and interval-valued neutrosophic sets. In *Fuzzy Multi-criteria Decision-Making Using Neutrosophic Sets* (pp. 45–62). Springer International Publishing. DOI: 10.1007/978-3-030-00045-5_3

Bharti, R. K., Suganthi, D., Abirami, S. K., Kumar, R. A., Gayathri, B., & Kayathri, S. (2022, December). Optimal Extreme Learning Machine based Traffic Congestion Control System in Vehicular Network. In 2022 6th International Conference on Electronics, Communication and Aerospace Technology (pp. 597-603). IEEE.

Bharti, R. K., Suganthi, D., Abirami, S. K., Kumar, R. A., Gayathri, B., & Kayathri, S. (2022, December). Optimal Extreme Learning Machine-based Traffic Congestion Control System in Vehicular Network. In *2022 6th International Conference on Electronics, Communication and Aerospace Technology* (pp. 597-603). IEEE.

Bibri, S E., & Krogstie, J. (2017, March 6). Smart sustainable cities of the future: An extensive interdisciplinary literature review. Elsevier BV, 31, 183-212. https://doi.org/DOI: 10.1016/j.scs.2017.02.016

Bodendorf, F., & Franke, J. (2021). A machine learning approach to estimate product costs in the early product design phase: A use case from the automotive industry. *Procedia CIRP, 100*, 643–648. DOI: 10.1016/j.procir.2021.05.137

Bohlender, G., & Kulisch, U. (2010). *Definition of the Arithmetic Operations and Comparison Relations for an Interval Arithmetic Standard*. Institut fur Angewandte und Numerische Mathematik, Universitat Karlsurhe.

Bojjagani, S., Reddy, Y. P., Anuradha, T., Rao, P. V., Reddy, B. R., & Khan, M. K. (2022). Secure authentication and key management protocol for deployment of internet of vehicles (IoV) concerning intelligent transport systems. *IEEE Transactions on Intelligent Transportation Systems*, *23*(12), 24698–24713. DOI: 10.1109/TITS.2022.3207593

Bos, J., Ducas, L., Kiltz, E., Lepoint, T., Lyubashevsky, V., Schanck, J. M., & Stehlé, D. (2018, April). CRYSTALS-Kyber: a CCA-secure module-lattice-based KEM. In: *IEEE European Symposium on Security and Privacy (EuroS&P)*. IEEE. DOI: 10.1109/EuroSP.2018.00032

Bouakouk, M. R., Abdelli, A., Mokdad, L., & Othman, J. B. (2022, May). Dealing with complex routing requirements using an MCDM based approach. In *2022 International Wireless Communications and Mobile Computing (IWCMC)* (pp. 1256-1261). IEEE. DOI: 10.1109/IWCMC55113.2022.9825024

Boyle, L N., Chin, H C., & Debnath, A K. (2013, March 1). Sustainable, safe, smart—three key elements of Singapore's evolving transport policies. Elsevier BV, 27, 20-31. https://doi.org/DOI: 10.1016/j.tranpol.2012.11.017

Brahim, M. B., Drira, W., & Filali, F. (2014, November). Roadside units placement within city-scaled area in vehicular ad-hoc networks. In *2014 International Conference on Connected Vehicles and Expo (ICCVE)* (pp. 1010-1016). IEEE.

Briguglio, G., & Crupi, V. (2024). Review on Sensors for Sustainable and Safe Maritime Mobility. *Journal of Marine Science and Engineering*, *12*(2), 353. Advance online publication. DOI: 10.3390/jmse12020353

Brown, S., Pyke, D F., & Steenhof, P A. (2010, March 17). Electric vehicles: The role and importance of standards in an emerging market. Elsevier BV, 38(7), 3797-3806. https://doi.org/DOI: 10.1016/j.enpol.2010.02.059

Bucaioni, A., & Pelliccione, P. (2020). Technical architectures for automotive systems, in: *IEEE International Conference on Software Architecture, ICSA, IEEE*.

Bučko, B., Michálek, M., Papierniková, K., & Zábovská, K. (2021). Smart Mobility and Aspects of Vehicle-to-Infrastructure: A Data Viewpoint. *Applied Sciences (Basel, Switzerland)*, *11*(22), 10514. Advance online publication. DOI: 10.3390/app112210514

Campisi, T., Severino, A., Al-rashid, M. A., & Pau, G. (2021). The Development of the Smart Cities in the Connected and Autonomous Vehicles (Cavs) Era: From Mobility Patterns to Scaling in Cities. *Infrastructures*, *6*(7), 100. Advance online publication. DOI: 10.3390/infrastructures6070100

Cao, Y., Zhao, Y., Colman-Meixner, C., Yu, X., & Zhang, J. (2017). Key on demand (KoD) for software-defined optical networks secured by quantum key distribution (QKD). *Optics Express*, *25*(22), 26453–26467.

Cao, Y., Zhao, Y., Wang, Q., Zhang, J., Ng, S. X., & Hanzo, L. (2022). The evolution of quantum key distribution networks: On the road to the qinternet. *IEEE Communications Surveys and Tutorials*, *24*(2), 839–894. DOI: 10.1109/COMST.2022.3144219

Cassandras, C G. (2017, January 1). Automating mobility in smart cities. Elsevier BV, 44, 1-8. https://doi.org/DOI: 10.1016/j.arcontrol.2017.10.001

Cavallaro, F. (2010). Fuzzy TOPSIS Approach for Assessing Thermal-Energy Storage in Concentrating Solar Power (CSP) Systems. *Applied Energy*, *87*(2), 496–503. DOI: 10.1016/j.apenergy.2009.07.009

Cavicchioli, R., Martoglia, R., & Verucchi, M. (2022). A Novel Real-Time Edge-Cloud Big Data Management and Analytics Framework for Smart Cities. *Journal of Universal Computer Science*, *28*(1), 3–26. Advance online publication. DOI: 10.3897/jucs.71645

Chaeikar, S. S., Jolfaei, A., & Mohammad, N. (2022). AI-enabled cryptographic key management model for secure communications in the internet of vehicles. *IEEE Transactions on Intelligent Transportation Systems*.

Chai, Z., & Ren, J. (2023). Multi-objective evolutionary optimization for spectrum allocation and power control in vehicular communications. *Wireless Personal Communications*, *129*(4), 2767–2789.

Chakraborty, A., Mondal, S. P., Alam, S., Pamucar, D., & Marinkovic, D. (2022). A new idea to evaluate networking problem and MCGDM problem in parametric interval valued pythagorean arena. *Discrete Dynamics in Nature and Society*, *2022*(1), 2022. DOI: 10.1155/2022/7369045

Chakraborty, S., & Zavadskas, E. K. (2014). Applications of WASPAS Method in Manufacturing Decision Making. *Informatica (Vilnius)*, *25*(4), 524–540. DOI: 10.15388/Informatica.2014.01

Chatterjee, S., & Roy, A. (2022). AI-based intelligent traffic management system for urban road networks. *Future Internet*, *14*(3), 74.

Chavhan, S., & Venkataram, P. (2019, July 4). Prediction based traffic management in a metropolitan area. Elsevier BV, 7(4), 447-466. https://doi.org/DOI: 10.1016/j.jtte.2018.05.003

Chen, S. P., & Eur, J. (2009). Parametric Nonlinear Programming for Analyzing Fuzzy Queues with Infinite Capacity, 157.

Cheng, E. W. L., & Li, H. (2001). Information Priority-Setting for Better Resource Allocation Using Analytic Hierarchy Process (AHP). *Information Management & Computer Security*, *9*(2), 61–70. DOI: 10.1108/09685220110388827

Chen, J., Wang, H., & Zhang, L. (2020). Game-theoretic reinforcement learning for congestion control in VANETs. *IEEE Transactions on Vehicular Technology*, *69*(5), 5406–5418.

Chen, L., Zhang, X., & Wang, Y. (2021). Multi-agent reinforcement learning for adaptive routing in VANETs. *IEEE Transactions on Intelligent Transportation Systems*, *22*(4), 2987–2997.

Chen, S. J., & Hwang, C. L. (1992). *Fuzzy Multiple Attribute Decision Making: Methods and Applications. Lecture Notes in Economics and Mathematical Systems, 375*. Springer., DOI: 10.1007/978-3-642-46768-4

Chen, T.-Y., Jiang, X., Tang, S.-B., Zhou, L., Yuan, X., Zhou, H., Wang, J., Liu, Y., Chen, L.-K., Liu, W.-Y., Zhang, H.-F., Cui, K., Liang, H., Li, X.-G., Mao, Y., Wang, L.-J., Feng, S.-B., Chen, Q., Zhang, Q., & Pan, J.-W. (2021). Implementation of a 46-node quantum metropolitan area network. *NPJ Quantum Information*, *7*(1), 1–6. DOI: 10.1038/s41534-021-00474-3

Chen, X., & Wang, Y. (2021). A machine learning-based adaptive traffic signal control approach for large-scale traffic networks. *Journal of Advanced Transportation*, *2021*, 6682917.

Chettibi, S. (2020). Combination of HF set and MCDM for stable clustering in VANETs. *IET Intelligent Transport Systems*, *14*(3), 190–195. DOI: 10.1049/iet-its.2019.0283

Chiu, S., & Chand, S. (1992). Adaptive Traffic Signal Control Using Fuzzy Logic. *Proceedings of the Second IEEE International Conference on Fuzzy Systems, 3*, 1371-1376. https://doi.org/DOI: 10.1109/FUZZY.1992.258699

Cho, Y., Park, M., & Paek, J. (2022, October). Broadcasting in the vehicular ad-hoc network: Survey on recent advances. In 2022 13th International Conference on Information and Communication Technology Convergence (ICTC) (pp. 813-816). IEEE.

Choukroun, Y., & Wolf, L. (2022). Error correction code transformer. *Advances in Neural Information Processing Systems*.

Chunduru, A., & Lenka, S. (2024). A Secure Method of Communication Through BB84 Protocol in Quantum Key Distribution. Scalable Computing. *Practice and Experience.*, *25*(1), 21–33. DOI: 10.12694/scpe.v25i1.2152

Clement Joe Anand, M., Martin, N., Clementking, A., Rani, S., Priyadharshini, S. S., & Siva, S. "Decision making on optimal selection of advertising agencies using machine learning," In 2023 International Conference on Information Management (ICIM), IEEE, 2023. https://ieeexplore.ieee.org/document/10145172

Clement Joe Anand, M., & Bharatraj, J. (2017). Theory of Triangular Fuzzy Number. *Proceedings of NCATM, 2017*, 80–83.

Clement Joe Anand, M., & Bharatraj, J. (2018). Gaussian qualitative trigonometric functions in a fuzzy circle. *Advances in Fuzzy Systems, 2018*, 1–9. DOI: 10.1155/2018/8623465

Clement Joe Anand, M., Moorthy, C. B., Sivamani, S., Indrakumar, S., Kalaiarasi, K., & Barhoi, A. (2023). Fuzzy intelligence inventory decision optimization model of sustainability and green technologies for mixed uncertainties of carbon emission. In *In: 2023 International Conference on Information Management (ICIM)*. IEEE., https://ieeexplore.ieee.org/document/10145085

Coifman, B., McCord, M R., Mishalani, R G., Iswalt, M., & Ji, Y. (2006, January 1). Roadway traffic monitoring from an unmanned aerial vehicle., 153(1), 11-11. https://doi.org/DOI: 10.1049/ip-its:20055014

Çolak, S., Lima, A., & González, M. C. (2016, March 15). Understanding congested travel in urban areas. *Nature Communications*, *7*(1), 10793. Advance online publication. DOI: 10.1038/ncomms10793 PMID: 26978719

Conway-Beaulieu, J., Athaide, A., Jalali, R., & El-Khatib, K. (2015, November 2).. . *Smartphone-based Architecture for Smart Cities.*, *55*, 79–83. DOI: 10.1145/2815347.2826698

Costandoiu, A., & Leba, M. (2019, February 18). Convergence of V2X communication systems and next generation networks. *IOP Conference Series. Materials Science and Engineering*, *477*, 012052. DOI: 10.1088/1757-899X/477/1/012052

Cui, X., & Chen, G. (2023). The application of advanced computational algorithms used for cooperative communication transmission of vehicular networks: A proposed method. *PeerJ. Computer Science*, *9*, e1643. DOI: 10.7717/peerj-cs.1643 PMID: 38077548

Cui, Y., Huang, X., Wu, D., & Zheng, H. (2020). Machine learning-based resource allocation strategy for network slicing in vehicular networks. *Wireless Communications and Mobile Computing*, *2020*, 1–10.

Cvar, N., Trilar, J., Kos, A., Volk, M., & Duh, E. S. (2020). The Use of Iot Technology in Smart Cities and Smart Villages: Similarities, Differences, and Future Prospects. *Sensors (Basel)*, *20*(14), 3897. Advance online publication. DOI: 10.3390/s20143897 PMID: 32668714

D'Anvers, J. P., Karmakar, A., Sinha Roy, S., & Vercauteren, F. (2018). Saber: Module-LWR based key exchange, CPA-secure encryption and CCA-secure KEM. In *Progress in Cryptology–AFRICACRYPT 2018: 10th International Conference on Cryptology in Africa, Marrakesh, Morocco, May 7–9, 2018, Proceedings*. Springer International Publishing.

Dabiri, S., & Heaslip, K. (2018, January 1). Transport-domain applications of widely used data sources in the smart transportation: A survey. Cornell University. https://doi.org//arxiv.1803.10902DOI: 10.48550

Damaj, I. W., Yousafzai, J. K., & Mouftah, H. T. (2022). Future Trends in Connected and Autonomous Vehicles: Enabling Communications and Processing Technologies. *IEEE Access : Practical Innovations, Open Solutions*, *10*, 42334–42345. Advance online publication. DOI: 10.1109/ACCESS.2022.3168320

Darzi, S., & Yavuz, A. A. (2024). PQC meets ML or AI: Exploring the Synergy of Machine Learning and Post-quantum Cryptography. *Authorea Preprints*.

Das, A., & Rad, P. (2020). Opportunities and challenges in explainable artificial intelligence (xai): A survey. arXiv preprint arXiv:2006.11371.

Dasanayaka, N., Hasan, K. F., Wang, C., & Feng, Y. (2020, January 1). Enhancing Vulnerable Road User Safety: A Survey of Existing Practices and Consideration for Using Mobile Devices for V2X Connections. Cornell University. https://doi.org/ DOI: 10.48550/arXiv.2010

Deng, B., Nan, J., Cao, W., & Wang, W. (2023). A Survey on Integration of Network Communication into Vehicle Real-Time Motion Control. *IEEE Communications Surveys and Tutorials*, *25*(4), 2755–2790. Advance online publication. DOI: 10.1109/COMST.2023.3295384

Deng, Z., Li, Y., & Wu, C. (2022). IoT-based traffic management in smart cities: Current challenges and future directions. *IEEE Communications Magazine*, *60*(4), 18–24.

Devadoss, A. V., & Anand, M. C. J. (2014). *Felix., A.: A CETD matrix approach to analyze the dimensions of the personality of a person. In: 2014 International Conference on Computational Science and Computational Intelligence*. IEEE., https://ieeexplore.ieee.org/abstract/document/6822301

Devadoss, A. V., Anand, M. C. J., & Felix, A. (2014). A CETD matrix approach to analyze the dimensions of the personality of the person. *2014 International Conference on Computational Science and Computational Intelligence*, DOI: 10.1109/CSCI.2014.92

Devadoss, A. V., Anand, M. C. J., & Felix, A. (2014, March). A CETD matrix approach to analyze the dimensions of personality of person. In *2014 International Conference on Computational Science and Computational Intelligence* (Vol. 2, pp. 40-45). IEEE.

Dey, J., & Dutta, R. (2023). Progress in multivariate cryptography: Systematic review, challenges, and research directions. *ACM Computing Surveys*, *55*(12), 1–34. DOI: 10.1145/3571071

Dhinesh Kumar, R., & Rammohan, A. (2023). Revolutionizing Intelligent Transportation Systems with Cellular Vehicle-to-Everything (C-V2X) Technology: Current Trends, Use Cases, Emerging Technologies, Standardization Bodies, Industry Analytics and Future Directions. *Vehicular Communications*, *43*, 100638. Advance online publication. DOI: 10.1016/j.vehcom.2023.100638

Dibaei, M., Zheng, X., Jiang, K., Abbas, R., Liu, S., Zhang, Y., Xiang, Y., & Yu, S. (2020). Attacks and Defences on Intelligent Connected Vehicles: A Survey. *Digital Communications and Networks*, *6*(4), 399–421. Advance online publication. DOI: 10.1016/j.dcan.2020.04.007

Dion, F., Rakha, H., & Kang, Y. S. (2004). Comparison of Delay Estimates at Under-Saturated and Over-Saturated Pre-Timed Signalized Intersections. *Transportation Research Part B: Methodological*, *38*(2), 99–122. DOI: 10.1016/S0191-2615(03)00003-1

do Vale Saraiva, T., Girardi, R., & Campos, C. A. V. (2021, September). Machine learning applied to network traffic for vehicular applications to predict the vehicle context. In 2021 IEEE 94th Vehicular Technology Conference (VTC2021-Fall) (pp. 1-5). IEEE. DOI: 10.1109/VTC2021-Fall52928.2021.9625538

Dong, S., Wang, P., & Abbas, K. (2021). A survey on deep learning and its applications. *Computer Science Review*, *40*, 100379. DOI: 10.1016/j.cosrev.2021.100379

Doosti-Aref, A., Neelam, S. G., Sahu, P. R., Zhu, X., Basar, E., Coleri, S., & Arslan, H. (2023). Performance Analysis of OTSM under Hardware Impairments in Millimeter-Wave Vehicular Communication Networks. *arXiv preprint arXiv:2309.04161*.

Druml, N., Ryabokon, A., Schorn, R., Koszescha, J., Ozols, K., Levinskis, A., Novickis, R., ... (2021). "Programmable Systems for Intelligence in Automobiles (PRYSTINE): Final Results after Year 3." In *Proceedings - 2021 24th Euromicro Conference on Digital System Design, DSD 2021*, DOI: 10.1109/DSD53832.2021.00049

Dzemydienė, D., Burinskienė, A., & Miliauskas, A. (2021). Integration of multi-criteria decision support with infrastructure of smart services for sustainable multi-modal transportation of freights. *Sustainability (Basel), 13*(9), 4675. DOI: 10.3390/su13094675

El-Rewini, Z., Sadatsharan, K., Selvaraj, D. F., Plathottam, S. J., & Ranganathan, P. (2020). Cybersecurity challenges in vehicular communications. *Vehicular Communications, 23*, 100214. DOI: 10.1016/j.vehcom.2019.100214

ElSahly, O., & Abdelfatah, A. (2022). A Systematic Review of Traffic Incident Detection Algorithms. *Sustainability (Basel), 14*(22), 14859. Advance online publication. DOI: 10.3390/su142214859

Erlang, A. K. (1917). Solution of some problems in the theory of probabilities of significance in automatic telephone exchanges. *Post Office Electrical Engineer's Journal, 10*, 189–197.

ETSI, G. (2021, March). 015. Quantum Key Distribution (QKD); Control Interface for Software Defined Networks.

Evangeline, C. S., & Kumaravelu, V. B. (2020). AHP-FVIKOR based access network selection in vehicular communications. In *Inventive Communication and Computational Technologies* [Springer Singapore.]. *Proceedings of ICICCT, 2019*, 829–837.

Evangeline, C. S., & Kumaravelu, V. B. (2022). A two-phase fuzzy based access network selection scheme for vehicular ad hoc networks. *Peer-to-Peer Networking and Applications, 15*(1), 107–133. DOI: 10.1007/s12083-021-01228-w PMID: 36188111

Eziama, E., Tepe, K., Balador, A., Nwizege, K. S., & Jaimes, L. M. (2018, December). Malicious node detection in vehicular ad-hoc network using machine learning and deep learning. In *2018 IEEE Globecom Workshops (GC Wkshps)* (pp. 1-6). IEEE.

Fang, Y., Huang, X., Qin, L., Zhang, Y., Zhang, W., Cheng, R., & Lin, X. (2020). A survey of community search over big graphs. *The VLDB Journal.*

Farman, H., Khan, Z., Jan, B., Boulila, W., Habib, S., & Koubaa, A. (2022). Smart Transportation in Developing Countries: An Internet-of-Things-Based Conceptual Framework for Traffic Control. *Wireless Communications and Mobile Computing, 2022*, 1–11. Advance online publication. DOI: 10.1155/2022/8219377

Fernando, K. C. Machine Learning in 6G-enabled Vehicular Network: A State-of-the-Art-Review.

Feukeu, E. A., & Mbuyu, S. (2023). Machine Learning Algorithm for a Link Adaptation strategy in a Vehicular Ad hoc Network. Inteligencia Artificial: Revista Iberoamericana de Inteligencia Artificial, 26(72).

Figueira J, Greco S and Ehrgott M (2005). Multiple Criteria Decision Analysis: State of the Art Surveys. *Springer, 78.*https://doi.org/DOI: 10.1007/b100605

Gandhi, M., Solanki, D S., Daptardar, R S., & Baloorkar, N S. (2020, December 1). Smart Control of Traffic Light Using Artificial Intelligence. https://doi.org/DOI: 10.1109/ICRAIE51050.2020.9358334

Gao, R., Liu, T., & Wang, Z. (2023). Hierarchical reinforcement learning for scalable congestion control in VANETs. *Computer Communications*, *191*, 24–33.

Gao, Y., Wang, H., & Sun, J. (2022). IoT-based emergency vehicle detection and prioritization using machine learning. *IEEE Access : Practical Innovations, Open Solutions*, *10*, 63774–63785.

Gautam, K., Jain, V. K., Verma, S. S., & Vyas, S. (2023, February). Vehicular Communication Strategy Using Machine Learning and Image Processing to Enhance Observations and Control on the Road Side Area. In *Proceedings of 3rd International Conference on Recent Trends in Machine Learning, IoT, Smart Cities and Applications: ICMISC 2022* (pp. 397-404). Singapore: Springer Nature Singapore. DOI: 10.1007/978-981-19-6088-8_34

Ghiasi, A., Hale, D A., Bared, J., Kondyli, A., & Ma, J. (2018, November 1). A Dynamic Signal Control Approach for Integrated Ramp and Mainline Metering. https://doi.org/DOI: 10.1109/ITSC.2018.8569567

Goh, M. (2002, March 1). Congestion management and electronic road pricing in Singapore. Elsevier BV, 10(1), 29-38. https://doi.org/DOI: 10.1016/S0966-6923(01)00036-9

Gohar, A., & Nencioni, G. (2021). The Role of 5g Technologies in a Smart City: The Case for Intelligent Transportation System. *Sustainability (Basel)*, *13*(9), 5188. Advance online publication. DOI: 10.3390/su13095188

Gökhan, A. K. E. L. (2022). Prioritization of the Theme Park satisfaction criteria with multi-criteria decision-making method: Level based weight assessment model. *Alphanumeric Journal*, *10*(2), 105–126. DOI: 10.17093/alphanumeric.1028864

Golbabaei, F., Yigitcanlar, T., & Bunker, J. (2021). The Role of Shared Autonomous Vehicle Systems in Delivering Smart Urban Mobility: A Systematic Review of the Literature. *International Journal of Sustainable Transportation*, *15*(10), 731–748. Advance online publication. DOI: 10.1080/15568318.2020.1798571

Gonçalves, M. C., Machado, T. R., Nara, E. O. B., Dias, I. C. P., & Vaz, L. V. (2023, November). Integrating Machine Learning for Predicting Future Automobile Prices: A Practical Solution for Enhanced Decision-Making in the Automotive Industry. In Conference on e-Business, e-Services and e-Society (pp. 91-103). Cham: Springer Nature Switzerland.

Gopal, K., & Prakash, R. (2012, September 28). Building the Business Case for Telematics Based Diagnostics at Mahindra Reva. *Multidisciplinary Digital Publishing Institute*, *5*(3), 722–729. DOI: 10.3390/wevj5030722

Grassl, M., Langenberg, B., Roetteler, M., & Steinwandt, R. (2016, February). Applying Grover's algorithm to AES: quantum resource estimates. In *International Workshop on Post-Quantum Cryptography*. Cham: Springer International Publishing. DOI: 10.1007/978-3-319-29360-8_3

Guidoni, D. L., Maia, G., Souza, F. S. H. D., Villas, L. A., & Loureiro, A A F. (2020, January 1). Vehicular Traffic Management Based on Traffic Engineering for Vehicular Ad Hoc Networks. *IEEE Access : Practical Innovations, Open Solutions*, *8*, 45167–45183. DOI: 10.1109/ACCESS.2020.2978700

Guo, H., & Yu, X. (2022). A survey on blockchain technology and its security. Blockchain: research and applications, 3(2), 100067.

Gupta, R., Sharma, V., & Malik, A. (2023). RL-driven priority management for mixed vehicular networks. *IEEE Access : Practical Innovations, Open Solutions*, *11*, 49256–49267.

Gyawali, S., & Qian, Y. (2019, May). Misbehavior detection using machine learning in vehicular communication networks. In *ICC 2019-2019 IEEE International Conference on Communications (ICC)* (pp. 1-6). IEEE. DOI: 10.1109/ICC.2019.8761300

Gyawali, S., Qian, Y., & Hu, R. Q. (2020). Machine learning and reputation-based misbehavior detection in vehicular communication networks. *IEEE Transactions on Vehicular Technology*, *69*(8), 8871–8885. DOI: 10.1109/TVT.2020.2996620

Hamza, A., & Kumar, B. (2020). A review paper on DES, AES, RSA encryption standards. *In:9th International Conference System Modeling and Advancement in Research Trends (SMART),* IEEE. DOI: 10.1109/SMART50582.2020.9336800

Haque, M A., & Hossain, M D. (2014, December 1). Technology survey of wireless communication for in-vehicle applications. https://doi.org/DOI: 10.1109/SKIMA.2014.7083523

Hasija, T., Ramkumar, K., Kaur, A., Mittal, S., & Singh, B. (2022). A survey on nist selected third round candidates for post quantum cryptography, in: *7th International Conference on Communication and Electronics Systems (ICCES)*, IEEE. DOI: 10.1109/ICCES54183.2022.9835864

Hejazi, H., & Bokor, L. (2021). A Survey on the Use-Cases and Deployment Efforts Toward Converged Internet of Things (IoT) and Vehicle-to-Everything (V2X) Environments. *Acta Technica Jaurinensis*, *15*(2), 58–73. Advance online publication. DOI: 10.14513/actatechjaur.00627

Hemmer, P., Kühl, N., & Schöffer, J. (2021). Utilizing Active Machine Learning for Quality Assurance: A case study of virtual car renderings in the automotive industry. arXiv preprint arXiv:2110.09023.

Hevey, D. (2018). Network analysis: A brief overview and tutorial. *Health Psychology and Behavioral Medicine*, *6*(1), 301–328.

He, Z.. (2021). Real-time traffic signal control for multiple intersections based on IoT. *IEEE Internet of Things Journal*, *8*(12), 9907–9918.

Hobbs, B. F., & Meier, P. (2000). *Energy Decisions and the Environment: A Guide to the Use of Multicriteria Methods. International Series in Operations Research & Management Science*. Springer., DOI: 10.1007/978-1-4615-4477-7

Hoffstein, J., Howgrave-Graham, N., Pipher, J., Silverman, J. H., & Whyte, W. (2003). NTRUSIGN: Digital signatures using the NTRU lattice. In *Cryptographers' track at the RSA conference*. Springer Berlin Heidelberg. DOI: 10.1007/3-540-36563-X_9

Hofheinz, D., Hövelmanns, K., & Kiltz, E. (2017). A modular analysis of the Fujisaki-Okamoto transformation. In: *Theory of Cryptography Conference*. Cham: Springer International Publishing. DOI: 10.1007/978-3-319-70500-2_12

Horrey, W J., Lesch, M F., Dainoff, M J., Robertson, M M., & Noy, Y I. (2011, December 24). On-Board Safety Monitoring Systems for Driving: Review, Knowledge Gaps, and Framework. Elsevier BV, 43(1), 49-58. https://doi.org/DOI: 10.1016/j.jsr.2011.11.004

Hossain, M. A., Md Noor, R., Yau, K. L. A., Azzuhri, S. R., Z'aba, M. R., Ahmedy, I., & Jabbarpour, M. R. (2021). Machine learning-based cooperative spectrum sensing in dynamic segmentation enabled cognitive radio vehicular network. *Energies*, *14*(4), 1169. DOI: 10.3390/en14041169

Huang, A., Bao, J., Zhang, J., & Zhang, T. (2012, August 1). Vehicle information monitoring system based on mobile phone. https://doi.org/DOI: 10.1109/ICAL.2012.6308216

Huang, Y., Zhang, L., & Li, W. (2021). Multi-agent reinforcement learning for dynamic task allocation in VANET-based emergency services. *IEEE Access : Practical Innovations, Open Solutions*, *9*, 89756–89766.

Hui, Y., & Dey, A. (2020). A study on directed neutrosophic social networks. Infinite Study.

Hülsing, A., Ning, K. C., Schwabe, P., Weber, F. J., & Zimmermann, P. R. (2021). Post-quantum wireguard. In: *IEEE Symposium on Security and Privacy (SP)*. IEEE. DOI: 10.1109/SP40001.2021.00030

Hussein, N. H., Yaw, C. T., Koh, S. P., Tiong, S. K., & Chong, K. H. (2022). A Comprehensive Survey on Vehicular Networking [IEEE Access.]. *Communications, Applications, Challenges, and Upcoming Research Directions*, *10*, 86127–86180.

Hwang, C. L., & Yoon, K. (1981). *Multiple Attribute Decision Making: Methods and Applications. Lecture Notes in Economics and Mathematical Systems, 186.* Springer., DOI: 10.1007/978-3-642-48318-9

Ihama, E. I., Akazue, M. I., Omede, E., & Ojie, D. (2023). A Framework for Smart City Model Enabled by Internet of Things (IoT). *International Journal of Computer Applications*, *185*(6), 6–11.

Ioniţă, S. (2017, October 1). Autonomous vehicles: From paradigms to technology. *IOP Conference Series. Materials Science and Engineering*, *252*, 012098–012098. DOI: 10.1088/1757-899X/252/1/012098

Ivanov, I., Maple, C., Watson, T., & Lee, S. (2018, March). Cyber security standards and issues in V2X communications for Internet of Vehicles. In *Living in the Internet of Things: Cybersecurity of the IoT-2018* (pp. 1–6). IET.

Jadaan, K., Zeater, S., & Abukhalil, Y. (2017, January 1). Connected Vehicles: An Innovative Transport Technology. Elsevier BV, 187, 641-648. https://doi.org/DOI: 10.1016/j.proeng.2017.04.425

Jahangeer, N. (2023). Fast multicriteria network selection scheme using hybrid of cat swarm optimization and TOPSIS algorithm for optimal handover in VANET. *Transactions on Emerging Telecommunications Technologies*, *34*(5), e4743. DOI: 10.1002/ett.4743

Jahn, O., Möhring, R. H., Schulz, A. S., & Stier-Moses, N. E. (2005, August 1). System-Optimal Routing of Traffic Flows with User Constraints in Networks with Congestion. *Operations Research*, *53*(4), 600–616. DOI: 10.1287/opre.1040.0197

Jain, S., Jain, V. K., & Mishra, S. (2024). Fuzzy-AHP-based optimal RSU deployment (Fuzzy-AHP-ORD) approach using road and traffic analysis in VANET. *Ad Hoc Networks*, •••, 103529.

Jalali, A., Azarderakhsh, R., Kermani, M. M., & Jao, D. (2017). Supersingular isogeny Diffie-Hellman key exchange on 64-bit ARM. *IEEE Transactions on Dependable and Secure Computing*.

Jambotkar, S., Guo, L., & Jia, Y. (2021, September 27). Adaptive Optimization of Autonomous Vehicle Computational Resources for Performance and Energy Improvement. https://doi.org/DOI: 10.1109/IROS51168.2021.9635828

Janwa, H., & Moreno, O. (1996). McEliece public key cryptosystems using algebraic-geometric codes. *Designs, Codes and Cryptography*, *8*(3), 293–307. DOI: 10.1023/A:1027351723034

Jaramillo-Alcazar, A., Govea, J., & Villegas-Ch, W. (2023). Advances in the Optimization of Vehicular Traffic in Smart Cities: Integration of Blockchain and Computer Vision for Sustainable Mobility. *Sustainability (Basel)*, *15*(22), 15736. Advance online publication. DOI: 10.3390/su152215736

Jayashree, S., & Kumar, S. S. (2023). A comprehensive survey on effective key management schemes for secured authentication in vehicular ad hoc network. *International Journal of Information and Computer Security*, *22*(3-4), 354–376.

Jiang, P., Li, P., Zhang, T., Huang, W., He, H., Nie, L., & Liu, Q. (2019, August). Budget constraint roadside units placement for traffic flows monitoring system with reliability in vehicular networks. In 2019 IEEE 21st International Conference on High Performance Computing and Communications; IEEE 17th International Conference on Smart City; IEEE 5th International Conference on Data Science and Systems (HPCC/SmartCity/DSS) (pp. 421-428). IEEE.

Jiang, P., Li, P., Zhang, T., Zhong, E., Jin, Y., He, H., & Liu, Q. (2018, September). Roadside units placement for traffic flows coverage requirement in vehicular networks. In *2018 IEEE International Conference on Smart Cloud (SmartCloud)* (pp. 145-152). IEEE.

Jojo, L. (2014). Predicting Traffic Congestion: A Queuing Perspective. [Open Journal of Modelling and Simulation.]. *Open Journal of Modelling and Simulation*, *2*(2), 57–66. DOI: 10.4236/ojmsi.2014.22008

Jokić, Ž., Božanić, D., & Pamučar, D. (2021). Selection of fire position of mortar units using LBWA and Fuzzy MABAC model. *Operational Research in Engineering Sciences: Theory and Applications*, *4*(1), 115–135.

Joshi, S. K., Aktas, D., Wengerowsky, S., Lončarić, M., Neumann, S. P., Liu, B., Scheidl, T., Lorenzo, G. C., Samec, Ž., Kling, L., Qiu, A., Razavi, M., Stipčević, M., Rarity, J. G., & Ursin, R. (2020). A trusted node–free eight-user metropolitan quantum communication network. *Science Advances*, *6*(36), 73–81. DOI: 10.1126/sciadv.aba0959 PMID: 32917585

Jurczenia, K., & Rak, J. (2022). A survey of vehicular network systems for road traffic management. *IEEE Access : Practical Innovations, Open Solutions, 10*, 42365–42385.

Kamarianakis, Y., & Prastacos, P. (2005). Space–time Modeling of Traffic Flow. *Computers & Geosciences, 31*(2), 119–133. DOI: 10.1016/j.cageo.2004.05.012

Karim, S. M., Habbal, A., Chaudhry, S. A., & Irshad, A. (2022). Architecture, protocols, and security in IoV: Taxonomy, analysis, challenges, and solutions. *Security and Communication Networks, 2022*, 1–19. DOI: 10.1155/2022/1131479

Kaur, K., & Rampersad, G. (2018, April 1). Trust in driverless cars: Investigating key factors influencing the adoption of driverless cars. Elsevier BV, 48, 87-96. https://doi.org/DOI: 10.1016/j.jengtecman.2018.04.006

Kavas-Torris, O., Cantas, M. R., Gelbal, Ş. Y., & Güvenç, L. (2023, January 1). Hardware-in-the-Loop and Road Testing of RLVW and GLOSA Connected Vehicle Applications. Cornell University. https://doi.org/DOI: 10.48550/arXiv.2306

Kaviyarasu, M., Aslam, M., Afzal, F., Mohammed Saeed, M., Mehmood, A., & Gul, S. (2024). The connectivity indices concept of neutrosophic graph and their application of computer network, highway system and transport network flow. *Scientific Reports, 14*(1), 4891.

Kayış, E. A. (2023). Demand forecasting using statistical and machine learning methods–a case studyin automotive industry (Master's thesis, İzmir Ekonomi Üniversitesi).

Khadka, S., Li, P. T., & Wang, Q. (2022). Developing Novel Performance Measures for Traffic Congestion Management and Operational Planning Based on Connected Vehicle Data. *Journal of Urban Planning and Development, 148*(2), 04022016. Advance online publication. DOI: 10.1061/(ASCE)UP.1943-5444.0000835

Khalil, O K. (2017, June 1). A Study on Road Accidents in Abu Dhabi Implementing a Vehicle Telematics System to Reduce Cost, Risk and Improve Safety. https://doi.org/DOI: 10.1109/DeSE.2017.41

Khan, I.. (2021). Smart traffic control system using IoT to improve vehicle flow and safety: A review. *IEEE Access : Practical Innovations, Open Solutions, 9*, 164366–164384.

Khan, M. A., Ectors, W., Bellemans, T., Janssens, D., & Wets, G. (2017). UAV-based traffic analysis: A universal guiding framework based on literature survey. *Transportation Research Procedia, 22*, 541–550.

Kihl, M., Linden, C., & Zhou, W. (2012). Adaptive message dissemination for congestion control in VANETs. *Journal of Communications and Networks (Seoul), 14*(6), 620–630.

Kokane, C. D., Mohadikar, G., Khapekar, S., Jadhao, B., Waykole, T., & Deotare, V. V. (2023). Machine Learning Approach for Intelligent Transport System in IOV-Based Vehicular Network Traffic for Smart Cities. International Journal of Intelligent Systems and Applications in Engineering, 11(11s), 06-16.

Köksalan, M., Wallenius, J., & Zionts, S. (2011). *Multiple Criteria Decision Making: From Early History to the 21st Century*. World Scientific., DOI: 10.1142/8042

Kuchta, R. (2014, January 1). Smart City Concept, Applications and Services. *OMICS Publishing Group, 03*(02). Advance online publication. DOI: 10.4172/2167-0919.1000117

Kumar, P. M.. (2020). An IoT-based smart traffic signal monitoring system using LoRa and cloud technologies. *Future Generation Computer Systems, 112*, 190–201.

Kumar, P., Singh, A., & Verma, R. (2022). Blockchain-enhanced reinforcement learning framework for secure VANET communication. *IEEE Internet of Things Journal, 9*(3), 1804–1815.

Kumar, S., Velliangiri, S., Karthikeyan, P., Kumari, S., & Khan, M. K. (2024). A survey on the blockchain techniques for the internet of vehicles security. *Transactions on Emerging Telecommunications Technologies, 35*(4), e4317. DOI: 10.1002/ett.4317

Kungumaraj, E., Lathanayagam, E., Saikia, U., Anand, M. C. J., Khanna, S. T., Martin, N., Tiwari, M., & Edalatpanah, S. A. (2024). Khanna, S.T., Martin, N., Tiwari, M., Edalatpanah, S. A.: Neutrosophic Topological Vector Spaces and its Properties.*International Journal of Neutrosophic Science, 23*(2), 63–76. DOI: 10.54216/IJNS.230206

Kungumaraj, E., Lathanayagam, E., Saikia, U., Anand, M., Khanna, S. T., Martin, N., & Edalatpanah, S. A. (2024). Neutrosophic Topological Vector Spaces and its Properties. [IJNS]. *International Journal of Neutrosophic Science, 23*(2).

Lazarus, J., Ugirumurera, J., Hinardi, S., Zhao, M., Shyu, F., Wang, Y., Yao, S., & Bayen, A M. (2018, November 1). A Decision Support System for Evaluating the Impacts of Routing Applications on Urban Mobility. https://doi.org/DOI: 10.1109/ITSC.2018.8569622

Leach, P., Mealling, M., & Salz, R. (2005). A universally unique identifier (uuid) urn namespace (No. rfc4122).

Lee, C., & Lee, H. (2001). Development and Evaluation of Fuzzy Logic-Based Traffic Signal Control. *Transportation Research Record: Journal of the Transportation Research Board, 1748*(1), 82–88. DOI: 10.3141/2663-11

Lee, E., Cen, X., Lo, H. K., & Ng, K. F. (2021, January 1). Designing zonal-based flexible bus services under stochastic demand. Cornell University. https://doi.org/ DOI: 10.48550/arXiv.2102

Li, H. (2016, January 1). Study on Green Transportation System of International Metropolises. Elsevier BV, 137, 762-771. https://doi.org/DOI: 10.1016/j.proeng.2016.01.314

Liang, Y., Liu, H., & Rajan, D. (2012, May). Optimal placement and configuration of roadside units in vehicular networks. In 2012 IEEE 75th Vehicular Technology Conference (VTC Spring) (pp. 1-6). IEEE.

Li, H., Zhao, J., & Xu, T. (2022). Edge computing-based reinforcement learning for real-time congestion control in VANETs. *Future Generation Computer Systems, 135*, 122–132.

Limbasiya, T., Teng, K. Z., Chattopadhyay, S., & Zhou, J. (2022). A Systematic Survey of Attack Detection and Prevention in Connected and Autonomous Vehicles. *Vehicular Communications, 37*, 100515. Advance online publication. DOI: 10.1016/j.vehcom.2022.100515

Li, P., Huang, C., & Liu, Q. (2014). BCDP: Budget constrained and delay-bounded placement for hybrid roadside units in vehicular ad hoc networks. *Sensors (Basel), 14*(12), 22564–22594. PMID: 25436656

Li, P., Liu, Q., Huang, C., Wang, J., & Jia, X. (2015, June). Delay-bounded minimal cost placement of roadside units in vehicular ad hoc networks. In *2015 IEEE International Conference on Communications (ICC)* (pp. 6589-6594). IEEE.

Liu, H. Q., Ding, S. J., Yang, L. C., & Yang, T. (2014). A connectivity-based strategy for roadside units placement in vehicular ad hoc networks. *International Journal of Hybrid Information Technology, 7*(1), 91–108.

Liu, Y.. (2023). Real-time traffic density estimation using deep learning and IoT sensors. *IEEE Sensors Journal, 23*(2), 1168–1178.

Liu, Y., Wang, H., & Chen, J. (2016). Clustering-based message dissemination for congestion mitigation in VANETs. *International Journal of Distributed Sensor Networks, 12*(9), 243–254.

Li, W., Wang, G. G., & Gandomi, A. H. (2021). A survey of learning-based intelligent optimization algorithms. *Archives of Computational Methods in Engineering, 28*(5), 3781–3799. DOI: 10.1007/s11831-021-09562-1

Li, X.. (2021). A comprehensive review of traffic signal control methods: Current status and future perspectives. *Transportation Research Part C, Emerging Technologies, 126*, 103057.

Li, Y., Wu, C., Zhong, L., & Yoshinaga, T. (2022, October). A communication-efficient distributed machine learning scheme in a vehicular network. In *Proceedings of the Conference on Research in Adaptive and Convergent Systems* (pp. 92-98). DOI: 10.1145/3538641.3561494

Li, Y., Zhao, J., & Xu, W. (2022). Exploring 6G for RL-based congestion control in vehicular networks. *IEEE Communications Magazine, 60*(5), 28–35.

Liya, X., Chuanhe, H., Peng, L., & Junyu, Z. (2013, December). A randomized algorithm for roadside units placement in vehicular ad hoc network. In 2013 IEEE 9th International Conference on Mobile Ad-hoc and Sensor Networks (pp. 193-197). IEEE.

Lopez Alduncin, M. B. (2018). A customer relationship management, machine learning case study in the automotive service industry.

Lundberg, H., Mowla, N. I., Abedin, S. F., Thar, K., Mahmood, A., Gidlund, M., & Raza, S. (2022). Experimental analysis of trustworthy in-vehicle intrusion detection system using explainable artificial intelligence (xai). *IEEE Access : Practical Innovations, Open Solutions, 10*, 102831–102841. DOI: 10.1109/ACCESS.2022.3208573

Lyu, H. M., Yin, Z. Y., Zhou, A., & Shen, S. L. (2023). MCDM-based flood risk assessment of metro systems in smart city development: A review. *Environmental Impact Assessment Review, 101*, 107154. DOI: 10.1016/j.eiar.2023.107154

Magableh, G., & Mumani, A. (2022). Simulation Based-MCDM Approach for Evaluating Traffic Solutions. *Promet (Zagreb), 34*(1), 117–133. DOI: 10.7307/ptt.v34i1.3842

Mahesh, J., Bodhisatwa, M., & Somnath, D. 2021. "Biometric-Based Secure Authentication for IoT-Enabled Devices and Applications." In *Security of Internet of Things Nodes*, DOI: 10.1201/9781003127598-4-4

Mahmood, A., Zen, H., & Hilles, S M S. (2018, January 1). Big Data and Privacy Issues for Connected Vehicles in Intelligent Transportation Systems. Springer Nature, 1-7. https://doi.org/DOI: 10.1007/978-3-319-63962-8_234-1

Mahrez, Z., Sabir, E., Badidi, E., Saad, W., & Sadik, M. (2021, June 11). Smart Urban Mobility: When Mobility Systems Meet Smart Data. *IEEE Transactions on Intelligent Transportation Systems, 23*(7), 6222–6239. DOI: 10.1109/TITS.2021.3084907

Malaca, P., Rocha, L. F., Gomes, D., Silva, J., & Veiga, G. (2019). Online inspection system based on machine learning techniques: Real case study of fabric textures classification for the automotive industry. *Journal of Intelligent Manufacturing, 30*, 351–361.

Malik, F., Shah, M A., & Biomechanics, A B A. (2018, September 1). Intelligent Transport System: An Important Aspect of Emergency Management in Smart Cities. https://doi.org/DOI: 10.23919/IConAC.2018.8749062

Malik, R. Q. (2021). A roadside unit positioning framework in the context of vehicle-to-infrastructure based on integrated AHP-entropy and group-VIKOR (Doctoral dissertation, Universiti Tun Hussein Onn Malaysia).

Malik, R. Q. (2021). *A roadside units positioning framework in the context of vehicle-to-infrastructure based on integrated AHP-entropy and group-VIKOR* (Doctoral dissertation, Universiti Tun Hussein Onn Malaysia).

Malik, S., Bandi, P., & Sun, W. (2021). "An Experimental Study of Denial of Service Attack against Platoon of Smart Vehicles." In *Proceedings - 2021 4th International Conference on Connected and Autonomous Driving, MetroCAD 2021*, DOI: 10.1109/MetroCAD51599.2021.00013

Malik, K. I., Arooj, H., Jamil, B., Munir, K., & Ijaz, H. (2023). MCDM-Based Optimal Gateway Selection in Mobile Fog Computing. *Technical Journal, 28*(04), 53–63.

Malik, R. Q., Zaidan, A. A., Zaidan, B. B., Ramli, K. N., Albahri, O. S., Kareem, Z. H., & Salih, M. M. (2022). Novel roadside unit positioning framework in the context of the vehicle-to-infrastructure communication system based on AHP—Entropy for weighting and borda—VIKOR for uniform ranking. *International Journal of Information Technology & Decision Making, 21*(04), 1233–1266.

Mamadou, A., & Chalhoub, G. (2022). MCDM-based RAT Selection method between LTE-D2D and IEEE 802.11.

Manshath, A., Kungumaraj, E., Lathanayagam, E., Anand, M. C., Martin, N., Muniyandy, E., & Indrakumar, S. (2024). Neutrosophic Integrals by Reduction Formula and Partial Fraction Methods for Indefinite Integrals. [IJNS]. *International Journal of Neutrosophic Science, 23*(1).

Mardle, S., Pascoe, S., & Herrero, I. (2004). Management Objective Importance in Fisheries: An Evaluation Using the Analytic Hierarchy Process (AHP). *Environmental Management, 33*(1), 1–11. DOI: 10.1007/s00267-003-3070-y PMID: 14994158

Martinelli, M., Martelli, P., & Gatto, A. (2022). "POLIQI: Milano quantum infrastructure," in ICOP.

Masood, A., Lakew, D. S., & Cho, S. (2020). Security and Privacy Challenges in Connected Vehicular Cloud Computing. *IEEE Communications Surveys and Tutorials*, *22*(4), 2725–2764. Advance online publication. DOI: 10.1109/COMST.2020.3012961

Massobrio, R., Toutouh, J., & Nesmachnow, S. (2020). Multi-Objective Evolutionary Algorithms for Smart Placement of Roadside Units in Vehicular Networks. In Evolutionary Multi-Objective System Design (pp. 85-114). Chapman and Hall/CRC.

Matsuda, K.. (2021). *"Field Demonstration of Real-time 14 Tb/s 220 m FSO Transmission with Class 1 Eye-safe 9-aperture Transmitter," Optical Fiber Communications Conference and Exhibition*. OFC.

Mavromatis, I., Tassi, A., Piechocki, R J., & Sooriyabandara, M. (2020, May 1). On Urban Traffic Flow Benefits of Connected and Automated Vehicles. https://doi.org/DOI: 10.1109/VTC2020-Spring48590.2020.9128758

Ma, Y., Zheng, Y. B., Wang, S. Y., Wong, Y. D., & Easa, S. M. (2024). Virtual-real-fusion simulation framework for evaluating and optimizing small-spatial-scale placement of cooperative roadside sensing units. *Computer-Aided Civil and Infrastructure Engineering*, *39*(5), 707–730.

Meenakshi, A., & Mythreyi, O. (2023). RETRACTED: Applications of Neutrosophic social network using max product networks. *Journal of Intelligent & Fuzzy Systems*, *45*(1), 407–420.

Meenakshi, A., Mythreyi, O., Bramila, M., Kannan, A., & Senbagamalar, J. (2023). Application of neutrosophic optimal network using operations. *Journal of Intelligent & Fuzzy Systems*, *45*(1), 421–433.

Mehar, S., Senouci, S. M., Kies, A., & Zoulikha, M. M. (2015, January). An optimized roadside units (RSU) placement for delay-sensitive applications in vehicular networks. In 2015 12th Annual IEEE consumer communications and networking conference (CCNC) (pp. 121-127). IEEE.

Meinzer, S., Jensen, U., Thamm, A., Hornegger, J., & Eskofier, B. M. (2017). Can machine learning techniques predict customer dissatisfaction? A feasibility study for the automotive industry. *Artificial Intelligence Research*, *6*(1), 80–90.

Mekala, M. S., Eyad, E., & Srivastava, G. (2023, October). Object-aware Multi-criteria Decision-Making Approach using the Heuristic data-driven Theory for Intelligent Transportation Systems. In *2023 IEEE 10th International Conference on Data Science and Advanced Analytics (DSAA)* (pp. 1-8). IEEE. DOI: 10.1109/DSAA60987.2023.10302554

Meneguette, R I., Grande, R E D., & Loureiro, A A F. (2018, January 1). Intelligent Transport System in Smart Cities. Springer International Publishing. https://doi.org/DOI: 10.1007/978-3-319-93332-0

Meng, L., Luo, J., & Wei, X. (2013, June 11). A Telematics Service System for Vehicle Safety Based on Vehicular Data Acquisition Module. https://doi.org/DOI: 10.1061/9780784413036.233

Miriam, M. R., Martin, N., & Clement Joe Anand, M. (2023). Inventory model promoting smart production system with zero defects. *International Journal of Applied and Computational Mathematics*, *9*(4), 47. DOI: 10.1007/s40819-023-01524-4

Mishra, A. R., Rani, P., & Gunasekaran, M. (2017). A Multi-Criteria Decision-Making Approach for Cloud Service Provider Selection Based on Fuzzy AHP and TOPSIS. *Journal of Enterprise Information Management*, *30*(4), 586–610. DOI: 10.1108/JEIM-12-2016-0201

Mittal, G., Sharma, M., Khatri, S K., & Johri, P. (2019, February 1). Optimizing Autonomous Vehicles to Prevent On-Road Congestion. https://doi.org/DOI: 10.1109/AICAI.2019.8701385

Modi, M., Agarwal, G., Patil, V., Khare, A., Shukla, S., & Sankhala, A. (2019). Minimization of traffic congestion by using queuing theory. *Int. J. Sci. Technol. Res*, *8*(10), 532–541.

Mohaghar, A., Heydarzadeh Moghaddam, H., & Ghasemi, R. (2023). Developing a Model to Optimize Maximum Coverage of Roadside Units Placement in Vehicular Ad–hoc Networks for Intelligent Transportation Systems. *Journal of Industrial Management Perspective*, *13*(2), 211–240.

Mohammad Asif, H. (2022). Cooperative spectrum sensing based on machine learning in cognitive radio vehicular network/Mohammad Asif Hossain (Doctoral dissertation, Universiti Malaya).

Msakni, M. K., Risan, A., & Schütz, P. (2023). Using machine learning prediction models for quality control: A case study from the automotive industry. *Computational Management Science*, *20*(1), 14. PMID: 36942085

Mudhivarthi, B. R., Thakur, P., & Singh, G. (2023). Aspects of Cyber Security in Autonomous and Connected Vehicles. *Applied Sciences (Basel, Switzerland)*, *13*(5), 3014. Advance online publication. DOI: 10.3390/app13053014

Mundra, A., Mundra, S., Srivastava, J. S., & Gupta, P. (2020). Optimized deep neural network for cryptanalysis of DES. *Journal of Intelligent & Fuzzy Systems*, *38*(5), 5921–5931. DOI: 10.3233/JIFS-179679

Munhoz, P A M. S. A., Dias, F. D. C., Chinelli, C. K., Guedes, A. L. A., Santos, J A N. D., Silva, W. D. S. E., & Soares, C A P. (2020, December 21). Smart Mobility: The Main Drivers for Increasing the Intelligence of Urban Mobility. *Sustainability (Basel), 12*(24), 10675–10675. DOI: 10.3390/su122410675

Musa, S. (2016, January 1). Smart Cities - A Roadmap for Development. *OMICS Publishing Group, 05*(03). Advance online publication. DOI: 10.4172/2167-0919.1000144

Mustafa, I., & Huang, C. Y. (2020). Lightweight cryptographic URLLC for 5G-V2X. *In:International Symposium on Networks, Computers and Communications (ISNCC),* IEEE.

Naderi, M., Chakareski, J., & Ghanbari, M. (2023). Hierarchical Q-learning-enabled neutrosophic AHP scheme in candidate relay set size adaption in vehicular networks. *Computer Networks*, *235*, 109968. DOI: 10.1016/j.comnet.2023.109968

Naderi, M., & Ghanbari, M. (2023). Adaptively prioritizing candidate forwarding set in opportunistic routing in VANETs. *Ad Hoc Networks*, *140*, 103048. DOI: 10.1016/j.adhoc.2022.103048

Naeem, S., Ali, A., Anam, S., & Ahmed, M. M. (2023). An unsupervised machine learning algorithms: Comprehensive review. *International Journal of Computing and Digital Systems*.

Nagaraj, C. T., Anand, M. C. J., Priyadharshini, S. S., & Aparna, P. (2023, December). GCNXG: Detecting Fraudulent Activities in Financial Networks: A Graph Analytics and Machine Learning Fusion. In International Conference on Renewable Energy, Green Computing, and Sustainable Development (pp. 17-32). Cham: Springer Nature Switzerland.

Nandy, T., Idris, M. Y. I. B., Noor, R. M., Ahmedy, I., & Bhattacharyya, S. (2020, December). *A multiple-criteria decision analysis clustering and cluster head selection algorithm in vehicular network. In 2020 IEEE 8th R10 Humanitarian Technology Conference (R10-HTC)*. IEEE.

Nasir, M. K., & Al Mamun, S. Application of Graph theory and Error Correcting Code in VANETs.

Nastjuk, I., Herrenkind, B., Marrone, M., Brendel, A. B., & Kolbe, L. M. (2020). What Drives the Acceptance of Autonomous Driving? An Investigation of Acceptance Factors from an End-User's Perspective. *Technological Forecasting and Social Change*, *161*, 120319. Advance online publication. DOI: 10.1016/j.techfore.2020.120319

Nath, A., Rather, Z., Mitra, I., & Sahana, L. (2023). Multi-criteria Approach for Identification and Ranking of Key Interventions for Seamless Adoption of Electric Vehicle Charging Infrastructure. *IEEE Transactions on Vehicular Technology*, *72*(7), 8697–8708. DOI: 10.1109/TVT.2023.3252273

Ndashimye, E., Sarkar, N. I., & Ray, S. K. (2021). A Multi-criteria based handover algorithm for vehicle-to-infrastructure communications. *Computer Networks*, *185*, 107652. DOI: 10.1016/j.comnet.2020.107652

Nejatollahi, H., Dutt, N., Ray, S., Regazzoni, F., Banerjee, I., & Cammarota, R. (2019). Post-quantum lattice-based cryptography implementations: A survey. *ACM Computing Surveys (CSUR)*.

Nguyen, D. D., Rohács, J., Rohács, D., & Boros, A. (2020). Intelligent Total Transportation Management System for Future Smart Cities. *Applied Sciences (Basel, Switzerland)*, *10*(24), 8933. Advance online publication. DOI: 10.3390/app10248933

Nguyen, K. T., Laurent, M., & Oualha, N. (2015). Survey on secure communication protocols for the Internet of Things. *Ad Hoc Networks*, *32*, 17–31. DOI: 10.1016/j.adhoc.2015.01.006

Nguyen, V., Hwang, R., Lin, P., Vyas, A., & Nguyen, V. (2022, September 12). Toward the Age of Intelligent Vehicular Networks for Connected and Autonomous Vehicles in 6G. *IEEE Network*, *37*(3), 44–51. DOI: 10.1109/MNET.010.2100509

Niittymäki, J., & Pursula, M. (2000). Signal Control Using Fuzzy Logic. *Fuzzy Sets and Systems*, *116*(1), 11–22. DOI: 10.1016/S0165-0114(99)00034-2

Nikitas, A., Kougias, I., Alyavina, E., & Tchouamou, E. N. (2017, November 30). How Can Autonomous and Connected Vehicles, Electromobility, BRT, Hyperloop, Shared Use Mobility and Mobility-As-A-Service Shape Transport Futures for the Context of Smart Cities? *Urban Science (Basel, Switzerland)*, *1*(4), 36–36. DOI: 10.3390/urbansci1040036

Noor-A-Rahim, M., Liu, Z., Lee, H., Khyam, M. O., He, J., Pesch, D., Moessner, K., Saad, W., & Poor, H. V. (2022, May 20). 6G for Vehicle-to-Everything (V2X) Communications: Enabling Technologies, Challenges, and Opportunities. *Proceedings of the IEEE*, *110*(6), 712–734. DOI: 10.1109/JPROC.2022.3173031

Nuñez, Y., Lovisolo, L., da Silva Mello, L., Ramos, G., Leonor, N. R., Faria, S., & Caldeirinha, R. F. (2023, November). Path Loss Prediction for Vehicular-to-Infrastructure Communication Using Machine Learning Techniques. In *2023 IEEE Virtual Conference on Communications (VCC)* (pp. 270-275). IEEE. DOI: 10.1109/VCC60689.2023.10474798

Ogunleye, O S., & Belle, J P V. (2014, May 1). Exploring the success, failure and factors influencing m-Government implementation in developing countries. https://doi.org/DOI: 10.1109/ISTAFRICA.2014.6880656

Okpeh, A. A., Okpe, J. M., Tekanyi, A., & Tanko, N. (2023). Cooperative Multi-criteria Handover Management For Reliable V2i Communication And Improved Qos In Overlayed Heterogeneous Network Environment. *Annals of the Faculty of Engineering Hunedoara-International Journal of Engineering*, *21*(4).

Okrepilov, V. V., Kovalenko, B. B., Getmanova, G. V., & Turovskaj, M. S. 2022. "Modern Trends in Artificial Intelligence in the Transport System." In *Transportation Research Procedia*, DOI: 10.1016/j.trpro.2022.01.038

Okumura, B., James, M., Kanzawa, Y., Derry, M., Sakai, K., Nishi, T., & Prokhorov, D. (2016, March 1). Challenges in Perception and Decision Making for Intelligent Automotive Vehicles: A Case Study. *IEEE Transactions on Intelligent Vehicles*, *1*(1), 20–32. DOI: 10.1109/TIV.2016.2551545

Pamucar, D., & Ecer, F. (2020). Prioritizing the weights of the evaluation criteria under fuzziness: The fuzzy full consistency method–FUCOM-F. Facta Universitatis, Series. *Mechanical Engineering (New York, N.Y.)*, *18*(3), 419–437.

Pandey, K., Gupta, A. K., Perumalla, K. R., & Dhillon, H. S. (2023, May). Vehicular Communication Networks with Platooned Vehicles: Modeling and Analysis. In *ICC 2023-IEEE International Conference on Communications* (pp. 3995-4000). IEEE.

Pandey, K., Perumalla, K. R., Gupta, A. K., & Dhillon, H. S. (2023). Fundamentals of vehicular communication networks with vehicle platoons. *IEEE Transactions on Wireless Communications*, *22*(12), 8634–8649. DOI: 10.1109/TWC.2023.3264662

Pandey, R., Singh, K., & Kumar, A. (2022). Blockchain-augmented RL frameworks for secure VANETs. *IEEE Transactions on Network and Service Management*, *19*(1), 340–353.

Pandya, D., & Mer, H. (2020, December). Inter-Vehicular Communication for Intelligent Collision Avoidance Using Machine Learning: An Overview. In *International Conference on Advanced Informatics for Computing Research* (pp. 151-159). Singapore: Springer Singapore.

Pan, Y., & Stark, R. (2022). An interpretable machine learning approach for engineering change management decision support in automotive industry. *Computers in Industry*, *138*, 103633.

Papageorgiou, M., Diakaki, C., Dinopoulou, V., Kotsialos, A., & Wang, Y. (2003). Review of Road Traffic Control Strategies. *Proceedings of the IEEE*, *91*(12), 2043–2067. DOI: 10.1109/JPROC.2003.819610

Parham, M., & Pouyan, A. A. (2020). An Effective Privacy-Aware Sybil Attack Detection Scheme for Secure Communication in Vehicular Ad Hoc Network. *Wireless Personal Communications*, *113*(2), 1149–1182. Advance online publication. DOI: 10.1007/s11277-020-07272-8

Park, Y.. (2022). Traffic signal control using reinforcement learning and IoT sensors. *IEEE Transactions on Intelligent Transportation Systems*, *23*(6), 8971–8982.

Partovi, Z., Zarei, M., & Rahmani, A. M. (2023). Data-centric approaches in the internet of vehicles: A systematic review on techniques, open issues, and future directions. *International Journal of Communication Systems*, *36*(3), e5383. DOI: 10.1002/dac.5383

Patel, M. R., Vashi, M. P., & Bhatt, B. V. (2017). SMART-multi-criteria decision-making technique for use in planning activities. New Horizons in Civil Engineering (NHCE 2017), 1-6.

Patel, S. R., & Ajmeri, M. (2020). Machine learning approach for transportation services using vehicular ad hoc network. [IJTIIR]. *Int J Trend Innov Res*, *2*(2).

Patil, S. D., & Ragha, L. (2020). Adaptive fuzzy-based message dissemination and micro-artificial bee colony algorithm optimised routing scheme for vehicular ad hoc network. *IET Communications*, *14*(6), 994–1004. DOI: 10.1049/iet-com.2019.0388

Patra, M., Mishra, S., & Murthy, C. S. R. (2014, May). An analytic hierarchy process-based approach for optimal roadside unit placement in vehicular ad hoc networks. In 2014 IEEE 79th Vehicular Technology Conference (VTC Spring) (pp. 1-5). IEEE.

Patro, S. K., & Smarandache, F. (2016). The neutrosophic statistical distribution, more problems, more solutions. Infinite Study.

Peres, R. S., Barata, J., Leitao, P., & Garcia, G. (2019). Multistage quality control using machine learning in the automotive industry. *IEEE Access : Practical Innovations, Open Solutions*, *7*, 79908–79916.

Perla, B., & Ramesh, D. A. (2022). Development of Roughness Index Model For Urban Roads Using Machine Learning Techniques and Prioritizing Using MCDM Techniques. *Available atSSRN* 4160025. DOI: 10.2139/ssrn.4160025

Ponnambalam, S. G.. (2022). IoT-based traffic monitoring and signal control system. *Internet of Things : Engineering Cyber Physical Human Systems*, *18*, 100377.

Prabha, S. K., Anand, M. C. J., Vidhya, V., Nagarajan, G., Saikia, U., Martin, N., Kumary, M. S., & Tiwari, M. (2024). Vidhya, V., Nagarajan, G., Saikia, U., Martin, N., Kumari, M.S., Tiwari, M.: Sorting Out Interval Valued Neutrosophic Fuzzy Shortest Cycle Route Problem by Reduced Matrix Method.*International Journal of Neutrosophic Science*, *23*(2), 91–103. DOI: 10.54216/IJNS.230208

Prabha, S. K., Anand, M., Vidhya, V., Nagarajan, G., Saikia, U., Martin, N., & Tiwari, M. (2024). Sorting Out Interval Valued Neutrosophic Fuzzy Shortest Cycle Route Problem by Reduced Matrix Method. [IJNS]. *International Journal of Neutrosophic Science*, *23*(2).

Pradhan, R., Agarwal, A., & De, T. (2022). A multi-criteria route selection method for vehicles using genetic algorithms based on driver's preference. *Wireless Personal Communications*, *125*(3), 2477–2496. DOI: 10.1007/s11277-022-09670-6

Prakash, J., Murali, L., Manikandan, N., Nagaprasad, N., & Ramaswamy, K. (2024). A vehicular network-based intelligent transport system for smart cities using machine learning algorithms. *Scientific Reports*, *14*(1), 468. DOI: 10.1038/s41598-023-50906-7 PMID: 38172149

Prevost, S., & Kettani, H. (2019, October 23). On Data Privacy in Modern Personal Vehicles. https://doi.org/DOI: 10.1145/3372938.3372940

Pu, C. (2023). Blockchain-Based Trust Management Using Multi-Criteria Decision-Making Model for VANETs. *Authorea Preprints*.

Purkait, R., & Tripathi, S. (2020). Fuzzy logic based multi-criteria intelligent forward routing in VANET. *Wireless Personal Communications*, *111*(3), 1871–1897. DOI: 10.1007/s11277-019-06962-2

Qiao, Y.. (2023). A review of IoT-based traffic management for smart cities. *Sustainable Cities and Society*, *95*, 104614.

R, S S., & Rajendran, L. (2021, January 27). Real-Time Adaptive Traffic Control System For Smart Cities. https://doi.org/DOI: 10.1109/ICCCI50826.2021.9402597

Rahman, T., Liu, P., & Qiu, J. (2018). Q-learning for dynamic congestion control in vehicular networks. *Ad Hoc Networks*, *72*, 13–25.

Rajesh, K., Rathod, S., Kundale, J., Rathod, N., Anand, M. C.J., Saikia, U., Tiwari, M,, Martin, N., "A Study on Interval Valued Temporal Neutrosophic Fuzzy Sets", International Journal of Neutrosophic Science, Volume 23, Issue 1, PP: 341-349, (2024). https://doi.org/DOI: 10.54216/IJNS.230129

Rajesh, K., Rathod, S., Kundale, J., Rathod, N., Anand, M., Saikia, U., & Martin, N. (2024). A Study on Interval Valued Temporal Neutrosophic Fuzzy Sets. [IJNS]. *International Journal of Neutrosophic Science*, *23*(1).

Raj, J., Prabhu, V. V., Krishnakumar, V., & Anand, M. C. J. (2023). Solar Powered Charging -of Fuzzy Logic Controller (FLC) Strategy with Battery Management System (BMS) Method Used for Electric Vehicle (EV). *International Journal of Fuzzy Systems*, *25*(7), 2876–2888. DOI: 10.1007/s40815-023-01537-7

Ram Seshadri, A. (2022). Machine Learning Driven Simulation in the Automotive Industry.

Rana, S., Parast, F. K., Kelly, B., Wang, Y., & Kent, K. B. (2023). A comprehensive survey of cryptography key management systems. *Journal of Information Security and Applications*, *78*, 103607. DOI: 10.1016/j.jisa.2023.103607

Raouf, I., Khan, A., Khalid, S., Sohail, M., Azad, M. M., & Kim, H. S. (2022). Sensor-Based Prognostic Health Management of Advanced Driver Assistance System for Autonomous Vehicles: A Recent Survey. *Mathematics*, *10*(18), 3233. Advance online publication. DOI: 10.3390/math10183233

Rashad, H., & Mohamed, M. (2021). Neutrosophic theory and its application in various queueing models: Case studies. *Neutrosophic Sets and Systems*, *42*(1), 8.

Rathore, R. S., Hewage, C., Kaiwartya, O., & Lloret, J. (2022). In-Vehicle Communication Cyber Security: Challenges and Solutions. *Sensors (Basel)*, *22*(17), 6679. Advance online publication. DOI: 10.3390/s22176679 PMID: 36081138

Razakova, D., Shmatko, L., Shatila, K., & Davtyan, A. (2023, January 1). Analysis for the sustainability of intelligent transport systems in the United States. *EDP Sciences*, *389*, 05002–05002. DOI: 10.1051/e3sconf/202338905002

Rinott, Y., Shoham, T., & Kalai, G. (2022). Statistical aspects of the quantum supremacy demonstration. *Statistical Science*, *37*(3). Advance online publication. DOI: 10.1214/21-STS836

Rocco, G., Pipino, C., & Pagano, C. (2023). An Overview of Urban Mobility: Revolutionizing with Innovative Smart Parking Systems. *Sustainability (Basel)*, *15*(17), 13174. Advance online publication. DOI: 10.3390/su151713174

Rogers, D., Wallace, S., Emrich, J., Chaput, S., & Noens, R. (2000, March 6). Evolving Telematics Systems and Standards. https://doi.org/DOI: 10.4271/2000-01-0814

Roy, K. S., & Kalita, H. K. (2019). A survey on post-quantum cryptography for constrained devices. *International Journal of Applied Engineering Research: IJAER*, *14*(11), 2608–2615.

Roy, S.. (2021). An IoT-based smart traffic management system for smart cities. *Journal of Urban Technology*, *28*(1-2), 97–115.

Saaty, T. L. (2008). Decision Making with the Analytic Hierarchy Process. *International Journal of Services Sciences*, *1*(1), 83–98. DOI: 10.1504/IJSSCI.2008.017590

Sadaf, M., Iqbal, Z., Javed, A. R., Saba, I., Krichen, M., Majeed, S., & Raza, A. (2023). Connected and Automated Vehicles: Infrastructure, Applications, Security, Critical Challenges, and Future Aspects. *Technologies*, *11*(5), 117. Advance online publication. DOI: 10.3390/technologies11050117

Sadhukhan, P. (2017, September 1). An IoT-based E-parking system for smart cities. https://doi.org/DOI: 10.1109/ICACCI.2017.8125982

Said, H., Nicoletti, T., & Perez-Hernandez, P. (2014, November 18). Utilizing Telematics Data to Support Effective Equipment Fleet-Management Decisions: Utilization Rate and Hazard Functions. American Society of Civil Engineers, 30(1). https://doi.org/DOI: 10.1061/(ASCE)CP.1943-5487.0000444

Sakaguchi, K., Fukatsu, R., Yu, T., Fukuda, E., Mahler, K., Heath, R. W., Fujii, T., Takahashi, K., Khoryaev, A., Nagata, S., & Shimizu, T. (2020, January 1). Towards mmWave V2X in 5G and Beyond to Support Automated Driving. Cornell University. https://doi.org/DOI: 10.48550/arXiv.2011

Sakr, A., El-Afifi, M. I., & Team, P. (2023). Intelligent traffic management systems: a review. Nile journal of communication and computer science, 5(1), 42-56.

Saleem, M. A., Li, X., Mahmood, K., Shamshad, S., Alenazi, M. J., Das, A. K. (2024). A Cost-Efficient Anonymous Authenticated and Key Agreement Scheme for V2I-Based Vehicular Ad-Hoc Networks. *IEEE Transactions on Intelligent Transportation Systems.*

Santoro, A., Lampinen, A., Mathewson, K., Lillicrap, T., & Raposo, D. (2021). Symbolic behaviour in artificial intelligence. *arXiv preprint arXiv:2102.03406.*

Sawant, S. V., & Rudra, B. (2023). Analysis of magnitude of threats for v2x authentication schemes under quantum powered adversary, in: *International Conference on Information and Communication Technology for Intelligent Systems,* Springer. DOI: 10.1007/978-981-99-3761-5_43

Schettler, M., Buse, D. S., Zubow, A., & Dressler, F. (2020, December). How to train your ITS? integrating machine learning with vehicular network simulation. In *2020 IEEE Vehicular Networking Conference (VNC)* (pp. 1-4). IEEE.

Schettler, M., Buse, D. S., Zubow, A., & Dressler, F. (2020, December). How to train your ITS? Integrating machine learning with vehicular network simulation. In *2020 IEEE Vehicular Networking Conference (VNC)* (pp. 1-4). IEEE. DOI: 10.1109/VNC51378.2020.9318324

Schwabe, P., Stebila, D., & Wiggers, T. (2020, October). Post-quantum TLS without handshake signatures. In *Proceedings of the 2020 ACM SIGSAC Conference on Computer and Communications Security.*

Sewalkar, P., & Seitz, J. (2019, January 17). Vehicle-to-Pedestrian Communication for Vulnerable Road Users: Survey, Design Considerations, and Challenges. *Sensors (Basel)*, *19*(2), 358–358. DOI: 10.3390/s19020358 PMID: 30658392

Shah, A., Hussain, T., & Ali, I. (2021). Optimal Shortest Path Selection Base on Multi-Criteria Decision Making in Vehicular Ad Hoc Network.

Shamsan, A. H., & Faridi, A. R. (2018). "SDN-Assisted Iot Architecture: A Review." In *2018 4th International Conference on Computing Communication and Automation, ICCCA 2018*, DOI: 10.1109/CCAA.2018.8777339

Sharma, M., & Khanna, H. (2018). Intelligent and Secure Vehicular Network using Machine Learning. *JETIR-International Journal of Emerging Technologies and Innovative Research (*www. jetir. org*), ISSN*, 2349-5162.

Sharma, M., & Khanna, H. (2018). Intelligent and Secure Vehicular Network using Machine Learning. JETIR-International Journal of Emerging Technologies and Innovative Research ISSN, 2349-5162.

Shilpa, B., Radha, R., & Movva, P. (2021). Comparative analysis of wireless communication technologies for IoT applications. In Artificial Intelligence and Technologies: Select Proceedings of ICRTAC-AIT 2020 (pp. 383-394). Singapore: Springer Singapore.

Shi, Q., & Abdel-Aty, M. (2021). Big data applications in real-time traffic operation and safety monitoring and improvement on urban expressways. *Transportation Research Part C, Emerging Technologies*, *125*, 102940.

Singh, M., & Kim, S. (2017, November 1). Security analysis of intelligent vehicles: Challenges and scope., 8, 13-14. https://doi.org/DOI: 10.1109/ISOCC.2017.8368805

Singh, M., & Kim, S. 2017, Introduce reward-based intelligent vehicles communication using blockchain, International SoC Design Conference (ISOCC), Seoul, Korea (South), pp. 15-16

Singh, A. K., Pamula, R., Jain, P. K., & Srivastava, G. (2023). An efficient vehicular-relay selection scheme for vehicular communication. *Soft Computing*, *27*(6), 3443–3459. DOI: 10.1007/s00500-021-06106-4

Singh, B., & Gupta, A. (2015, April 1). Recent trends in intelligent transportation systems: a review. SciELO, 9(2), 30-34. https://doi.org/DOI: 10.1590/2238-1031.jtl.v9n2a6

Singh, B., Kumar, R., & Singh, V. P. (2022). Reinforcement learning in robotic applications: A comprehensive survey. *Artificial Intelligence Review*, *55*(2), 945–990. DOI: 10.1007/s10462-021-09997-9

Singh, P., Sharma, K., & Gupta, S. (2021). 5G-enabled reinforcement learning for adaptive VANET communication. *IEEE Vehicular Technology Magazine*, *16*(1), 28–37.

Singh, P., & Singh, R. K. (2020). A smart traffic management system using IoT. *Procedia Computer Science*, *167*, 2061–2070.

Sinha, S., & Saha, S. (2023). An improved communication strategy in vehicular ad hoc networks: Adaptive game theoretic modelling approach. *International Journal of Mobile Communications*, *21*(2), 273–293. DOI: 10.1504/IJMC.2023.128824

Sivabalan, S., & Minu, R. I. (2022, December). Statistical Sales Forecasting Using Machine Learning Forecasting Methods for Automotive Industry. In International Conference on Communication and Intelligent Systems (pp. 311-319). Singapore: Springer Nature Singapore.

Smarandache, F. (2002). Neutrosophy and Neutrosophic Logic. First International Conference on Neutrosophy, Neutrosophic Logic, Set, Probability, and Statistics, University of New Mexico, Gallup, NM 87301, USA.

Smarandache, F. (2005). Neutrosophic set is a generalization of the intuitionistic fuzzy sets, 24, 287-297. Inter. J. Pure Appl. Math.

Smarandache, F. (1999). *A Unifying Field in Logics: Neutrosophic Logic. Neutrosophy, Neutrosophic Set, Neutrosophic Probability*. American Research Press.

Soleymanian, M., Weinberg, C. B., & Zhu, T. (2019, January 1). Sensor Data and Behavioral Tracking: Does Usage-Based Auto Insurance Benefit Drivers? *Marketing Science*, *38*(1), 21–43. DOI: 10.1287/mksc.2018.1126

Solke, N. S., Shah, P., Sekhar, R., & Singh, T. P. (2022). Machine learning-based predictive modeling and control of lean manufacturing in automotive parts manufacturing industry. *Global Journal of Flexible Systems Managment*, *23*(1), 89–112.

Souza, A M D., Yokoyama, R S., Botega, L C., Meneguette, R I., & Villas, L A. (2015, October 1). Scorpion: A Solution Using Cooperative Rerouting to Prevent Congestion and Improve Traffic Condition. https://doi.org/DOI: 10.1109/CIT/IUCC/DASC/PICOM.2015.71

Souza, A. M. D., Brennand, C A R. L., Yokoyama, R. S., Donato, E., Madeira, E. R. M., & Villas, L. A. (2017, April 1). Traffic management systems: A classification, review, challenges, and future perspectives. *International Journal of Distributed Sensor Networks*, *13*(4), 155014771668361–155014771668361. DOI: 10.1177/1550147716683612

Sriram, S., Ramachandran, M., Chinnasamy, S., & Mathivanan, G. (2022). A review on multi-criteria decision-making and its application. REST Journal on Emerging trends in Modelling and Manufacturing, 7(4), 101-107.

Sudha, S., Martin, N., Anand, M. C. J., Palanimani, P. G., Thirunamakkani, T., & Ranjitha, B. (2023). MACBETH-MAIRCA Plithogenic Decision-Making on Feasible Strategies of Extended Producer's Responsibility towards Environmental Sustainability. Infinite Study.

Sudha, S., Nivetha, M., Clement Joe Anand, M., Palanimani, P. G., Thirunamakkani, T., & Ranjitha, B. (2023). MACBETH-MAIRCA Plithogenic Decision-Making on Feasible Strategies of Extended Producer's Responsibility towards Environmental Sustainability. *International Journal of Neutrosophic Science*, *22*(2), 114–130. DOI: 10.54216/IJNS.220210

Sun, Y., Wang, F., & Zhang, Q. (2023). Hybrid reinforcement learning for traffic prediction and control in VANETs. *IEEE Transactions on Intelligent Transportation Systems*, *24*(3), 1654–1665.

Suresh, S. (2023). Performance Evaluation of Vehicular Communication in Simulated Environment: A Comprehensive Evaluation of Vehicular Communication in LTE/5G Networks using a Simulated Environment.

Taleb, M A., & Farooque, O A. (2020, September 6). Towards a circular economy for sustainable development: An application of full cost accounting to municipal waste recyclables. Elsevier BV, 280, 124047-124047. https://doi.org/DOI: 10.1016/j.jclepro.2020.124047

Tamura, H., Nakano, K., Sengoku, M., & Shinoda, S. (2011). On applications of graph/network theory to problems in communication systems. ECTI Transactions on Computer and Information Technology (ECTI-CIT), 5(1), 15-21.

Tan, X., & Leon-Garcia, A. (2018, October 1). Autonomous Mobility and Energy Service Management in Future Smart Cities: An Overview. DOI: 10.1109/UV.2018.8642141

Tang, F., Kawamoto, Y., Kato, N., & Liu, J. (2019). Future intelligent and secure vehicular network toward 6G: Machine-learning approaches. *Proceedings of the IEEE*, *108*(2), 292–307. DOI: 10.1109/JPROC.2019.2954595

Tang, F., Mao, B., Kato, N., & Gui, G. (2021). Comprehensive survey on machine learning in vehicular network: Technology, applications and challenges. *IEEE Communications Surveys and Tutorials*, *23*(3), 2027–2057. DOI: 10.1109/COMST.2021.3089688

Tang, F., Wen, C., Zhao, M., & Kato, N. (2022). Machine learning for space–air–ground integrated network assisted vehicular network: A novel network architecture for vehicles. *IEEE Vehicular Technology Magazine*, *17*(3), 34–44. DOI: 10.1109/MVT.2022.3188405

Tang, Q., & Teague, V. (2024). Public-Key Cryptography–PKC 2024. In: *Proceedings of the 27th IACR International Conference on Practice and Theory of Public-Key Cryptography,* Sydney, NSW, Australia.

Tan, Y., Wu, Y., & Yu, H. (2020). Adaptive traffic signal control with deep reinforcement learning. *IEEE Transactions on Intelligent Transportation Systems*, *21*(6), 2375–2387.

Taslimasa, H., Dadkhah, S., Neto, E. C. P., Xiong, P., Ray, S., & Ghorbani, A. A. (2023). Security issues in Internet of Vehicles (IoV): A comprehensive survey. *Internet of Things : Engineering Cyber Physical Human Systems*, *22*, 100809. DOI: 10.1016/j.iot.2023.100809

Ta, V. T., & Dvir, A. (2020). A Secure Road Traffic Congestion Detection and Notification Concept Based on V2I Communications. *Vehicular Communications*, *25*, 100283. Advance online publication. DOI: 10.1016/j.vehcom.2020.100283

Taylor, S. J., Ahmad, F., Nguyen, H. N., Shaikh, S. A., Bryans, J., & Wartnaby, C. E. (2023). A comparative analysis of multi-criteria decision methods for secure beacon selection in vehicular platoons. *Transactions on Emerging Telecommunications Technologies*, *34*(11), e4841. DOI: 10.1002/ett.4841

Tayyaba, S. K., Khattak, H. A., Almogren, A., Shah, M. A., Din, I. U., Alkhalifa, I., & Guizani, M. (2020). 5G vehicular network resource management for improving radio access through machine learning. *IEEE Access : Practical Innovations, Open Solutions*, *8*, 6792–6800. DOI: 10.1109/ACCESS.2020.2964697

Teixeira, P., Sargento, S., Rito, P., Luis, M., & Castro, F. (2023). A Sensing, Communication and Computing Approach for Vulnerable Road Users Safety. *IEEE Access : Practical Innovations, Open Solutions*, *11*, 4914–4930. Advance online publication. DOI: 10.1109/ACCESS.2023.3235863

Theissler, A., Pérez-Velázquez, J., Kettelgerdes, M., & Elger, G. (2021). Predictive maintenance enabled by machine learning: Use cases and challenges in the automotive industry. *Reliability Engineering & System Safety*, *215*, 107864.

Tiwari, J., Dubey, D., Prakash, A., & Tripathi, R. (2022, December). A trustworthy and cooperative MAC protocol for cognitive vehicular networks. In *2022 IEEE 9th Uttar Pradesh Section International Conference on Electrical, Electronics and Computer Engineering (UPCON)* (pp. 1-6). IEEE. DOI: 10.1109/UPCON56432.2022.9986422

Tiwari, P. K., Tripathi, A., Srivastava, G., Prakash, S., & Shukla, N. K. (2022). A Novel Framework for Recommendation of Various Communication Networks Using Multi-criteria. In *VLSI, Microwave and Wireless Technologies: Select Proceedings of ICVMWT 2021* (pp. 635-645). Singapore: Springer Nature Singapore.

Tiwari, J., Prakash, A., & Tripathi, R. (2023). A metaheuristic MAC protocol for safety applications in cognitive vehicular networks. *Vehicular Communications*, *44*, 100686. DOI: 10.1016/j.vehcom.2023.100686

Tsai, C. W., Chin, F. L., Chao, H. C., & Vasilakos, A. V. (2015). Big Data Analytics: A Survey. *Journal of Big Data*, *2*(1), 21. Advance online publication. DOI: 10.1186/s40537-015-0030-3 PMID: 26191487

Uddin, M., Imran, M., & Qazi, A. (2018). Multi-Criteria Decision Making Based on Fuzzy Logic for Urban Transportation System. *International Journal of Transportation Engineering*, *6*(3), 257–270. DOI: 10.22119/IJTE.2018.88293

Ugwuishiwu, C. H., Orji, U. E., Ugwu, C. I., & Asogwa, C. N. (2020). An overview of quantum cryptography and shor's algorithm. *International Journal of Advanced Trends in Computer Science and Engineering*.

Upadhyay, P., Marriboina, V., Goyal, S. J., Kumar, S., El Sayed, M. E.-K., Ibrahim, A., Alhussan, A. A., & Khafaga, D. S. (2023). An Improved Deep Reinforcement Learning Routing Technique for Collision-Free VANET. *Scientific Reports*, *13*(1), 21796. Advance online publication. DOI: 10.1038/s41598-023-48956-y PMID: 38066104

Vahidi, A., & Eskandarian, A. (2003). Research Advances in Intelligent Collision Avoidance and Adaptive Cruise Control. *IEEE Transactions on Intelligent Transportation Systems*, *4*(3), 143–153. DOI: 10.1109/TITS.2003.821292

Vaidya, B., & Mouftah, H T. (2019, March 28). Connected Autonomous Electric Vehicles as Enablers for Low-Carbon Future. IntechOpen. https://doi.org/DOI: 10.5772/intechopen.84287

Vakula, D., & Raviteja, B. (2017, December 1). Smart public transport for smart cities. https://doi.org/DOI: 10.1109/ISS1.2017.8389288

Varalakshmi, A., Kumar, S. S., Shanmugapriya, M. M., Mohanapriya, G., & Clement Joe Anand, M. (2022). Markers location monitoring on images from an infrared camera using optimal fuzzy inference system. *International Journal of Fuzzy Systems*.

Veiga, D. T. T. F., de Miranda, A. W., Junior., Silva, L. N., Cavalcante, M. S., & dos Santos, M. D. C. (2023). Applying Machine Learning on automotive customer quality data to improve user experience and increase industry competitiveness (No. 2022-36-0070). SAE Technical Paper.

Velsberg, O., Jonsson, K., Westergren, U. H., & Saarikko, T. (2021). IoT Triggers. How Municipalities Are Transforming to Smarter Cities through IoT Use. *Scandinavian Journal of Information Systems, 33*(1).

Vermesan, O., Bahr, R., Falcitelli, M., Brevi, D., Bosi, I., Dekusar, A., Velizhev, A., ... 2020. "IoT Technologies for Connected and Automated Driving Applications." In *Internet of Things - The Call of the Edge: Everything Intelligent Everywhere*, DOI: 10.1201/9781003338611-6

Vermesan, O., John, R., Pype, P., Daalderop, G., Kriegel, K., Mitic, G., Lorentz, V., Bahr, R., Sand, H. E., Bockrath, S., & Waldhör, S. (2021). Automotive Intelligence Embedded in Electric Connected Autonomous and Shared Vehicles Technology for Sustainable Green Mobility. *Frontiers in Future Transportation, 2*, 688482. Advance online publication. DOI: 10.3389/ffutr.2021.688482

Vitianingsih, A. V., Baharin, S. S. K., Othman, O., & Suraji, A. (2022). Empirical study of a spatial analysis for prone road traffic accident classification based on MCDM method. *International Journal of Advanced Computer Science and Applications, 13*(5), 665–679. DOI: 10.14569/IJACSA.2022.0130578

Wang, Q., Yuan, Y., Yang, X., & Huang, Z. (2021, January 20). Adaptive and multi-path progression signal control under connected vehicle environment. Elsevier BV, 124, 102965-102965. https://doi.org/DOI: 10.1016/j.trc.2021.102965

Wang, H., Liu, Y., & Zhou, J. (2019). Deep reinforcement learning for spectrum management in VANETs. *Wireless Communications and Mobile Computing, 2019*, 1–11.

Wang, J., & Chen, Q. (2021). A Traffic Prediction Model Based on Multiple Factors. *The Journal of Supercomputing, 77*(3), 2928–2960. Advance online publication. DOI: 10.1007/s11227-020-03373-0

Wang, J., Shao, Y., Ge, Y., & Yu, R. (2019, January 15). A Survey of Vehicle to Everything (V2X) Testing. *Sensors (Basel), 19*(2), 334–334. DOI: 10.3390/s19020334 PMID: 30650658

Wang, N., Zhang, Y., Ahn, C. K., & Xu, Q. (2022). Autonomous Pilot of Unmanned Surface Vehicles: Bridging Path Planning and Tracking. *IEEE Transactions on Vehicular Technology, 71*(3), 2358–2374. Advance online publication. DOI: 10.1109/TVT.2021.3136670

Wang, S., Chen, K., & Zhang, X. (2023). Multi-agent RL for adaptive traffic signal optimization in smart cities. *IEEE Internet of Things Journal, 10*(2), 1568–1578.

Wang, W., Szefer, J., & Niederhagen, R. (2018). FPGA-based Niederreiter cryptosystem using binary Goppa codes. In *Post-Quantum Cryptography: 9th International Conference, PQCrypto 2018, Fort Lauderdale, FL, USA, April 9-11, 2018* [Springer International Publishing.]. *Proceedings*, *9*, 77–98.

Wang, Y., Hu, J., & Guo, S. (2020). An IoT-based approach for intelligent traffic monitoring and prediction [English Edition]. *Journal of Traffic and Transportation Engineering*, *7*(5), 714–729.

Wang, Y., Qian, J., & Xu, F. (2018). Reinforcement learning-based routing protocol for dynamic vehicular networks. *Wireless Networks*, *24*(3), 723–734.

Wijesekara, P. A. D. S. N., & Gunawardena, S. (2023, July). A Machine Learning-Aided Network Contention-Aware Link Lifetime-and Delay-Based Hybrid Routing Framework for Software-Defined Vehicular Networks. In Telecom (Vol. 4, No. 3, pp. 393-458). MDPI. DOI: 10.3390/telecom4030023

Xie, L., Chen, M., & Zhao, Y. (2019). Reinforcement learning for priority-based message dissemination in VANETs. *IEEE Transactions on Mobile Computing*, *18*(12), 2907–2918.

Xie, Z. (2020, February 1). Research on Bus Passenger Flow Analysis Method Based on Intelligent Transportation. *IOP Conference Series. Earth and Environmental Science*, *440*(4), 042005–042005. DOI: 10.1088/1755-1315/440/4/042005

Xue, J., Zhang, T., Wu, W., Zhou, H., & Shen, X. (2022, December). Sparse big data for vehicular network traffic flow estimation: A machine learning approach. In GLOBECOM 2022-2022 IEEE Global Communications Conference (pp. 4959-4963). IEEE. DOI: 10.1109/GLOBECOM48099.2022.10001710

Xu, G., Liu, H., & Li, M. (2020). Intelligent traffic management system based on the Internet of Things. *IEEE Access : Practical Innovations, Open Solutions*, *8*, 10826–10836.

Xu, T., Zhang, H., & Li, W. (2021). Edge-assisted RL frameworks for real-time congestion control in VANETs. *Future Generation Computer Systems*, *120*, 15–25.

Yacheur, B. Y., Ahmed, T., & Mosbah, M. (2023). Efficient DRL-Based Selection Strategy in Hybrid Vehicular Networks. *IEEE Transactions on Network and Service Management*, *20*(3), 2400–2411. DOI: 10.1109/TNSM.2023.3300653

Yang, F., Vereshchaka, A., Lepri, B., & Dong, W. (2021, August 24). Optimizing City-Scale Traffic Through Modeling Observations of Vehicle Movements. *IEEE Transactions on Intelligent Transportation Systems*, *23*(8), 10492–10503. DOI: 10.1109/TITS.2021.3094758

Yang, Z., Zolanvari, M., & Jain, R. (2023). A survey of important issues in quantum computing and communications. *IEEE Communications Surveys and Tutorials*, *25*(2), 1059–1094. DOI: 10.1109/COMST.2023.3254481

Yannis, G., Golias, J., & Antoniou, C. (2006). Effects of Real-Time Traffic Information on Driver Behaviour and Traffic Flow. *Transportation Research Part C, Emerging Technologies*, *14*(5), 402–420. DOI: 10.1016/j.trc.2006.09.001

Yao, B., & Feng, T. (2018). Machine learning in automotive industry. *Advances in Mechanical Engineering*, *10*(10), 1687814018805787.

Yao, Y., Shu, F., Li, Z., Cheng, X., & Wu, L. (2023). Secure transmission scheme based on joint radar and communication in mobile vehicular networks. *IEEE Transactions on Intelligent Transportation Systems*, *24*(9), 10027–10037. DOI: 10.1109/TITS.2023.3271452

Ye, L., Wu, F., & Zhang, R. (2020). Deep reinforcement learning for adaptive traffic signal control in urban areas. *IEEE Transactions on Intelligent Transportation Systems*, *21*(2), 665–674.

Yigit Katanalp, B., Eren, E., & Alver, Y. (2023). An integrated solution to identify pedestrian-vehicle accident prone locations: GIS-based multicriteria decision approach. *Journal of Transportation Safety & Security*, *15*(2), 137–176. DOI: 10.1080/19439962.2022.2048760

Yuan, W., Du, Z., Meng, X., Liu, F., & Masouros, C. (2023). Integrated Sensing and Communication for Vehicular Networks. In *Integrated Sensing and Communications* (pp. 419–444). Springer Nature Singapore. DOI: 10.1007/978-981-99-2501-8_15

Yue, W., Li, C., Mao, G., Cheng, N., & Zhou, D. (2021). Evolution of Road Traffic Congestion Control: A Survey from Perspective of Sensing, Communication, and Computation. *China Communications*, *18*(12), 151–177. Advance online publication. DOI: 10.23919/JCC.2021.12.010

Yusuf, A., Syed, A. K., & Souissi, R. (2024). Vehicle-to-Everything (V2X) in the Autonomous Vehicles Domain – A Technical Review of Communication, Sensor, and AI Technologies for Road User Safety. *Transportation Research Interdisciplinary Perspectives*, *23*, 100980. Advance online publication. DOI: 10.1016/j.trip.2023.100980

Zeina, M. B. (2020). Erlang service queueing model with neutrosophic parameters. *International Journal of Neutrosophic Science*, *6*(2), 106–112.

Zeina, M. B. (2020). Neutrosophic event-based queueing model. *International Journal of Neutrosophic Science*, *6*(1), 48–55.

Zeng, F., Pang, C., & Tang, H. (2023). Sensors on the Internet of Things Systems for Urban Disaster Management: A Systematic Literature Review. *Sensors (Basel)*, *23*(17), 7475. Advance online publication. DOI: 10.3390/s23177475 PMID: 37687929

Zeng, S., Zhang, N., Zhang, C., Su, W., & Carlos, L. A. (2022). Social network multiple-criteria decision-making approach for evaluating unmanned ground delivery vehicles under the Pythagorean fuzzy environment. *Technological Forecasting and Social Change*, *175*, 121414. DOI: 10.1016/j.techfore.2021.121414

Zeydan, E., Turk, Y., Aksoy, B., & Ozturk, S. B. (2022). Recent advances in post-quantum cryptography for networks: A survey. In: *Seventh International Conference OnMobileAnd Secure Services(MobiSecServ)*. IEEE. DOI: 10.1109/MobiSecServ50855.2022.9727214

Zhang, C., & Lu, Y. (2021). Study on artificial intelligence: The state of the art and future prospects. *Journal of Industrial Information Integration*, *23*, 100224. DOI: 10.1016/j.jii.2021.100224

Zhang, H., Wang, W., & Zhou, B. (2021). Density-based adaptive traffic signal control using IoT sensors. *Sensors (Basel)*, *21*(4), 1305. PMID: 33670358

Zhang, J.. (2022). Smart traffic management systems: An overview and their implementation in connected and autonomous vehicles. *IEEE Vehicular Technology Magazine*, *17*(1), 35–43.

Zhang, J., & Letaief, K. B. (2019, October 28). Mobile Edge Intelligence and Computing for the Internet of Vehicles. *Proceedings of the IEEE*, *108*(2), 246–261. DOI: 10.1109/JPROC.2019.2947490

Zhang, Q., Ayoub, O., & Gatto, A.. (2022). Joint routing, channel, and key-rate assignment for resourcie-efficient qkd networking. *IEEE Globecom Workshops*.

Zhang, Q., Zhao, L., & Gao, M. (2020). Predictive congestion control in VANETs using reinforcement learning and analytics. *IEEE Systems Journal*, *14*(3), 3058–3067.

Zhang, S., Chen, J., Lyu, F., Cheng, N., Shi, W., & Shen, X. (2018). Vehicular communication networks in the automated driving era. *IEEE Communications Magazine*, *56*(9), 26–32.

Zhao, Z., Xiang, T., Bi, Y., Zhi, M., & Huang, Z. (2023, July). A Novel Multi-Criteria Contribution Evaluation Scheme for Federated Learning in Internet of Vehicles. In *2023 15th International Conference on Communication Software and Networks (ICCSN)* (pp. 319-325). IEEE. DOI: 10.1109/ICCSN57992.2023.10297351

Zhao, W., Li, Q., & Huang, S. (2023). Resource allocation optimization in VANETs using deep reinforcement learning. *IEEE Communications Surveys and Tutorials*, *25*(1), 165–182.

Zhou, N. (2021). "The Future Integrated Traffic Management System to Optimize the Automobile Traffic." In *Journal of Physics: Conference Series*, DOI: 10.1088/1742-6596/1972/1/012097

Zhou, M., Wu, F., & Chen, T. (2020). Collaborative multi-agent reinforcement learning for efficient message dissemination in VANETs. *IEEE Access : Practical Innovations, Open Solutions*, *8*, 190453–190463.

Zhou, S., Wei, C., Song, C., Pan, X., Chang, W., & Yang, L. (2023). Short-Term Traffic Flow Prediction of the Smart City Using 5G Internet of Vehicles Based on Edge Computing. *IEEE Transactions on Intelligent Transportation Systems*, *24*(2), 1–10. Advance online publication. DOI: 10.1109/TITS.2022.3147845

Zhou, T., Ma, S., Yu, D., Li, M., & Hang, T. (2020, July 22). Development of Reliable, High Performance WLCSP for BSI CMOS Image Sensor for Automotive Application. *Sensors (Basel)*, *20*(15), 4077–4077. DOI: 10.3390/s20154077 PMID: 32707858

Zhou, X., & Tang, X. (2011). Research and implementation of RSA algorithm for encryption and decryption. In: *Proceedings of 2011 6th international forum on strategic*. IEEE. DOI: 10.1109/IFOST.2011.6021216

Zidi, S., Alaya, B., Moulahi, T., Al-Shargabi, A., & El Khediri, S. (2024). Fault Prediction and Recovery Using Machine Learning Techniques and the HTM Algorithm in Vehicular Network Environment. *IEEE Open Journal of Intelligent Transportation Systems*, *5*, 132–145. DOI: 10.1109/OJITS.2023.3347484

Žižović, M., & Pamucar, D. (2019). New model for determining criteria weights: Level Based Weight Assessment (LBWA) model. *Decision Making: Applications in Management and Engineering*, *2*(2), 126–137.

About the Contributors

Truong Khang Nguyen is currently the Head of the Department of Computer Engineering, Faculty of Information Technology at Van Lang University, Ho Chi Minh City. He leads the Intelligent Systems and Networks Research Group and is an IEEE Senior Member. His research includes reconfigurable antennas for 4G/5G/6G, photoconductive antennas for terahertz sources, and nano antennas for energy conversion. Dr. Nguyen holds a B.Sc. in Computational Physics from HCM Vietnam National University and a Ph.D. in Electrical and Computer Engineering from Ajou University, Korea. He has published over 100 ISI/Scopus papers, authored 3 book chapters, and holds 2 patents in Korea and the USPTO. His accolades include the UNESCO Research Grant Award and the NAWA Foreign Scientist Exchange Award.

Devasis Pradhan currently working as an Assistant Professor Grade 1 & Dean Research and Development,Acharya Institute of Technology, Bengaluru, Karnataka from 2017 onwards. His current research includes the effectiveness of 5G Green Communications, mmWave antenna design, UWB antennas, and its implementation. With the experience of 17+ Years of Academics and Research field he has published 80 + research papers (Journal & Conference Paper) and 4 Authored Books and 3 Edited Books (CRC,ISTE, Bentham Science) with a reputed publishing house. He is a co-editor in the Editorial Board & Peer-review of 8 International Journals and a committee member in the much-reputed organization. He has authored and co-authored 18 Book Chapters (IGI, CRC, TaylorSeries, Bentham, Wiley etc;). He has been a Session Chair, Technical Committee Member and Reviewer for reputed Internal Conferences such as IEEE, Grenz Society, Springer, etc; He has received 7 National Awards in the field of Academic and Research work from various governing bodies associated with the government of India. He is also an active member of ISTE, IEEE, ATMS, and other professional associations toward professional growth.

***

Abhiram A is currently pursuing a B.Tech degree in Electronics and Computer Engineering at Amrita Vishwa Vidyapeetham, Amritapuri Campus, Kollam, Kerala. He has completed numerous mini projects and participated in an Engine Control Unit tuning workshop at the National Institute of Technology, Calicut. Abhiram is actively building skills in Android development and is currently working on a major project focused on connected cars and IoT. His research interests include the Internet of Things, Computer Networking, and app development.

Manokamana Agarwal is an Assistant Professor of Mathematics at Silver Oak College of Engineering and Technology, Silver Oak University, Ahmedabad. She is also a dedicated research scholar at Charusat University, Changa, focusing on Special Functions. Her research interests include Special Functions, Fractional Calculus, and Fuzzy Logic and it's application. Outside of her academic life, Manokamana enjoys singing and car riding, reflecting her diverse interests and dynamic personality. Her commitment to both her profession and personal hobbies showcases her well-rounded character and passion for continuous learning.

M. Clement Joe Anand is working as an Assistant Professor in the Department of Mathematics, Mount Carmel College (Autonomous), Bengaluru, Karnataka. He has 12 years of teaching and research experience and received an honorary award from the University of Mexico, USA by the Board of Neutrosophic Science International Association. A total of 50 research articles were published indexed in Scopus and SCI databases. He has two Indian patents. He has completed a short-term course at the Indian Institute of Technology, Madras. He has received a doctoral degree from the University of Madras, Chennai. He is a reviewer of Scopus, SCI-indexed journals: 1. International Journal of Applied and Computational Mathematics (Scopus) 2. International Journal of Fuzzy Systems (Scopus, SCI) 3.Journal of Mathematics and Computer Science (Scopus, SCIE) 4.Iraqi Journal of Computer Science and Mathematics (Scopus) 5.Journal of Applied Mathematics and Informatics (Scopus) 6.Journal of Computer Science (Scopus, SCI) 7.IEEE Conferences (Scopus) He has received a grant from the Science and Engineering Research Board (SERB)-DST, Government of India.

Anjana V A completed her MSc in Mathematics from Mount Carmel College (Autonomous), Bengaluru. She is a passionate and active researcher in the fields of Fuzzy Logic and Statistics. Her research paper, "Optimizing Electricity Distribution Through Neutrosophic Transportation Problem Using Contingency Technique," was accepted and published by IEEE at the ICPEICES-2024 conference. She has also co-authored two more research papers titled "Enhancing Stock Market Forecasting with Fuzzy Logic Relation-Based System Identification" and "A Method to

Reduce the Negative Nature in Intuitionistic Fuzzy Assignment Problems in the Automation Sector" for the same conference. Additionally, she presented a paper on "Maximization of Fuzzy Transportation Problem (Maxi-FuzTP) and Comparative Discussions Using Triangular and Trapezoidal Fuzzy Numbers" at ICMMIS 2024, held in Uthangarai, where she received the Best Paper Award.

A. Manikandan completed his B.E in Electronics and Communication Engineering from Madurai Kamaraj University in 2004, Master of Engineering in Communication Systems from Anna University in 2006. He received his Ph.D from Anna University at 2018. He has 18 years of teaching experience and guided many UG & PG Projects. He is a Life member of IETE & ISTE. He has published papers in 13 International Journals, 14 international conferences, and 6 national conferences. Currently, he is working in M2M communication and IoT. He is currently associated with the Amrita School of Engineering, Amrita Vishwa Vidhyapeetham, Kollam, kerala, India.

Chandan A. Chavadi is a Dean and Professor with a Ph.D. in Management. He has published and edited a book widely. His research on retail and consumer behaviour has been recognized with awards and grants. He launched a journal and faculty programs, and played a key role in institutional development.

Ravindran K, who holds degrees in M.B.A, M.S.W, M.Phil., and Ph.D., presently serves at Presidency College (Autonomous), Bangalore, India. With 14 years of expertise in teaching and research, his focus lies in Marketing. He boasts a publication record of over 40 research papers in esteemed national and international journals. Noteworthy achievements include multiple faculty and teaching competency awards, as well as two patents in marketing. Additionally, he guides Ph.D. research scholars, with one student successfully attaining a doctoral degree under his mentorship. He also contributes as a reviewer for prestigious journals.

Sakshi Taaresh Khanna has received a B.Sc. (Hons.) Computer Science from University of Delhi, in 2006, M.C.A. from Guru Gobind Singh Indraprastha University in 2009, and pursuing Ph.D. degree in Information Technology from Amity University Uttar Pradesh, India. She is currently working as an Assistant Professor in Department of Computer Science Ram Lal Anand College, University of Delhi. She has qualified UGC-NET in Computer Science and Applications in June 2009. She has published six research articles in reputed journals and published two books. The memberships in various bodies like ACM and IEEE.

Sheela Maharajpet is a distinguished academician and a dedicated educator with over nine years of teaching experience. She is currently serving as an Assistant

Professor in the Department of Master of Computer Applications at the prestigious Acharya Institute of Technology, Soladevanahalli, Bangalore. She holds a Master of Computer Applications (MCA) degree from Kuvempu University, Jnana Sahyadri, Shankaraghatta, Shimogga. Her academic journey is marked by a commitment to excellence and a passion for continuous learning. She has an impressive portfolio of research, having published numerous papers in both national and international journals and conferences. In addition to her publications, Prof. Maharajpet has demonstrated her innovative prowess by filing a patent and copyright, showcasing her ability to translate theoretical research into practical applications. Her research interests lie at the intersection of Artificial Intelligence and Machine Learning.

C Manoj Kumar As a researcher dedicated to advancing the fields of nanofluids and mathematical modeelling. Through this domine i am exploring the use of artificial neural networks (ANNs) to optimize heat transfer parame-ters for nanofluids, a topic of substantial interest for applications in microelectronics cooling, biomedical devices, and energy systems.

Shaik Khaja Mohiddin, working as Associate Professor, at KL University Vaddeswaram, Andhra Pradesh. He has a teaching experience of more than 18 years. He compted his M.Tech from JNTU Kakinada in 2010, and Ph.D from Acharya Nagarjuna Unviersity in the domain of cloud computing working on Cloud Forensics in 2020. He thaught many subjects for both UG and PG students. His analytical way and algorithmic appraoch to solve a given problem in a real time makes him unique compared with that of others. He has published many research papers indexed in SCI, Scopus, Web of Sciecne and UGC Journals. He is acting as editor for many Elsevier and Scopus indexed Journals. He has guidede many project at both UG and PG level and guding three research scholars.

Quoc-Huy Ngo Im is a lecturer at Van Lang University, Vietnam. I has a background in communication networks and wireless technology, particularly in device-to-device communication in 6G networks. I has published several research papers in international conferences and scientific journals. His research interests include network performance optimization, data security in wireless networks, and AI applications in communications.

Pradeep Pandian is currently pursuing a Ph.D. in Social Media Analytics at Presidency Research Centre, Bengaluru City University. He holds an MBA from PES University, Bengaluru, and possesses 6 years of teaching and 3 years of industry experience. Additionally, he is qualified in NTA-NET and KSET.

A.P. Pushpalatha received her Master degree in Mathematics in 1995 and received her Ph.D degree in 2011 from Madurai Kamaraj University, Madurai, India. She is currently working as a Professor and Head of Mathematics Department, Velammal College of Engineering and Technology, Madurai, India. Previously she worked as an Associate Professor in Thiagarajar College of Engineering, Madurai-15 for 23 years. She has published nearly 35 research articles which include Scopus, Web of Science and UGC care list of journals. She has 29 years of teaching experience and 20 years of research experience. She acquired IJRULA Best Researcher award in 2018. She is the reviewer for the following Advances and Applications in Mathematical Sciences and Discrete Mathematics, Algorithms and applications journals. Her area of interest includes graph theory, Networking, discrete mathematics and fuzzy graphs. She is the memberships in various bodies like KMA and IJRULA and so on.

Vaishnav Raj M is currently pursuing a bachelor's degree in Electronics and Communication Engineering at Amrita Vishwa Vidyapeetham, Amritapuri Campus, Kollam, Kerala. At 21 years old and in the second year of the B.Tech program, Vaishnav has shown a keen interest in research and innovation. This research paper marks his first contribution to the academic community. In addition to his academic pursuits, Vaishnav has actively participated in hackathons, demonstrating his ability to apply theoretical knowledge to practical problems. Vaishnav's research interests include Computer networking and Communication. Despite being early in his academic journey, he is enthusiastic about contributing to advancements in these fields and looks forward to further opportunities to engage in research and development.

R. Rajeshkanna is an accomplished academic with over 16 years of experience in the field of computer science and information technology. He holds a Ph.D. from Bharathiar University, Coimbatore, and an M.Phil from Vinayaka Missions University, Salem. He also earned his MCA from Periyar University, Salem. Currently, Dr. Rajeshkanna serves as an Assistant Professor in the Department of Computer Science at CHRIST (Deemed to be University), Bangalore, a position he has held since July 2022. Prior to this, he was the Professor and Head of the Department of Information Technology at Dr. N.G.P Arts & Science College (Autonomous), Coimbatore. His tenure at this institution, spanning from January 2020 to June 2022, followed a series of progressive roles, including Associate Professor and Head of the Department, and earlier, as an Assistant Professor. Dr. Rajeshkanna has made significant academic contributions through his research and scholarly activities. He has published 35 papers in reputed international journals with high impact factors and has contributed 16 book chapters through IGI Global Publications.

Chitra Ramaprakash did her PhD from Bangalore University in 2015. She is currently working as an assistant Professor in Department of Mathematics, Dayananda Sagar College of Engineering, Bengaluru (Autonomous), since 2021. She has a very vast experience in handling Graph Theory courses for undergrads of foreign Universities. Her main research work focuses on Graph theory and algorithms. She has 17 years of teaching experience and 8 years of research experience.During this tenure she has published 8 research articles, 2 patents. These research articles were also presented in National and international Graph theory conferences. She has also attended various workshops on different domains of Graph theory and currently she is working on coloring and DominationDomination along with certain concepts of Machine learning, deep learning and Artificial intelligence, Mathematical modeling etc

B Ranjitha working as a Assistant Professor in the Department of Mathematics, School of Liberal Arts and Sciences, Mohan Babu University, Tirupati, Andhra Pradesh. Previously, she was an Assistant Professor at KMM Institute of Technology and Science, Tirupati. She has 18 years of teaching and 4 years research experience. She qualified AP-SET 2012. She published 6 papers in National journals. She is a member of AMTI, Chennai. She attended conferences and Faculty Development programs In reputed institutions. She completed 3 NPTEL courses. She published 5 patents. She completed 21 Coursera certification courses.

Y. Monikarchana working as a Assistant Professor (SL) in the Department of Mathematics, School of Liberal Arts and Sciences, Mohan Babu University, Tirupati, Andhra Pradesh. She has 15 years of teaching and 10 years research experience. She published 10 papers in International journals. She is a member of AMTI, Chennai. She attended and presented papers in national and International conferences and Faculty Development programs In reputed institutions. She completed 3 NPTEL courses. She published one patent. She completed 21 Coursera certification courses.

Bhawana Rudra is currently Assistant Professor at National Institute of Technology, Karnataka - Surathkal, India (NITK) in the Department of Information Technology. She has been honoured Ph.D. degree from Indian Institute of Information Technology, Allahabad, India (IIIT - Allahabad). Her research interest include, Cryptography and Cyber Security, Network Security, Internet of Things, Quantum Computing and Cryptography, Blockchain Security, Future Internet and its Security, Computer Networks and Android Security Issues.

Chiranjeevi S, from Bengaluru, is an exceptional MCA student at Acharya Institute of Technology, having consistently topped his department with an overall average of 92.04%. He has presented papers at prestigious conferences, including the IEEE International Conference and an International Conference on Management and Technology. His project "Density-Based Traffic Management System" earned first place at a departmental exhibition.

Sarvesh V. Sawant received Bachelor's Degree and Master's Degree in Computer Science Engineering National Institute of Technology, Goa, India (NIT Goa) in 2014 and 2017 respectively. Currently, he is pursuing his Ph.D. in National Institute of Technology, Karnataka - Surathkal, India (NITK) in the Department of Information Technology. His research interest include, Information Security, Cryptography, Cybersecurity and Blockchain.

Parul Sharda, is currently serving as an Associate Professor under Statue21 of Medi Caps University, Indore. She has served as a College Code-28 Associate Professor and Head of Management Department at Indore Management Institute, IMI Group of Institutions, Indore. She is an enthusiastic academician with over 15 years of teaching experience in different college and Universities. She is a registered guide at Medi-Caps university Indore. She is also having a corporate experience of two years with Reliance Industries, Navi Mumbai. She is currently holding the position of MP State Convener- Research Foundation of India. Also, she is a MP State President of WICCI (Women Indian Chamber of Commerce and Industry) - Financial Literacy and Management Council. Dr Sharda had also been a trainer at MSME-Development Institute and ATDC (Apparel Training and Design Center) for budding entrepreneurs. She is a faculty of International Business, Economics and Entrepreneurship, and an author of Two Book's "Macro Economics-An Indian Perspective" and "Minimalism-A New Lifestyle Statement" for Management Students and "   -I" for BALLB - Ist Sem Students. Dr Sharda has published more than 50 research Papers in Varied National/International, UGC Care, Scopus and ABDC Indexed Journals She is been awarded with Teacher with Huge Potential by Research Foundation of India and World Federation of Science and Technology in 2019, Best Researcher Award in the area of Management by RULA foundation Trichy, Tamil Nadu and Best Presentation award for Sustainable Development Goal-9 by Global Peace and Sustainable Development Summit- 2021. Also, she had received Global Empowerment Award-2021 for dedicating her service as a "Zestful Academician with Enormous Aptitudes" in the research area of International Business, Economics and Entrepreneurship. She is a thoughtful leader and visionary with a disruptor outlook and has proven examples of success with established national, international and global educational platforms. She has also received the SDG Ambassador Award in 2024 for raising awareness among the youth and academic community related to sustainable development goals. She is a columnist at Free Press and News Track Live, and believes that words are the powerful weapons for teachers, which can be used to bring a positive change in the society. Dr Parul has also organized various Workshops, Training Programs, SDPs and FDPs for faculties and students' overall enhancement. She has shared her knowledge and expertise as a resource person and guest speaker in various institutions including IIT-Indore (MPVS), Parul University, Vadodra (Gujrat), Imperial College Bargarh, Odisha, GLA University Mathura, AIDASCO-Republic of Serbia, Witty Gossip Association, Medi-Caps University Indore, Oriental University Indore, Acropolis Institute of Management and Research Indore, Shri Vaishnav College of Commerce, Indore, St. Paul Institute of Professional Studies, Indore and many more.

Neeraj Kumar Sharma is working as Associate Professor in the Department of Computer Science in Ram Lal Anand College, University of Delhi. He has completed his graduation in the B.Sc.(Hons.) Mathematics from University of Delhi in 1992, Masters in Computer Application from Indira Gandhi National Open University Delhi in 1999. He has completed Ph.D. in Computer Science from University of Delhi in 2014. He has authored two booklets of for IGNOU, MCA and various research articles in journals and conferences. His research interests include Artificial Intelligence Techniques, Multi-agent Systems, Trust and Reputation Systems, e-Commerce, Algorithms, Software Engineering.

Monika Sirothiya received her PhD degree in 'Marketing of compressed biogas in India' from Bangalore University in 2022. She has a master's degree in international business management and a bachelor's degree in biotechnology. She has made significant contributions to several research publications in which primary and secondary data were analyzed utilizing statistical packages such as SPSS, Amos, and R. Published papers in Scopus, UGC Care, and ABDC indexed Journals. She currently holds positions as co-founder and research analyst at the Research Rescue consultancy.

Manish Kumar Thakur, an accomplished academician and seasoned professional, holds a PhD in Computer Applications from Visvesvaraya Technological University. With a robust academic background, including an MCA from Visvesvaraya Technological University and an MTech in Information Technology from Karnataka State Open University, he has seamlessly blended theoretical knowledge with practical expertise. Dr. Thakur's research focuses on machine learning, data analytics, artificial intelligence, and cloud computing. His noteworthy contributions include the development of the “Alive” integrated

LMS platform and significant work on image recognition and content evaluation using machine learning techniques. His publications in prestigious journals and presentations at international conferences reflect his commitment to advancing technology and education. As the Assistant Director of Operations at Acharya Institutes, Bangalore, Dr. Thakur has demonstrated exceptional leadership in campus infrastructure, project management, and innovative educational solutions. His dedication to mentorship has earned him accolades, including the "Best Mentor Award" by IBM-India. Dr. Thakur continues to inspire and lead in both academic and professional spheres, driving technological innovation and educational excellence.

Rakesh Thoppaen Suresh Babu is a Senior Technical Architect specializing in Data Science at Hexaware Technologies in the USA. With 13 years of industry experience, he is a Senior IEEE Member recognized for his expertise in data science, Industrial IoT, and cybersecurity. Rakesh has successfully led data-driven transformation programs utilizing deep learning, OCR, and generative AI to enhance profitability, optimize efficiency, streamline analytics operations, form analytics teams, and develop analytics infrastructure strategies. He holds a Master of Science in Computer Science, specializing in unstructured data analysis and object detection. Rakesh has spearheaded the establishment of an AI Center of Excellence (COE) for a major semiconductor client and has assisted numerous clients in identifying, developing, and productionizing AI use cases and capabilities. As a co-founder of BayofTech Pvt Ltd, he has contributed to creating Industry 4.0 solutions for manufacturing companies, focusing on IIoT and AI-based hardware.In addition to his professional achievements, Rakesh has served as a judge in various technology forums and peer-reviewed publications such as TheSciPub. He holds a pending US patent for an innovative authentication process designed to mitigate vulnerabilities caused by HID. His research interests encompass AI-computer vision, IIoT, and Cybersecurity. Rakesh can be contacted at rakesh.thoppaen@gmail.com or rakesht@ hexaware.com.

Mohit Tiwari I'm an Assistant Professor at Bharati Vidyapeeth's College of Engineering,Delhi with a B.Tech. in Computer Science and an MBA in Operations & Research. With over 20 years of experience in Software Development, Networking, and IT Infrastructure Management, I oversee the IT Maintenance Cell and conduct research to stay updated with emerging IT trends. My diverse background enriches my teaching and research, benefiting both students and my contributions to the field.

Quang-Nhat Tran received the B.S and M.Sc degrees in computer science and information technology, Mendeleev University, Moscow, Russia, in 2008. He is currently the Deputy Head of Computer Networking and IoT Department, Faculty of Information Technology, Van Lang University, Ho Chi Minh City (HCMC), Vietnam. He is also working toward the Ph.D. degree at the Institute of Fundamental and Applied Sciences, Duy Tan University, HCMC, Vietnam

Index

A

C

E

F

I

K

L

M

N

O

P

Q

R

S

T

V

W

www.ingramcontent.com/pod-product-compliance
Lightning Source LLC
LaVergne TN
LVHW080341011026
849161LV00009B/2151

* 9 7 9 8 3 6 9 3 6 4 2 2 2 *